Delmar's Inspection & Maintenance Series

A Technician's Guide to

Advanced Automotive Emissions Systems

By Rick Escalambre

Delmar Publishers

I(T)P™ An International Thomson Publishing Company

Albany • Bonn • Boston • Cincinnati • Detroit • London • Madrid • Melbourne
Mexico City • New York • Pacific Grove • Paris • San Francisco • Singapore • Tokyo
Toronto • Washington

Rick would like to thank his wife, Pat, for her hours of work helping him to research, develop, and write this guide.

NOTICE TO THE READER

Publisher does not warrant or guarantee any of the projects described herein or perform any independent analysis in connection with any of the project information contained herein. Publisher does not assume, and expressly disclaims, any obligation to obtain and include information other than that provided to it by the manufacturer.

The reader is expressly warned to consider and adopt all safety precautions that might be indicated by the activities herein and to avoid all potential hazards. By following the instructions contained herein, the reader willingly assumes all risks in connection s with such instructions.

The publisher makes no representation or warranties of any kind, including but not limited to, the warranties of fitness for particular purpose or merchantability, nor are any such representations implied with respect to the material set forth herein, and the publisher takes no responsibility with respect to such material. The publisher shall not be liable for any special, consequential, or exemplary damages resulting, in whole or part, from the readers' use of, or reliance upon, this material.

Delmar Staff:
Publisher: Dale R. Bennie
Administrative Editor: Vernon Anthony
Editor: Jack Erjavec
Production Manager: Melanie Hope

Printed in the United States of America

For more information, contact:

Delmar Publishers
3 Columbia Circle, Box 15015
Albany, New York 12212-5015
1-800-347-7707

International Thomson Publishing Europe
Berkshire House 168 - 173
High Holborn
London WC1V7AA
England

Thomas Nelson Australia
102 Dodds Street
south Melbourne, 3205
Victoria, Australia

Nelson Canada
1120 Birchmount Road
Scarborough, Ontario
Canada M1K5G4

International Thomson Editores
Campos Eliseos 385, Piso 7
Col Polanco
11560 Mexico D F Mexico

International Thomson Publishing GmbH
Konigswinterer Strasse 418
53227 Bonn
Germany

International Thomson Publishing Asia
221 Henderson Road
#05 - 10 Henderson Building
Singapore 0315

International Thomson Publishing - Japan
Hirakawacho Kyowa Building, 3F
2-2-1 Hirakawacho
Chiyoad-ku, Tokyo 102
Japan

2 3 4 5 6 7 8 9 10 X X X 01 00 99 98 97 96 95

ry of Congress Cataloging-in-Publication Data

nbre, Rick
chnician's Guide to Advanced Automotive Emissions Systems / by Rick Escalambre.
cm
s index.
8273-7154-3
obiles--Pollution control devices--Maintenance and repair--Handbooks, manuals, etc. I. Title.
3 1995
8--dc20 94-46450
CIP

Contents

Preface

A Technician's Guide to Advanced Automotive Emissions Systems is written to provide the technician with a look into the causes of automotive emissions, the various methods of emissions inspection/maintenance testing, the diagnosis of failed vehicles, and enhanced on-board computer systems.

Chapter One provides the technician with an understanding of the causes and effects of automotive emissions. In addition, it covers the information necessary to understand why the EPA mandates emissions testing in many states. Emphasis is on reducing the number of vehicles failing an enhanced emissions test.

Chapters Two through Four examine the rules and regulations of the EPA-mandated emissions systems testing, IM240 program requirements, and alternative emissions inspection/maintenance programs. This provides the technician with insights about how a vehicle is tested and how it can fail an enhanced emissions test. Then the technician can better communicate with the customer about how and why the vehicle failed.

Chapters Five through Eight cover on-board computer systems. To fully understand the diagnosis of advanced emissions systems, the technician must have a good working knowledge of the on-board computer. Chapter Seven deals specifically with the use of scan tools and how to use their many features. These chapters provide the foundation for Chapters Eleven and Twelve.

Chapter Nine is designed to introduce the technician to automotive Lab Scopes which have become vital to proper diagnostics. Chapter Ten is designed to assist the technician with diagnosing emissions-related failures. The technician will learn the parts of an IM240 trace, how to establish a baseline reading using the tools and equipment available in the shop, and how to recreate the conditions under which the vehicle failed.

Chapter Eleven covers how to effectively diagnose on-board computer systems using a systematic approach. The technician will learn what the computer does and does not know. Scan tool diagnosis is based on the information covered in Chapters Five through Eight.

Chapter Twelve covers OBD-II (On-Board Diagnostics Second Generation). OBD-II is an enhanced emissions system that relates directly to all of the information covered in this book. Automotive emissions, emissions systems inspection and maintenance programs, on-board computers, scan tool diagnosis, and lab scopes are a vital part of OBD-II.

CHAPTER 1

Causes and Effects of Automotive Emissions

An ideal engine (if it existed) would burn its fuel completely. The combustion byproducts (exhaust emissions) of this ideal engine would be carbon dioxide (CO_2) and water (H_2O). (See Figure 1-1.)

The automotive engine needs three ingredients for the combustion process:

1. *Fuel* — In the automotive engine, gasoline is the most common fuel in use. The addition of ethanol, methanol, LPG, etc. does not change this requirement for combustion.
2. *Oxygen* — Fuel by itself cannot burn unless it is combined with enough oxygen to support combustion. In the automotive engine, oxygen is supplied through the induction system. The fuel is atomized by a carburetor or fuel injector(s). The proportion of oxygen to fuel is called the air/fuel ratio.

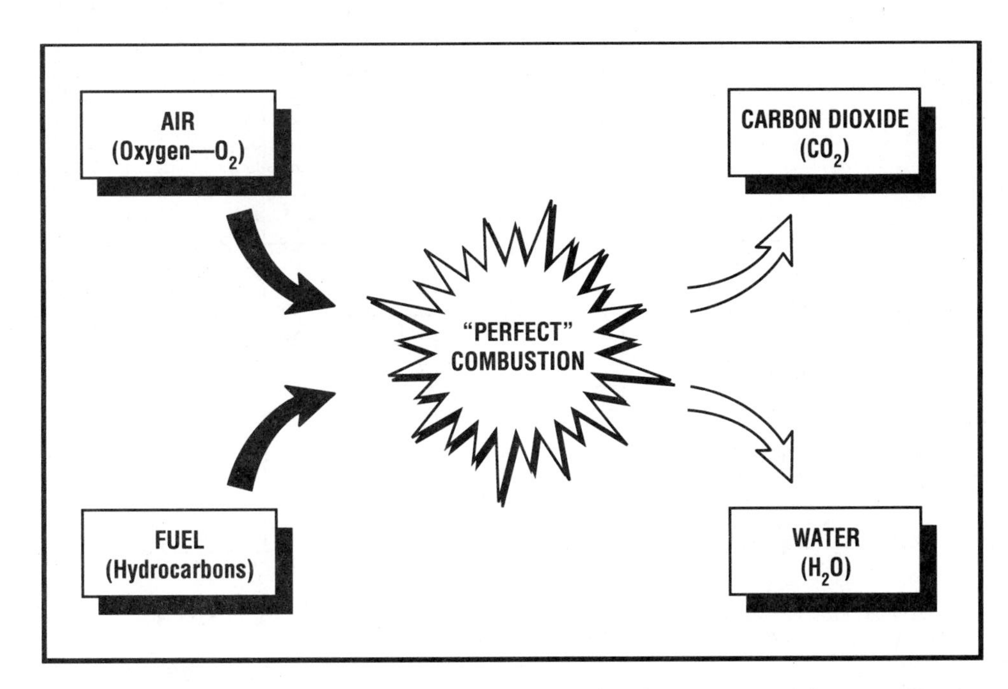

Figure 1-1 An *ideal* engine does not produce pollutants (courtesy Environmental Systems Products, Inc.).

3. *Heat* — The compression of fuel and oxygen does not generate enough heat to ignite the air/fuel mixture. Instead, the heat necessary for the burning process comes from the ignition system.

For the internal combustion engine to develop power from the combustion process, the following is required:

1. *Volumetric Efficiency* — An engine must be capable of filling each cylinder with a sufficient amount of oxygen and fuel. The intake and exhaust systems must be balanced to handle the flow. The camshaft and crankshaft must be in time for correct operation of the pistons and the intake and exhaust valves.
2. *Air/Fuel* — It is important to have the correct air/fuel ratio for the driving conditions (engine load). The air/fuel ratio will vary from very rich (9:1) to very lean (18:1).
3. *Compression* — The ability of an engine to contain and compress the air/fuel in a cylinder is a vital part of the combustion process. Any loss of compression affects the engine power.
4. *Ignition* — Once the air/fuel is compressed, ignition must occur. The spark must travel to the correct cylinders, in the correct order, at the right time, with sufficient voltage and duration.

Unfortunately, nothing is perfect. Real-world engines require regular preventive maintenance (Figure 1-2), non-tampered emission control systems, quality fuel, and immediate attention when a component or system has failed.

Failure to meet these requirements results in the release of undesirable crankcase, evaporative, and exhaust emissions being emitted into the air we breathe.

The key undesirable emissions are hydrocarbons (HC), carbon monoxide (CO), and oxides of nitrogen (NO_x). The major emissions byproducts of our real-world engines are shown in Figure 1-3.

Scheduled Maintenance Services

GASOLINE ENGINES WITH HEAVY DUTY EMISSIONS— MAINTENANCE SCHEDULE I†

Item No.	Service	If your driving conditions meet those specified on page 7-5, use Maintenance Schedule I (+).				If your driving conditions do NOT meet those specified on page 7-5, use Maintenance Schedule II (•).															
	Miles (000)	3	6	9	12	15	18	21	24	27	30	33	36	39	42	45	48	51	54	57	60
	Kilometers (000)	5	10	15	20	25	30	35	40	45	50	55	60	65	70	75	80	85	90	95	100
1	Engine Oil Change*—Every 3 Months, or	+	+	+	+	+	+	+	+	+	+	+	+	+	+	+	+	+	+	+	+
	Oil Filter Change*—Every 3 Months, or	+	+	+	+	+	+	+	+	+	+	+	+	+	+	+	+	+	+	+	+
2	Chassis Lubrication—Every 12 Months, or	+	+	+	+	+	+	+	+	+	+	+	+	+	+	+	+	+	+	+	+
3	Clutch Fork Ball Stud Lubrication										+										+
5	Cooling System Service*—Every 24 Months or								+								+				
6	Air Cleaner Element Replacement▲*								+								+				
7	Front Wheel Bearing Repack				+				+				+				+				
8	Transmission Service—See page 7-19																				
10	Fuel Filter Replacement*				+				+				+				+				+
11	Spark Plugs Replacement*									+									+		
12	Spark Plug Wire Inspection*																				+
13	EGR System Inspection*																				+
15	Engine Timing Check▲*								+								+				
16	Fuel Tank, Cap and Lines Inspection*																				+
17	Thermostatically Controlled Air Cleaner Inspection▲*								+								+				
18	Engine Accessory Drive Belt(s) Inspection*				+				+				+				+				+
19	Evaporative Control System Inspection*																				+
20	Shields and Underhood Insulation Inspection▲■				+				+				+				+				+
21	Air Intake System Inspection▲■								+								+				
22	Thermostatically Controlled Engine Cooling Fan Check ▲■— Every 12 Months or				+				+				+				+				+
24	Tire and Wheel Rotation—See page 7-21																				
25	Drive Axle Service	+	+	+	+	+	+	+	+	+	+	+	+	+	+	+	+	+	+	+	+
26	Brake Systems Inspection—See page 7-21																				

FOOTNOTES:
* An Emission Control Service
▲ Also a Noise Emission Control Service
■ Applicable only to vehicles sold in the United States
† To determine the emissions classification of your entine refer to page 7-4.
T0339

THE SERVICES SHOWN ON THIS CHART UP TO 60,000 MILES (100 000 km) ARE TO BE DONE AFTER 60,000 MILES AT THE SAME INTERVALS.
T0340

Figure 1-2 Regular maintenance is vital to low emissions (courtesy GM).

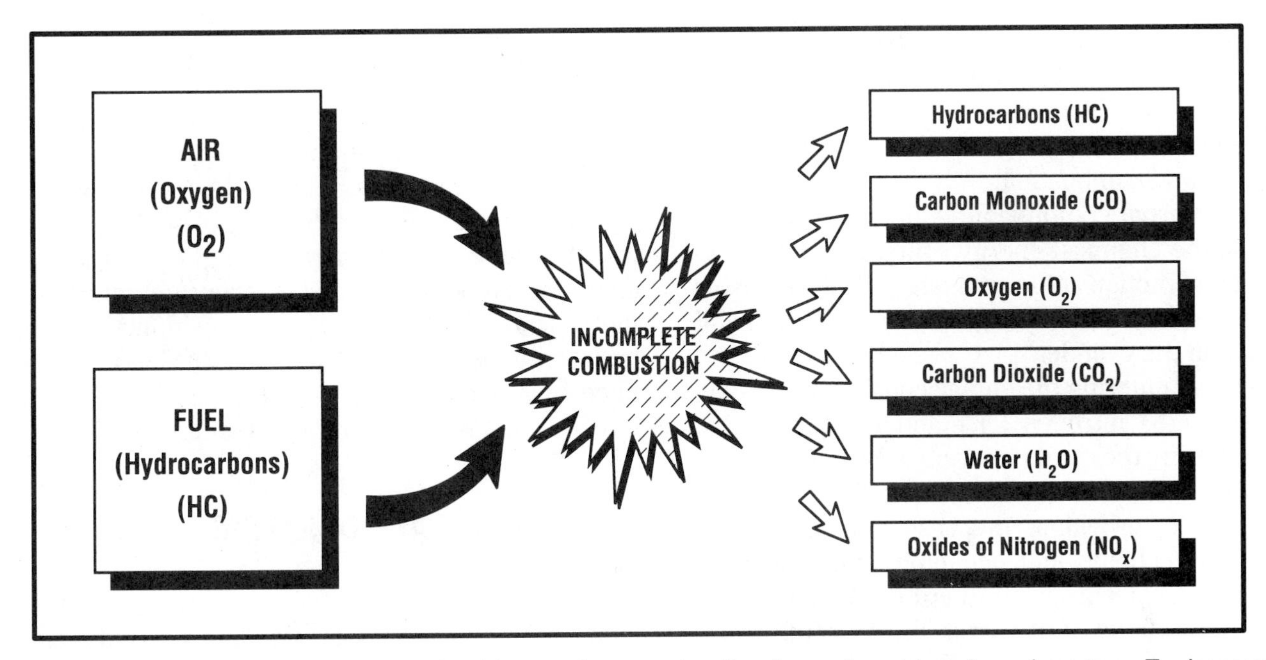

Figure 1-3 Emissions are the result of incomplete combustion in *real-world* engines (courtesy Environmental Systems Products, Inc.).

Hydrocarbons (HC) consist of gasoline that did not burn in the combustion process. Hydrocarbons can be measured in *parts per million* (ppm) or *grams per mile* (g/mi). A low HC level is desirable. On vehicles equipped with a catalytic converter, it is possible to detect no hydrocarbons at the tailpipe. High HC levels result from poor combustion due to ignition, compression, or timing abnormalities. As the flame front moves across the cylinder and contacts the cooler cylinder walls, the flame is quenched and the result is unburned fuel in the cylinder. Also, a very lean or very rich air/fuel mixture can raise the level of hydrocarbons.

Carbon monoxide (CO) is created when there is not enough oxygen present during combustion. When there is an oxygen deficiency in the combustion process, some of the carbon in the gasoline will join with some of the oxygen in the air, on a one-to-one basis, to form carbon monoxide (CO). Carbon monoxide can be measured as a *percentage* (%) or in *grams per mile* (g/mi). High CO is typically the result of a rich air/fuel mixture.

Carbon monoxide interferes with the oxygen-carrying capacity of the blood. Exposure aggravates angina and other aspects of coronary heart disease and decreases exercise tolerance in persons with cardiovascular problems. Infants, fetuses, elderly persons, and individuals with respiratory diseases are also particularly susceptible to CO poisoning.

Oxides of nitrogen (NO_x) can begin to form when combustion chamber temperatures exceed 1600°F. Under these conditions, nitrogen combines with oxygen to form a family of new compounds called oxides of nitrogen or NO_x. The "N" stands for the nitrogen, the "O" for oxygen, and the "x" means that the amount of oxygen combined with nitrogen varies for each of the nitrogen compounds. Oxides of nitrogen can be measured in *parts per million* (ppm) or *grams per mile* (g/mi). Lean air/fuel ratios, advanced spark timing, carbon, malfunctioning EGR systems, etc. can cause combustion chamber temperatures to rise and create NO_x.

Oxides of nitrogen include nitrogen dioxide (NO_2) and nitric oxide (NO). These gases irritate the lungs, lower resistance to respiratory infections, and contribute to the development of emphysema, bronchitis, and pneumonia. NO_x contributes to ground-level ozone formation and can also react chemically in the air to form nitric acid.

THE CHEMISTRY OF COMBUSTION

Automotive emissions controls are becoming increasingly more numerous and sophisticated. Automotive manufacturers are attempting to cope with new systems and designs required by federal emissions standards, as well as the complex interrelationship of the three basic automotive emissions under a variety of driving conditions. To effectively troubleshoot the emissions system, the technician must have a good understanding of the chemistry of combustion.

An internal combustion engine is designed to release the chemical energy contained in fuel and turn it into motion at the crankshaft. The chemistry of combustion requires fuel to be mixed with air and burned in the cylinder.

The air entering the engine consists of 21% oxygen (O_2), 78% nitrogen (N_2), and 1% other gases. Fuel entering the combustion chamber consists primarily of hydrocarbons (HC). The ratio of air to fuel is critical to providing the best compromise between driveability, fuel economy, and emissions control. This occurs when there is 14.7 pounds of air to each pound of fuel, or a 14.7:1 air/fuel ratio. Ideally, when combustion occurs the following chemical reaction should take place:

$$O_2 + N_2 + HC = CO_2 + H_2O + N_2 + O_2$$

In this ideal scenario, all of the hydrocarbons are completely oxidized, chemically changing into carbon dioxide (CO_2) and water vapor (H_2O). The nitrogen (N_2) passes through the combustion process unchanged. A small percentage of oxygen (O_2) should also be present.

Under real-world conditions, the quench area in the combustion chamber, variations in the air/fuel mixture, and high combustion temperatures due to engine load result in the following chemical reaction in the engine:

$$O_2 + N_2 + HC = CO_2 + H_2O + N_2 + O_2 + HC + CO + NO_x$$

In this scenario, some hydrocarbons (HC) do not burn at all. Other hydrocarbons oxidize only partially, forming carbon monoxide (CO) instead of carbon dioxide (CO_2). Additionally, the high combustion temperature causes the nitrogen to oxidize, forming oxides of nitrogen (NO_x). During the exhaust stroke, unburned hydrocarbons, carbon monoxide, and nitrogen oxides are forced out of the engine.

To solve these problems, many manufacturers use an air injection system and a catalytic converter system to reduce HC and CO levels. They also use some form of exhaust gas recirculation to reduce NO_x that supplies spent (inert) exhaust gas through the intake manifold into the cylinder to reduce the combustion chamber temperatures.

This process creates a conflict between the control of hydrocarbons and the control of nitrogen oxides. Higher compression ratios, higher combustion chamber temperatures, and lean air/fuel ratios promote cleaner burning of the fuel, resulting in fewer hydrocarbons. Unfortunately, most things that reduce hydrocarbons also result in the higher combustion chamber temperatures that create NO_x.

HISTORY OF AUTOMOTIVE EMISSIONS

Smog and air pollution are exceedingly complex subjects that are not completely understood and are highly controversial. Both are recognized as major environmental problems.

Most urban and industrial areas experience air pollution from time to time in varying degrees of intensity, from barely discernible to highly objectionable. The amount and type of air pollution at any given location at any particular time is influenced by such factors as:

1. Population density.
2. Industrial activity.
3. Traffic density.
4. Fuel consumption.
5. Climate, weather, and topography.

There are many different types and sources of air pollution (Figure 1-4), including smoke from ordinary burning, gases from industrial processes, dust from many sources, and photochemical smog. The visible substances that contribute to air pollution are minute particles consisting of either liquids or solids suspended in the air. They may consist of a great variety of compounds such as hydrocarbons, lead salts, and many chemical compounds of iron, ammonia, sulfur, calcium, and others. This is the type of air pollution commonly seen in industrial areas and large cities.

Other major sources of HC are: industrial processing, petroleum refining, fuel handling, space heating, and refuse burning. Nationwide, nitrogen oxides originate primarily from stationary sources,

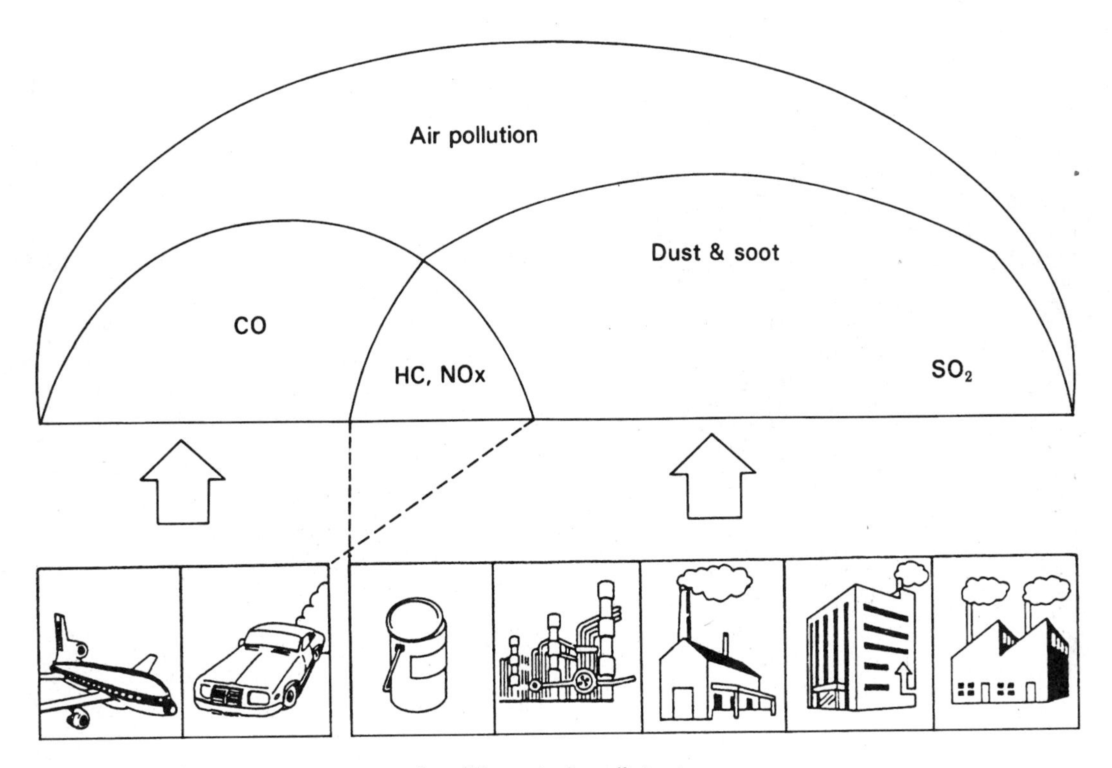

Figure 1-4 A variety of sources create the different air pollutants.

although automobiles are also important contributors. In many localities, electric generating plants and industrial burning are the principal sources, with space heating and refuse burning also adding sizeable quantities. Nature itself produces large quantities of these pollutants. Natural sources are responsible for large amounts of hydrocarbons, such as the terpene produced by certain types of growing trees (conifers). Similarly, a large quantity of methane hydrocarbons in the atmosphere comes from the decomposition of vegetable matter.

A relationship between automobile emissions and air pollution was first suggested in the U.S. by studies that began during the 1940s in California. Air pollution characterized by plant damage, eye and throat irritation, stressed rubber cracking, and decreased visibility were detected in Los Angeles as early as 1943.

By 1948, the California legislature established air pollution control districts empowered to curb emissions sources. Initial efforts addressed the reduction of industrial particulate emissions and resulted in improved visibility, but eye irritation and other smog symptoms remained.

A unique type of air pollution known as *photochemical* smog was first recognized in the Los Angeles Basin of California during the last half of the 1940s. The atmospheric requirements for smog formation are sunshine and still air. When the concentration of HC in the atmosphere becomes sufficiently high and NO_x is present in the correct ratio, the action of sunshine causes them to react chemically, forming photochemical smog.

When these conditions occur simultaneously with a thermal inversion (where warmer air above prevents upward movement of cooler air near the ground), smog can accumulate under this *lid,* all the way down to the ground within a few hours. The first effect of a smog buildup is reduced visibility. As the buildup of smog approaches ground level, its irritating effects on the eyes, nose, and throat are sensed.

As the phenomenon of smog gradually increased in intensity and frequency of occurrence in the Los Angeles area, it became the object of much investigation and research. In the early 1950s, it was learned that motor vehicles were major sources of these unburned hydrocarbons and oxides of nitrogen. Automobile manufacturers began an extensive research program to help California reduce its air pollution problem.

The question to be investigated was the precise source from which each air pollutant was emitted

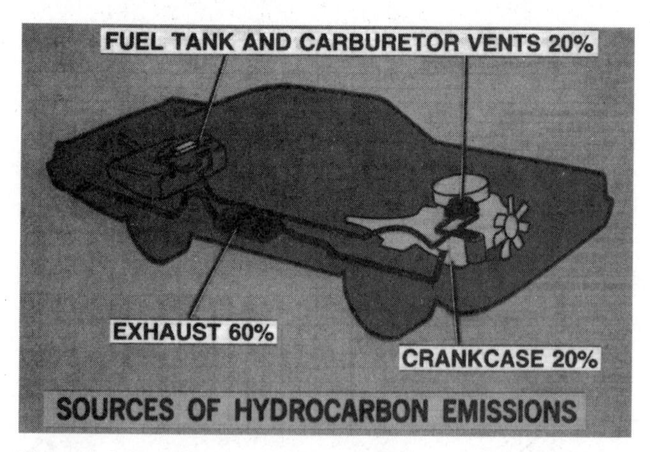

Figure 1-5 Three main sources of hydrocarbon emissions (courtesy Chrysler Corp.).

from an automobile (Figure 1-5). It was found that of the total hydrocarbons emitted:

- 60% escaped from the engine and out the tailpipe.
- 20% escaped from the engine crankcase.
- 20% evaporated through vents in the fuel tank filler cap and the carburetor.

Nitrogen oxides in several forms were found in exhaust gases. Carbon monoxide (CO), a toxic gas but not an ingredient of photochemical smog, also was found in exhaust gases.

During the early 1950s, the theory became accepted that the Los Angeles smog was formed by a complex chemical reaction that took place in the atmosphere. The U.S. government became involved with air pollution in 1955 by empowering the Department of Health, Education, and Welfare to provide technical assistance for resolution of problems related to the chemical reaction called air pollution.

The California legislature enacted air quality standards for oxidants and carbon monoxide (CO) in 1959, and the California Motor Vehicle Pollution Control Board was created in 1960 to implement these standards. Its function was to establish standards for vehicle exhaust and evaporative emissions and to certify that vehicles sold in California were meeting these standards.

To meet the air quality goals by 1970, it was estimated that an 80% reduction in motor vehicle HC emissions and a 60% reduction in motor vehicle CO emissions would be required. In 1961, the first emissions control requirement addressed crankcase blowby emissions and was required to be factory-installed on all new motor vehicles sold in California. In 1963, these devices were installed nationally on all new automobiles.

This activity was further enhanced by the Clean Air Act of 1963, which was structured to stimulate state and local air pollution control activity. In 1965, amendments to the Clean Air Act specifically authorized the writing of national standards for emissions. In 1966, all new motor vehicles sold in California were required to have such devices installed at the factory. The exhaust controls had to limit unburned hydrocarbons to 275 ppm and carbon monoxide to a concentration of 1.5% by volume.

These exhaust emissions standards were measured by the California Air Resources Board (CARB) using a seven-cycle operating mode test. The vehicle was stored overnight and then pushed onto a dynamometer to assure a cold start that allowed for testing the carburetor choke circuit. A tape-recorded set of instructions prompted the driver to accelerate and decelerate the engine to specified speeds. This assured that each vehicle was driven through the same modes of cold and hot engine driving conditions. The exhaust pollutants were collected and analyzed, and then the measurements were weighted and averaged by the computer for the seven driving modes.

In 1968, Congress enacted the Pure Air Act which recognized the work of the California Air Resources Board by adopting the CARB limits of 275 ppm for HC and 1.5% CO for engines of over 140-cubic-inch displacement.

The history of motor vehicle emissions control in the U.S. has involved both progressively reduced standards and improved test procedures. Beginning in 1968, the certification practice began to measure the mass emissions rates in grams per mile (g/mi). This was done to compensate for the size differences between engines. Using the original 275 ppm standard for unburned hydrocarbons, a 500-cubic-inch engine would emit approximately twice the amount of unburned hydrocarbons as a 250-cubic-inch engine. Both engines would still be within the established standards measured in concentration for HC and CO. The new limits for 1970 models were set at 2.2 g/mi of the total hydrocarbons (THC) and 23 g/mi CO for light-duty vehicles. In addition, motor vehicle evaporative emissions limits were certified at 6 grams per test (g/test) for 1971 models and 2 g/test for 1972 models (Figure 1-6).

MODEL YEAR	VEHICLE CATEGORY [a]	TAILPIPE HYDROCARBON	CARBON MONOXIDE	OXIDES OF NITROGEN	EVAPORATIVE HYDROCARBON	PARTICULATE	FUEL ECONOMY
1970 [b]	LDV, LDT	2.2 g/mi	23 g/mi	-	-	-	-
1971	LDV, LDT	2.2 g/mi	23 g/mi	-	6 g/test [c]	-	-
1972 [d]	LDV, LDT	3.4 g/mi	39 g/mi	-	2 g/test	-	-
1973	LDV, LDT	3.4 g/mi	39 g/mi	3.0 g/mi	2 g/test	-	-
1974 [e]	HDGV, HDDV	-	40 g/bhphr	-	-	-	-
1975 [f]	LDV	1.5 g/mi	15 g/mi	3.1 g/mi	2 g/test	-	-
	LDT	2.0 g/mi	20 g/mi	3.1 g/mi	2 g/test	-	-
1977	LDV	1.5 g/mi	15 g/mi	2.0 g/mi	2 g/test	-	-
1978	LDV	1.5 g/mi	15 g/mi	2.0 g/mi	6 g/test [g]	-	18 mpg
	LDT	2.0 g/mi	20 g/mi	3.1 g/mi	6 g/test	-	-
1979	LDV	1.5 g/mi	15 g/mi	2.0 g/mi	6 g/test	-	19 mpg
	LDT	1.7 g/mi	18 g/mi	2.3 g/mi	6 g/test	-	-
	HDGV, HDDV	1.5 g/bhphr	25 g/bhphr	-	-	-	-
1980	LDV	0.41 g/mi	7.0 g/mi	2.0 g/mi	6 g/test	-	20 mpg
1981	LDV	0.41 g/mi	3.4 g/mi	1.0 g/mi	2 g/test	-	22 mpg
	LDT	1.7 g/mi	18 g/mi	2.0 g/mi	2 g/test	-	-
1982	LDV	0.41 g/mi	3.4 g/mi	1.0 g/mi	2 g/test	-	24 mpg
1983	LDV	0.41 g/mi	3.4 g/mi	1.0 g/mi	2 g/test	-	26 mpg
1984	LDV	0.41 g/mi	3.4 g/mi	1.0 g/mi	2 g/test	-	27 mpg
	LDT	0.8 g/mi	10 g/mi	2.3 g/mi	2 g/test	-	-
1985	LDV	0.41 g/mi	3.4 g/mi	1.0 g/mi	2 g/test	0.6 g/mi	27.5 mpg
	LDT	0.8 g/mi	10 g/mi	2.3 g/mi	2 g/test	0.6 g/mi	-
[h]	HDGV (<14k)	2.5 g/bhphr	40.1 g/bhphr	-	3 g/test	-	-
[h]	(>14K)				4 g/test	-	-
[h]	HDDV	1.3 g/bhphr	15.5 g/bhphr	-	-	-	-
1987	LDV	0.41 g/mi	3.4 g/mi	1.0 g/mi	2 g/test	0.2 g/mi	27.5 mpg
	LDT	0.8 g/mi	10 g/mi	1.2 g/mi	2 g/test	0.26 g/mi	-
	HDGV (<14K)	1.3 g/bhphr	15.5 g/bhphr	6.0 g/bhphr	3 g/test	-	-
	(>14K)	2.5 g/bhphr	40 g/bhphr	6.0 g/bhphr	4 g/test	-	-
	HDDV	1.3 g/bhphr	15.5 g/bhphr	6.0 g/bhphr	-	-	-
1988	HDGV (<14k)	1.1 g/bhphr	14.4 g/bhphr	6.0 g/bhphr	3 g/test	-	-
	(>14K)	1.9 g/bhphr	37.1 g/bhphr	6.0 g/bhphr	4 g/test	-	-
	HDDV	1.3 g/bhphr	15.5 g/bhphr	6.0 g/bhphr	-	0.6 g/bhphr	-
1991	HDGV (<14K)	1.1 g/bhphr	14.4 g/bhphr	5.0 g/bhphr	3 g/test	-	-
	HDDV	1.3 g/bhphr	15.5 g/bhphr	5.0 g/bhphr	-	0.25 g/bhphr	-
1993	HDDV	1.3 g/bhphr	15.5 g/bhphr	5.0 g/bhphr	-	0.1(bus) g/bhphr	-
1994 [i]	LDV	0.25 g/mi	3.4 g/mi	0.4 g/mi	2 g/test	0.08 g/mi	27.5 mpg
[i]	LDT	0.32 g/mi	4.4 g/mi	0.7 g/mi	2 g/test	0.26 g/mi	-
	HDDV	1.3 g/bhphr	15.5 g/bhphr	5.0 g/bhphr	-	0.1 g/bhphr	-
1995 [j]	LDT	0.32 g/mi	4.4 g/mi	0.7 g/mi	2 g/test	0.08 g/mi	-
1998	HDDV	1.3 g/bhphr	15.5 g/bhphr	4.0 g/bhphr	-	0.1 g/bhphr	-

note: a_LDV = light-duty passenger car, LDT = light-duty truck, HDGV = heavy-duty gasoline truck/bus, HDDV = heavy-duty diesel truck/bus, g/mi = grams per mile, g/bhphr = grams per brake horsepower hour, mpg = miles per gallon; b_7-mode steady-state driving procedure; c_carbon trap technique; d_CVS-72 transient driving procedure; e_13-mode steady-state driving procedure; f_CVS-75 transient driving procedure; g_SHED technique; h_transient driving procedure; i_phased in from 1994 to 1996; j_PM phased in from 1995 to 1997

Figure 1-6 United States Motor Vehicle Emissions and Fuel Economy Standards (courtesy US EPA).

California's adoption of tighter emissions standards and controls has forced automotive manufacturers to build three types of vehicles to meet certification standards (Figure 1-7):

1. Vehicles for sale in all 50 states.
2. Vehicles for sale in 49 states and/or Canada.
3. Vehicles for sale in California.

These systems differ in engine applications, engine-transmission combinations, carburetor type and size, distributor spark advance curves, emissions controls, emissions standards, and tune-up specifications. A typical emissions decal is shown in Figure 1-8, page 10. The technician has to accurately identify the system in question in order to make the correct diagnosis and repair.

Studies conducted during the 1970 and since have revealed that large percentages of vehicles on roadways are exceeding the emissions limits for which they were designed. This is primarily because of inadequate maintenance by vehicle owners, and sometimes because owners or their mechanics intentionally disable the emission control systems.

One particularly effective way to control this problem has been to establish inspection and maintenance (I/M) programs. These programs were designed to identify vehicles in need of maintenance and to assure the effectiveness of their emission control systems.

Inspection and maintenance (I/M) and anti-tampering (ATP) programs were established to identify such vehicles and to require their repair at the owner's expense. Since 1970, manufacturers and dealers have been subject to a $10,000 fine per violation for tampering with a motor vehicle's emission control system. The 1977 Clean Air Act Amendments extended the tampering prohibition to include automotive repair and service facilities and fleet operators. Such a violation can result in a civil penalty of up to $2,500 for each motor vehicle tampered with. In addition, the 1990 amendment to the Clean Air Act contained laws and sanctions prohibiting vehicle owners from tampering with emission control systems.

Beginning in 1978, evaporative emissions were measured using the Sealed Housing for Evaporative Determination (SHED) technique. During the test, the entire vehicle is enclosed in a shed to collect emissions from all evaporative sources. The current evaporative emissions standard is 2 g/test.

Although significant reductions in the levels of new motor vehicle emissions have been experienced, the impact of air quality improvements has decreased. This decrease is related to an increase in the number of vehicles in operation, poor maintenance, and a large increase in the number of roadway miles traveled. Because of these and other observations, the U.S. Congress amended the Clean Air Act in 1990 to require more stringent new motor vehicle emissions standards. These amendments altered the procedures for identification and repair of motor vehicles tampered with or poorly maintained, and mandated the use of cleaner, more environmentally-benign alternatives to conventional petroleum-based fuels.

The original source of hydrocarbons and carbon monoxide emissions from an automobile is, of course, the fuel on which the engine operates. Recent studies suggest that the elimination of certain elements from the fuel and changes in certain characteristics of the fuel will help reduce some undesirable emissions from the vehicle.

FEDERAL TEST PROCEDURE (FTP)

The FTP is used to determine the compliance of light-duty vehicles (LDV) and light-duty trucks (LDT) with federal emissions standards. As designed, the FTP is intended to represent typical driving patterns in primarily urban areas.

Preproduction vehicles are tested using the FTP as part of the motor vehicle certification process. The certification process is used to establish that each vehicle is designed to comply with emissions standards for its full useful life. The FTP is also used to test production line and in-use vehicles for compliance with appropriate emissions standards.

The FTP is more than just a driving cycle. It provides a way to measure, consistently and repetitively, concentrations of HC, NO_x, CO, and CO_2 emissions that occur when a vehicle is driven over a simulated urban driving trip. Exhaust emissions are measured by driving the vehicle (placed on a dynamometer) on a simulated urban driving trip under two conditions: with a cold start designed to

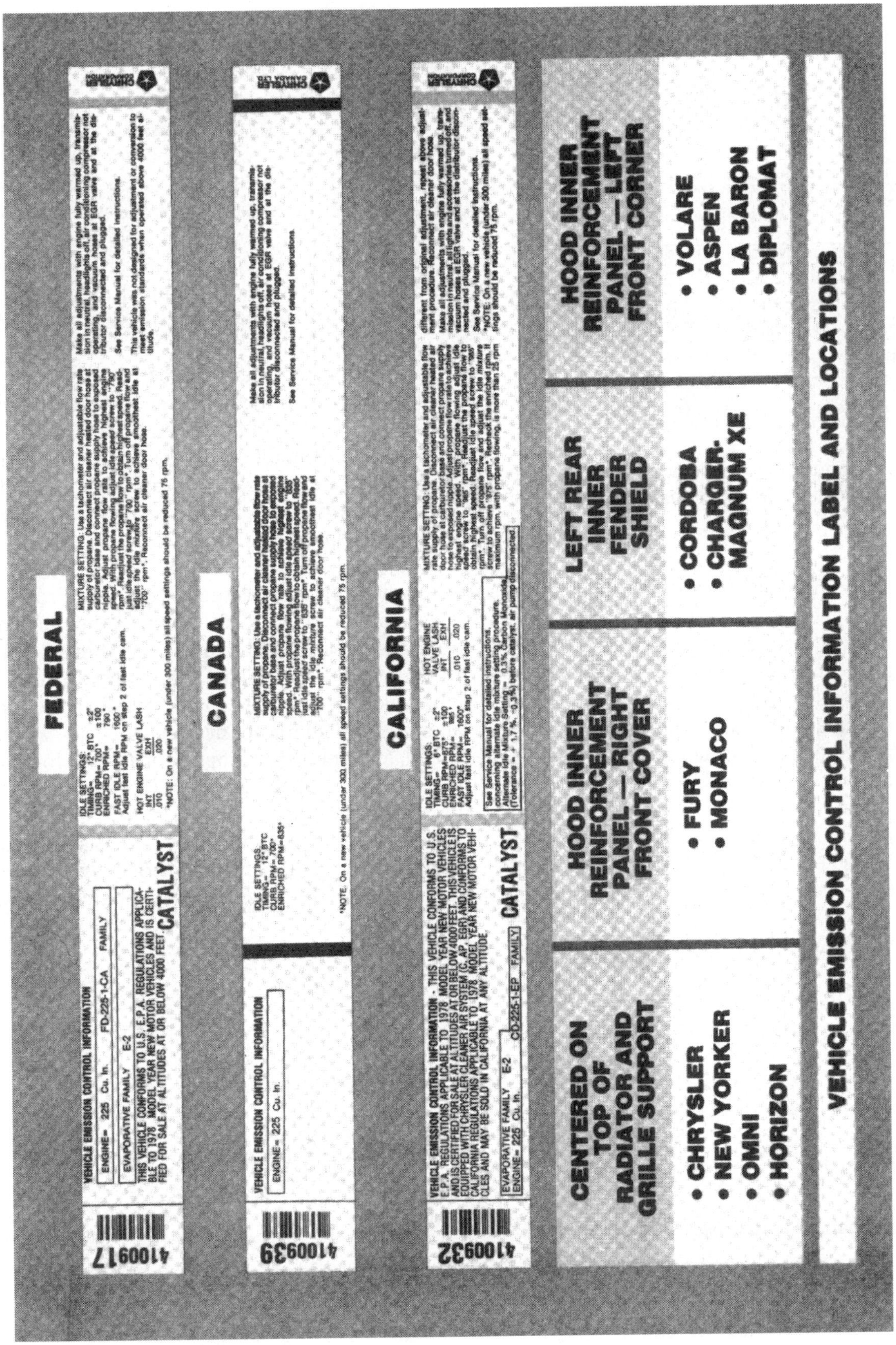

FEDERAL

4100917

VEHICLE EMISSION CONTROL INFORMATION

ENGINE= 225 Cu. In. FD-225-1-CA FAMILY

EVAPORATIVE FAMILY E-2

THIS VEHICLE CONFORMS TO U.S. E.P.A. REGULATIONS APPLICABLE TO 1978 MODEL YEAR NEW MOTOR VEHICLES AND IS CERTIFIED FOR SALE AT ALTITUDES AT OR BELOW 4000 FEET. CATALYST

IDLE SETTINGS:
TIMING= 12° BTC ±2°
CURB RPM= 700* ±100
ENRICHED RPM= 790*
FAST IDLE RPM= 1600*
Adjust fast idle RPM on step 2 of fast idle cam.

HOT ENGINE VALVE LASH
INT .010 EXH .020

MIXTURE SETTING: Use a tachometer and adjustable flow rate supply of propane. Disconnect air cleaner heated door hose at carburetor base and connect propane supply hose to exposed nipple. Adjust propane flow rate to achieve highest engine speed. With propane flowing adjust idle speed screw to "790" rpm*. Readjust the propane flow to obtain highest speed. Readjust idle speed screw to "790" rpm*. Turn off propane flow and adjust the idle mixture screw to achieve smoothest idle at "700" rpm*. Reconnect air cleaner door hose.

Make all adjustments with engine fully warmed up, transmission in neutral, headlights off, air conditioning compressor not operating, and vacuum hoses at EGR valve and at the distributor disconnected and plugged.

See Service Manual for detailed instructions.

This vehicle was not designed for adjustment or conversion to meet emission standards when operated above 4000 feet altitude.

*NOTE: On a new vehicle (under 300 miles) all speed settings should be reduced 75 rpm.

CHRYSLER CORPORATION

CANADA

4100939

VEHICLE EMISSION CONTROL INFORMATION

ENGINE= 225 Cu. In.

IDLE SETTINGS:
TIMING= 12° BTC
CURB RPM= 700*
ENRICHED RPM=835*

MIXTURE SETTING: Use a tachometer and adjustable flow rate supply of propane. Disconnect air cleaner heated door hose at carburetor base and connect propane supply hose to exposed nipple. Adjust propane flow rate to achieve highest engine speed. With propane flowing adjust idle speed screw to "835" rpm*. Readjust the propane flow to obtain highest speed. Readjust idle speed screw to "835" rpm*. Turn off propane flow and adjust the idle mixture screw to achieve smoothest idle at "700" rpm*. Reconnect air cleaner door hose.

Make all adjustments with engine fully warmed up, transmission in neutral, headlights off, air conditioning compressor not operating, and vacuum hoses at EGR valve and at the distributor disconnected and plugged.

See Service Manual for detailed instructions.

*NOTE. On a new vehicle (under 300 miles) all speed settings should be reduced 75 rpm.

CHRYSLER CANADA LTD.

CALIFORNIA

4100932

VEHICLE EMISSION CONTROL INFORMATION - THIS VEHICLE CONFORMS TO U.S. E.P.A. REGULATIONS APPLICABLE TO 1978 MODEL YEAR NEW MOTOR VEHICLES AND IS CERTIFIED FOR SALE AT ALTITUDES AT OR BELOW 4000 FEET. THIS VEHICLE IS EQUIPPED WITH CHRYSLER CLEANER AIR SYSTEM (C. AP. EGR) AND CONFORMS TO CALIFORNIA REGULATIONS APPLICABLE TO 1978 MODEL YEAR NEW MOTOR VEHICLES AND MAY BE SOLD IN CALIFORNIA AT ANY ALTITUDE.

EVAPORATIVE FAMILY E-2
ENGINE= 225 Cu. In. CD-225-1-EP FAMILY

CATALYST

IDLE SETTINGS:
TIMING= 6° BTC ±2°
CURB RPM=875* ±100
ENRICHED RPM= 985*
FAST IDLE RPM= 1600*
Adjust fast idle RPM on step 2 of fast idle cam.

HOT ENGINE VALVE LASH
INT .010 EXH .020

See Service Manual for detailed instructions, concerning alternate idle mixture setting procedure. Alternate Idle Mixture Setting = 0.3% Carbon Monoxide (Tolerance = + 1.7 %, -0.3%) before catalyst, air pump disconnected.

MIXTURE SETTING: Use a tachometer and adjustable flow rate supply of propane. Disconnect air cleaner heated air door hose at carburetor base and connect propane supply hose to exposed nipple. Adjust propane flow rate to achieve highest engine speed. With propane flowing adjust idle speed screw to "985" rpm*. Readjust the propane flow to obtain highest speed. Readjust idle speed screw to "985" rpm*. Turn off propane flow and adjust the idle mixture screw to achieve "875" rpm*. Recheck the enriched rpm. If maximum rpm, with propane flowing, is more than 25 rpm different from original adjustment, repeat above adjustment procedure. Reconnect air cleaner door hose.

Make all adjustments with engine fully warmed up, transmission in neutral, all lights and accessories turned off, and vacuum hoses at EGR valve and at the distributor disconnected and plugged.

See Service Manual for detailed instructions.

*NOTE: On a new vehicle (under 300 miles) all speed settings should be reduced 75 rpm.

CENTERED ON TOP OF RADIATOR AND GRILLE SUPPORT

- CHRYSLER
- NEW YORKER
- OMNI
- HORIZON

HOOD INNER REINFORCEMENT PANEL — RIGHT FRONT COVER

- FURY
- MONACO

LEFT REAR INNER FENDER SHIELD

- CORDOBA
- CHARGER-MAGNUM XE

HOOD INNER REINFORCEMENT PANEL — LEFT FRONT CORNER

- VOLARE
- ASPEN
- LA BARON
- DIPLOMAT

VEHICLE EMISSION CONTROL INFORMATION LABEL AND LOCATIONS

Figure 1-7 Emissions standards require manufacturers to build vehicles for three types of certifications (courtesy Chrysler Corp.).

VEHICLE EMISSION CONTROL INFORMATION - ACURA NSX

ENGINE FAMILY MHN3.0V5F3AX
EVAPORATIVE FAMILY 91FK
DISPLACEMENT 2977cm³
182CID

TWC HO2S
EGR SMPI

CATALYST

TUNE UP SPECIFICATIONS

TUNE UP CONDITIONS:
ENGINE AT NORMAL OPERATING TEMPERATURE
ALL ACCESSORIES TURNED OFF, COOLING FAN OFF
TRANSMISSION IN NEUTRAL
ADJUSTMENTS TO BE MADE IN ACCORDANCE WITH INDICATIONS GIVEN IN SHOP MANUAL.

IDLE SPEED	5 SPEED TRANSMISSION	800±50 rpm
	AUTOMATIC	750±50 rpm
IGNITION TIMING AT IDLE		15±2° BTDC
VALVE LASH	SETTING POINTS BETWEEN CAMSHAFT AND ROCKER ARM	IN. 0.17±0.02mm COLD
		EX. 0.19±0.02mm COLD

NO OTHER ADJUSTMENTS NEEDED.

THIS VEHICLE CONFORMS TO U.S. EPA AND STATE OF CALIFORNIA REGULATIONS APPLICABLE TO 1991 MODEL YEAR NEW MOTOR VEHICLES.

Figure 1-8 Tune-up specifications will vary depending on the certification type shown at the bottom of the decal.

represent a morning start-up after a long soak (a period of non-use) and with a hot start that takes place while the engine is still hot.

Evaporative emissions are determined using the Sealed Housing for Evaporative Determination (SHED) technique. The test measures evaporative emissions after heating the fuel tank to simulate heating by the sun (the diurnal test) and again after the car has been driven and parked with a hot engine (the hot soak test).

The FTP also encompasses all factors used for vehicle testing, such as fuel, vehicle preconditioning, ambient temperature and humidity, aerodynamic loss, and vehicle inertia simulations. Besides evaporative and exhaust emissions, the FTP is also used in evaluating fuel economy.

The FTP driving cycle was designed to simulate a vehicle operating over a road route in Los Angeles that represented a typical home-to-work commute. The original road route was developed by trial and error. It tried to match the engine operating mode distribution (based on manifold vacuum and RPM ranges) obtained in central Los Angeles using the same test vehicle with a variety of drivers and routes.

Using an instrumented 1964 Chevrolet, recordings were made of actual home-to-work commute trips by employees of the state of California's Vehicle Pollution Laboratory. By trial and error, a specific street route near the lab was found that matched the average speed/load distribution on the commute trips. The 12-mile route was called the *LA4* (Figure 1-9).

To develop an improved Federal Test Procedure (based on speed-time distributions rather than manifold vacuum and RPM ranges), six different drivers from the EPA's West Coast Laboratory drove a 1969 Chevrolet over the LA4 route. The six traces were analyzed for idle time, average speed, maximum speed, and number of stops per trip. The total time required for the six trips ranged from 35 to 40 minutes, with an average of 37.6 minutes. Of the six traces, one varied significantly and was discarded. The other five traces were surprisingly similar. Of those five, the trace with the actual time closest to

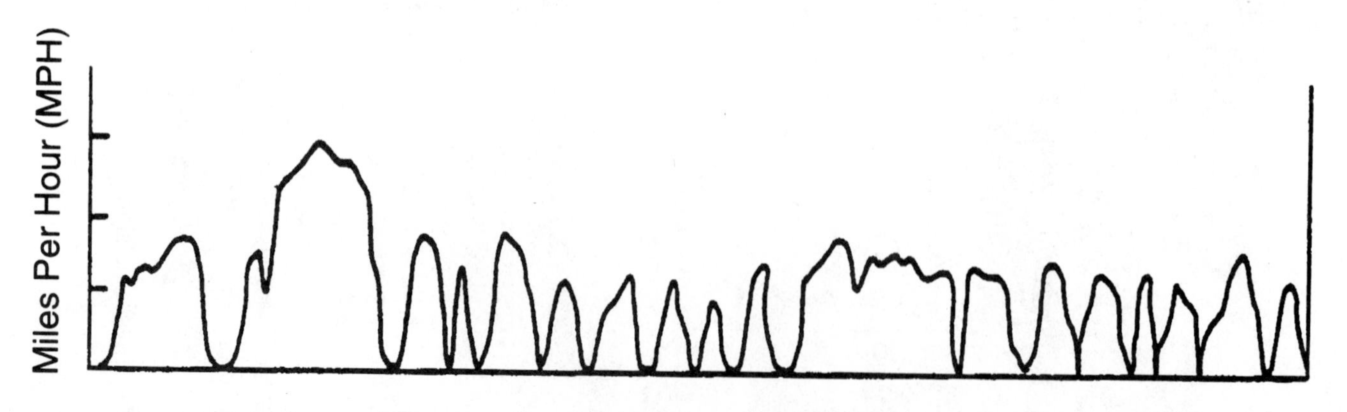

Figure 1-9 LA4 speed/load distribution.

the average was selected as the most representative speed-time trace. That trace contained 28 "hills" of non-zero speed activity separated by idle periods and had an average speed of 19.2 miles per hour.

Based on a 1969 report on driving patterns in Los Angeles, the average trip length was estimated to be 7.5 miles. Several hills were eliminated to shorten the test cycle to 7.5 miles while maintaining the same average speed. The shortened route, designated the *LA4-S3,* was 7.486 miles in length with an average speed of 19.8 mph. Slight modifications to some speed-time profiles have been made in cases where the acceleration or deceleration rate exceeded the 3.3 mph/s limit of the belt-driven chassis dynamometers. Mass emissions tests comparing the shortened cycle to the full cycle showed very high correlation. The final version of the cycle was designated the *LA4-S4* cycle and is 7.46 miles in length with an average speed of 19.6 mph.

This cycle is now commonly referred to as the *LA4* or the Urban Dynamometer Driving Schedule (UDDS). It has been the standard driving cycle for the certification of LDVs and LDTs since the 1972 model year. Beginning with the 1975 model year, the cycle was modified to repeat the initial 505 seconds of the cycle (Figure 1-10) following a 10-minute soak at the end of the cycle. This allows emissions to be collected on a hot start (when the engine is still warm) as well as after a cold start and during operation. The test then provides a more accurate reflection of typical customer service than running just one 7.46-mile cycle from a cold start.

The 1990 amendment to the Clean Air Act contains a large number of provisions to further improve ambient air quality. This amendment requires that the EPA review its regulations for the testing of motor vehicles and revise them if necessary to ensure that motor vehicles are tested under circumstances reflecting actual current driving conditions.

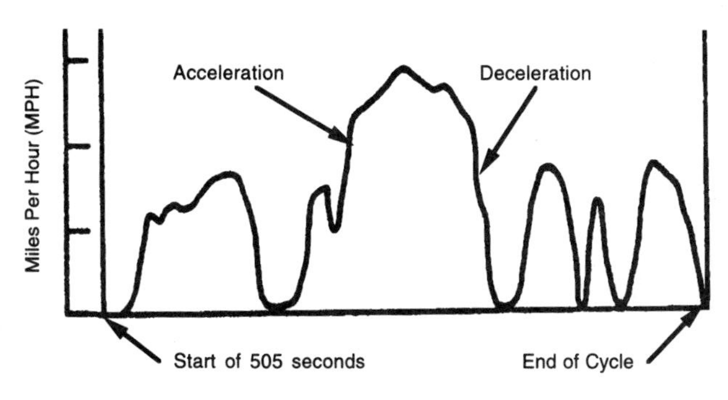

Figure 1-10 FTP 505-second drive cycle.

CONTROLLING AUTOMOTIVE EMISSIONS

Automotive emission control systems have been in use since 1961. Generally, California was the first to implement the use of many emissions systems. Federally-certified motor vehicles were usually one to two years behind.

Technicians must have a good working knowledge of the various emissions systems in order to maintain and troubleshoot them. The following is designed as an overview of the most common groups of emission control systems. This material will include the:

1. Basic purpose of each system.
2. Types of systems most commonly used.
3. Year introduced.
4. Types of emission(s) most affected by each system.
5. Failure symptoms.

Positive Crankcase Ventilation (PCV)

Crankcase emissions are responsible for 20% of the total emissions produced by an internal combustion engine. The main function of the crankcase ventilation system is to allow the crankcase to breathe. Blowby gases (HC) that escape by the piston rings into the crankcase are removed to prevent a pressure buildup. They are recirculated to the intake manifold and added to the incoming air/fuel mixture. As blowby gases are removed from the crankcase, fresh air is drawn in to prevent a vacuum. This system reduces the amount of fuel vapors entering the atmosphere, sludge in the crankcase, and varnish buildup on engine components.

Types of systems most commonly used:

1. *Type One* — Open system vented to the atmosphere
2. *Type Two* — Restricted (often used as a retrofit)
3. *Type Three* — Free flow (no PVC valve is used)
4. *Type Four* — Closed or sealed system

Year introduced:

1. California = 1961
2. Federal = 1963
3. 1968 required all new vehicles be equipped with a Type Four closed system.

Emissions controlled:

1. Hydrocarbons (HC)

Failure symptoms:

1. Excessive vapors escaping into the atmosphere.
2. Oil in the air cleaner.
3. *Rich Mixture* — Excessive vapors or plugged valve.
4. *Lean Mixture* — Cracked hose or bad valve.
5. Excessive amount of blue smoke at the tailpipe.

Air Injection Reaction System (AIR)

Exhaust emissions account for approximately 60% of the pollutants released into the atmosphere.

The Air Injection Reaction System (AIR) injects a controlled amount of fresh air into the exhaust stream to oxidize HC and CO emissions. This system might also supply fresh air to the catalytic converter to assist the oxidizing converter. On vehicles equipped with an oxygen sensor, this system will assist in bringing the sensor up to operating temperature in a shorter time.

Types of systems most commonly used:

1. Pump type
2. Pulse air

Year introduced:

1. Domestic = 1966
2. Import = 1970

Emissions controlled:

1. Hydrocarbons (HC)
2. Carbon monoxide (CO)

Failure symptoms:

1. High HC and CO readings, but engine performs okay.
2. Noisy pump, belt, or exhaust leaks.
3. Backfire, especially on deceleration.
4. Damaged catalytic converter due to lack of oxygen or excessive hydrocarbons.
5. Excessively lean oxygen sensor voltage.

Thermostatic Air Cleaner (TAC)

This system regulates incoming air temperature through all driving conditions. Heating of the incoming air charge will result in a faster warm-up and improved vaporization of the fuel during cold engine operation. By using heat rising from the exhaust manifold, it can maintain a steady inlet air temperature.

Types of systems most commonly used:

1. Vacuum motor and bi-metallic sensor
2. Thermostatic bulb

Year introduced:

1. Domestic = 1966, limited use until 1968
2. Import = 1970

Emissions controlled:

1. Hydrocarbons (HC)
2. Carbon monoxide (CO)

Failure symptoms:

1. Vacuum motor door fully open on cold engine results in poor driveability when the engine is cold.
2. Vacuum motor door fully closed on hot engine can cause pinging because of the high inlet temperatures.

Evaporative Emissions Controls (EVP or EEC)

Fuel vapors lost to the atmosphere account for 20% of the hydrocarbon emissions in our atmosphere today. This system is designed to collect and store fuel vapors from the carburetor float bowl and fuel tank when the engine is not running. With the vehicle running, this system purges stored vapors to the intake manifold to mix with the incoming air/fuel.

Types of systems most commonly used:

1. Crankcase storage
2. Charcoal canister

Year introduced:

1. California = 1970
2. Federal = 1971

Emissions controlled:

1. Hydrocarbons (HC)

Failure symptoms:

1. Excessive vapors escaping into the atmosphere.
2. Excessive fuel tank pressure.
3. No fuel pressure or a collapsed tank.
4. Hard starting of a hot engine.
5. *Rich Mixture* — Bad purge valve or loaded canister.

Catalytic Converters (OC = 2-Way)

The primary function of the converter is to clean up exhaust emissions. This is accomplished through the use of a catalyst made of platinum or palladium. The catalyst combined with the proper amount of oxygen creates a chemical reaction, thus converting HC and CO to harmless water (H_2O) and carbon dioxide (CO_2).

Types of systems most commonly used:

1. Pellet type
2. Monolithic type

Year introduced:

1. 1974 = Limited use
2. 1975 = Included California and many federal vehicles

Emissions controlled:

1. Hydrocarbons (HC)
2. Carbon monoxide (CO)

Failure symptoms:

1. High HC and CO readings and engine runs okay.
2. No start or hard to start, lack of power, and low engine vacuum due to a plugged converter.
3. Warning lamp on due to high converter temperatures.
4. Catalyst material blown out the tailpipe.

Exhaust Gas Recirculation (EGR)

This valve recirculates inert exhaust gases back into the intake manifold to dilute the incoming air charge. The recirculation of inert gases causes the air/fuel ratio to move to the rich side. Since a rich mixture tends to cool, this has the effect of lowering cylinder temperatures. EGR is necessary on a warm engine, during light to moderate engine load. The main differences between systems is when and how the valve is opened under load or blocked during cold engine operation.

Types of systems most commonly used:

1. *Floor Jet* — Intake manifold below throttle plates
2. *Diaphragm Type* — Ported or venturi vacuum
3. *Back Pressure Type* — Positive and negative
4. *Digital* — Computer-controlled solenoid(s)

Year introduced:

1. 1971 = California Chrysler
2. 1973 = Other applications

Emissions controlled:

1. Oxides of nitrogen (NO_x)

Failure symptoms:

1. High NO_x readings under load.
2. If stuck open, can cause rough idle or cold engine stumble.
3. If stuck closed, can cause pinging under load.

4. Lean air/fuel ratio under load if the valve does not open.

Spark Controls

In addition to the EGR valve, controlling ignition timing is another way to lower cylinder temperatures and decrease NO_x. Delaying or reducing spark advance lowers combustion chamber temperatures and pressures, which in turn reduce NO_x. Spark-controlled systems eliminate or modify vacuum advance during low speed operation. This can be done through solenoids or switches located between the carburetor vacuum port and the distributor vacuum advance. Total control of spark timing can now be done electronically by the computer using a network of input sensors.

Types of systems most commonly used:

1. Transmission-controlled switches
2. Vacuum delay valves
3. Coolant temperature override switches
4. Retrofit devices
5. Computer-controlled electronic spark

Year introduced:

1. 1966

Emissions controlled:

1. Oxides of nitrogen (NO_x)

Failure symptoms:

1. High NO_x or HC readings.
2. Lack of power, sluggish performance, and poor gas mileage.
3. Engine pinging and overheating.

Catalytic Converters (TWC = 3-Way)

In addition to the benefits of a two-way catalyst, a three-way catalyst also allows for the reduction of NO_x to harmless nitrogen and oxygen. This converter may be located in front of the two-way converter (monolithic) or in the top part of the two-way converter (dual bed). It may also be a separate converter located ahead of the two-way catalyst.

Types of systems most commonly used:

1. *Pellet Type* — Dual bed
2. *Monolithic Type* — Honeycomb

Year introduced:

1. 1977 = Volvo certified for California
2. 1978 = Other applications

Emissions controlled:

1. Hydrocarbons (HC)
2. Carbon monoxide (CO)
3. Oxides of nitrogen (NO_x)

Failure symptoms:

1. High HC and CO, or NO_x readings and engine runs okay.
2. No start, lack of power, and low engine vacuum due to a plugged converter.
3. Warning lamp on due to high converter temperatures.
4. Catalyst material blown out the tailpipe.

Technician's Role in Assuring Clean Air

Well-trained technicians can insure that more emissions reductions are realized through properly performed emissions-related maintenance. In addition to emissions-related repair knowledge, the technician should also be aware of the tampering laws.

Both federal and state governments have laws regarding tampering with vehicle emission control systems. Tampering is defined as removing, disconnecting, or modifying in any way that makes the device or system inoperable. Tampering might include:

1. Disconnecting vacuum lines.
2. Removing or rendering inoperative such components as the catalytic converter, air pump, or EGR.
3. Adjusting the vehicle outside the specifications set forth by the manufacturer.

4. Modifying a system through the use of components not certified for the vehicle.

The technician can help to insure effective operation of emission control systems by:

1. Emphasizing to the customer the importance of performing regularly scheduled preventive maintenance.
2. Performing a thorough visual inspection and looking for emissions devices that are missing, disconnected, or modified.

The following list includes some things to look for when performing a visual inspection of the various emission control systems:

1. *Positive Crankcase Ventilation* — Check to see if the PCV system is installed as required. Verify that the valve, required hoses, connections, flame arresters, etc. are present, routed properly, and in serviceable condition. Verify that the appropriate oil filler cap and dipstick are installed.
2. *Air Injection (AIR)* — Check the following systems:
   a. Air Pump Systems:
      Examine the air pump for a missing or disconnected belt, check valve(s), diverter valve(s), distribution hoses, and vacuum signal line(s).
   b. Pulse Air Systems:
      Inspect for the presence of required check valve(s) and air distribution manifolds.
   c. Air Pump and Pulse Air Systems:
      Check the air injection system for proper hose routing. Charred delivery hoses indicate defective check valves.
3. *Thermostatic Air Cleaner (TAC)* — Check to see that the required heat stoves, delivery pipes, vacuum hoses, and air cleaner components are present and installed properly. Verify that any required thermostatic vacuum switches are in place and the hoses are installed and in good condition.

   Check to see that the air cleaner lid is installed properly. Also, check for an oversized filter element and additional holes in the housing because a modified air cleaner system may affect the PCV and Fuel Evaporation systems.
4. *Fuel Evaporation System* — Check for the presence of the vapor storage canister or crankcase storage connections when required. Verify that required hoses, solenoids, etc. are present and connected properly. Check for the proper type of fuel tank cap. Check any non-OEM or auxiliary fuel tanks for compliance and the required number of evaporation canisters.
5. *Exhaust Gas Recirculation (EGR)* — Verify that the EGR valve is present and not modified or purposely defeated. Check to see that any thermal vacuum switches, pressure transducers, speed switches, computer-operated solenoids and vacuum regulating valves, etc. are present and are not bypassed or modified. Verify that vacuum hoses and wiring are installed and the hoses are not plugged. Verify that EGR passages are free and open.
6. *Spark Controls* — Use vacuum diagrams to confirm that vacuum hoses and wiring which connect to the distributor, carburetor, intake manifold, spark delay valves, thermal vacuum switches, etc. are in place and routed properly.
7. *Catalytic Converter* — Visually check for the presence of the catalytic converter(s). Check for external damage such as severe dents, removed or damaged heat shields, etc. Also, check for pellets or pieces of the converter in the tailpipe. If damage is found, report it to the vehicle owner. Check for the presence of any required air supply systems for the oxidizing section of the converter.

Progress in Reducing Vehicle Emissions

Extensive research and development have enabled automotive manufacturers to achieve substantial reductions in vehicle exhaust emissions. Now, there are effective control systems for all major sources of emissions from automobiles. Today, crankcase emissions and evaporative emissions (from the fuel tank and carburetor), which represented approximately 40 percent of automotive emissions, have been virtually eliminated.

The cornerstone of the Clean Air Act is the effort to attain and maintain National Ambient Air Quality Standards (NAAQS). Regulation of emissions from on-highway and stationary sources prior to enactment of the Clean Air Act Amendments (CAAA) of 1990 has resulted in significant emissions reductions from these sources. However, many air quality regions have failed to attain the

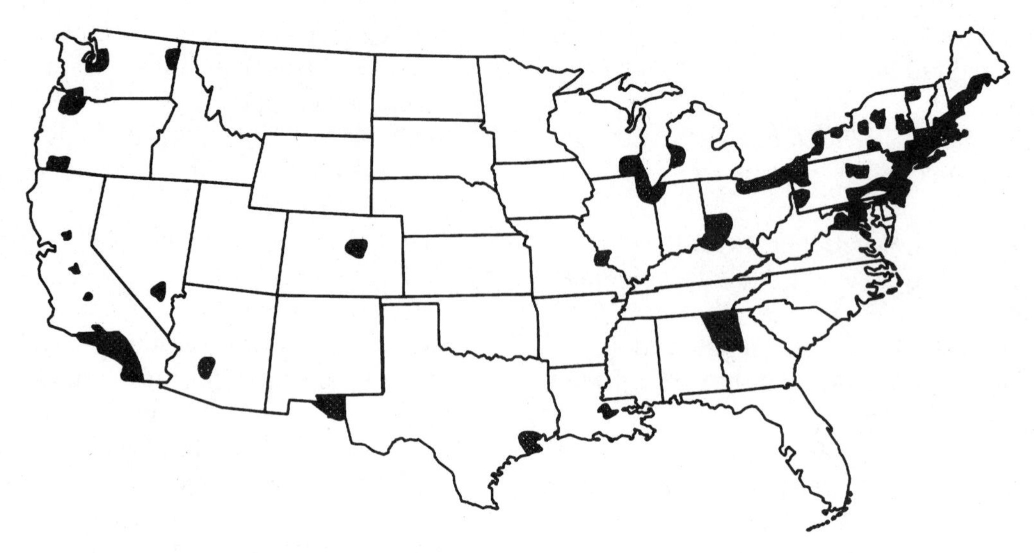

Figure 1-11 United States non-attainment areas.

NAAQS for ozone and CO. This is due to many factors, including the number of vehicles on the road and a corresponding increase in the number of miles driven by the in-use fleet.

While significant progress has been made in controlling motor vehicle emissions, as of August 1990, 96 air quality control regions still failed to meet the National Ambient Air Quality Standard (NAAQS) for ozone, and 41 regions failed to attain the NAAQS for CO. Areas across the country that cannot meet federal clean air standards are called *non-attainment areas* and are shown in Figure 1-11.

CHAPTER 2

Guidance for Inspection and Maintenance Programs

The control of motor vehicle emissions is crucial to the attainment of clean air. While motor vehicle manufacturers have made tremendous advances in emissions control technology, cars and trucks still account for about half of the ozone problem and nearly all of the carbon monoxide air pollution in the United States. 10 to 30% of the vehicles cause approximately 80% of the problem. Inspection and maintenance (I/M) programs are designed to reduce mobile source emissions by identifying those vehicles in need of engine or emissions control repairs.

The Environmental Protection Agency (EPA) has held oversight and policy development responsibility for I/M programs since the passage of the 1977 amendments to the Clean Air Act. I/M was then mandated for certain areas with long-term air quality problems. The 1990 amendments to the Clean Air Act expanded the role of I/M programs as an attainment strategy by requiring the EPA to develop different performance standards for basic and enhanced I/M programs.

On November 5, 1992, the EPA amended the Clean Air Act to include the inspection and maintenance program's *Final Rule*. This rule established standards and other requirements for basic and enhanced vehicle inspection and maintenance (I/M) programs.

The EPA revised the Final Rule in September 1994 and issued a supplemental guidance for I/M programs entitled: "Vehicle Repair, Technical Assistance, Performance Monitoring, and Technician Education and Certification." This document is intended to be a resource for state and local inspection and maintenance (I/M) planners and administrators. They, in turn, are asked to share it with automotive technician educators, citizen groups, repair industry professional groups, interested local press, individual technicians, etc.

The rest of this chapter covers the parts of this EPA document that will impact the customer, technician, and the entire automotive repair industry. It is important that these regulations be understood because an educated technician can help the consumer through the process. If the technician and the consumer are both educated about the program, the program should run much smoother.

I/M STATUTORY REQUIREMENTS

The Clean Air Act Amendments (CAAA) of 1990 require enhanced I/M in areas with the worst air quality problems and in the Northeast Ozone Transport Region. The Clean Air Act Amendments also require the EPA to issue guidance for I/M programs, which states must comply with in all respects. The guidance, which must be incorporated by each state into its State Implementation Plan (SIP), provides each state with the continued flexibility necessary to establish a reasonable and fair program for the affected consumer.

The Inspection/Maintenance Program Requirements issued on November 5, 1992 define the SIP requirements as follows:

"The SIP is to include a description of the technical assistance program to be implemented, a description of the procedures and criteria to be used in meeting the performance monitoring requirements of this section, and a description of the repair technician training resources available in the community."

Part of the CAAA requires guidance for enhanced I/M areas which includes a performance standard achievable by a program combining emissions testing and inspection criteria. This will include administration features necessary to reasonably assure that adequate management resources, tools, and practices are in place to attain and maintain the performance standard.

The CAAA further specifies that each enhanced I/M program will include, at a minimum:

1. Computerized emissions analyzers.
2. On-road testing devices.
3. Denial of waivers for warranted vehicles or repairs related to tampering.
4. A $450 expenditure to qualify for waivers for emissions-related repairs that are not covered by warranty.
5. Enforcement through registration denial unless an existing program with a different mechanism can be demonstrated to have greater effectiveness.
6. Annual inspection unless a state can demonstrate that decentralized testing is equally effective.
7. Inspection of the emission control diagnostic system.

I/M REGULATORY REQUIREMENTS

Public Awareness

The I/M Final Rule contains public awareness requirements that must be met in both basic and enhanced I/M areas, and requires that the SIP include a plan for informing the public about various I/M-related issues. It also requires that motorists failing I/M be provided with information on results of repairs performed by facilities in the area, as well as diagnostic information on the parts of the test that were failed by their vehicle. The rule states that the SIP must include a plan for informing the public, on an ongoing basis, throughout the life of the I/M program about:

1. The air quality problem.
2. The role of motor vehicles in the air quality problem.
3. How to maintain a vehicle in a low-emissions condition.
4. How to find a qualified repair technician.

Motorists that fail the I/M test in enhanced I/M areas will be given a list of repair facilities in the area and information on the results of repairs performed by those repair facilities. Motorists that fail the I/M test will also be provided with software-generated, interpretive diagnostic information based on the portions of the test that were failed.

Technical Assistance

The I/M Final Rule contains a technical assistance requirement, which must also be met in basic and enhanced I/M areas. This requirement states that the oversight agency must provide the repair industry with information and assistance related to vehicle inspection, diagnosis, and repair. There is also a requirement that each state provide a technician repair hotline. It reads as follows:

"The oversight agency shall provide the repair industry with information and assistance related to vehicle inspection, diagnosis, and repair."

1. The agency will regularly inform repair facilities of changes in the inspection program, training course schedules, common problems being found with particular engine families, diagnostic tips, etc.
2. The agency must provide a hotline service to assist repair technicians with specific repair problems, answer technical questions that arise in the repair process, and answer questions related to the legal requirements of state and federal law with regard to emissions control device tampering, engine switching, or similar issues.

Performance Monitoring

The performance monitoring requirement under the Final Rule is required in enhanced areas, although it is encouraged in basic areas as well. This requirement reads as follows:

1. In enhanced I/M program areas, the oversight agency will monitor the performance of individual motor vehicle repair facilities, and provide to the public, at the time of initial failure, a summary of the performance of local repair facilities that have repaired vehicles for retest.
2. Programs will provide feedback, including statistical and qualitative information to individual repair facilities on a regular basis (at least annually) regarding their success in repairing failed vehicles.
3. A prerequisite for a retest will be a completed repair form that indicates what repairs were performed, as well as any technician-recommended repairs that were not performed, and identification of the facility that performed the repairs.

Technician Education

The Inspection/Maintenance Program Requirements issued November 5, 1992 also states that the state shall assess the availability of adequate repair technician training in the I/M area and, if the required types of training are not currently available, insure that training is made available to all interested individuals in the community either through private or public facilities. This may involve working with local community colleges or vocational schools to add curricula to existing programs or to start new programs. It might also involve attracting private training providers to offer classes in the area. The training available should include:

1. Diagnosis and repair of malfunctions in computer-controlled, closed-loop vehicles.
2. The application of emissions control theory and diagnostic data to the diagnosis and repair of failures on the transient emissions test and the evaporative system functional checks.
3. Utilization of diagnostic information on systematic or repeated failures observed in the transient emissions test and the evaporative system functional checks.
4. General training on the various subsystems related to engine emission control systems.

Mandatory technician certification is not a requirement of the I/M rule but adequate training must be made available in both basic and enhanced I/M areas. The rule does not require the state to conduct training or require technician or facility certification. Instead, the rule allows each state significant flexibility in insuring these minimum levels of training are available.

Start-Up IM240 Emissions Standards in Enhanced I/M Areas

The I/M Final Rule allows areas to use higher IM240 emissions standards at the start of the enhanced I/M program. The EPA has issued technical guidance and has recommended start-up standards for various vehicle classes. These should be used during the calendar years 1995 and 1996. Looser start-up standards are recommended because when high-tech testing starts, the fleet will have a sizable accumulation of vehicles (generally older than 5 years) that have suffered problems which went undetected by the idle test or which escaped repair due to improper inspections. Looser start-up standards allow I/M programs to send only a manageable number of these cars for repair in the first two years. This should prevent repair facilities from being overwhelmed by these challenging repair cases and owners from becoming overly frustrated.

By 1997, repair facilities will have more experience repairing IM240 failed vehicles. In addition, more may have entered the I/M repair business, and much of the accumulated backlog of defective vehicles will have already been repaired, allowing the IM240 standards to be tightened for greater emissions reduction. The EPA advises states to retain as much flexibility as possible in revising inspection standards during 1995 and 1996, since local experience may indicate the advisability of either more or less stringent start-up standards than shown in the EPA's Technical Guidance.

Hardship Extension Option

The CAAA requires that any waiver in an enhanced I/M program be granted only if the owner has spent at least $450 on repairs, but does not define precisely what a waiver is. However, the I/M Final Rule defines a waiver as "a form of compliance with the program requirements that allows a motorist to comply without meeting the applicable test standards, as long as prescribed criteria are met." One of the prescribed criteria set forth by the EPA is:

"A time extension, not to exceed the period of the inspection frequency, may be granted to obtain needed repairs on a vehicle in the case of economic hardship when waiver requirements have not been met, but the extension may be granted only once for each vehicle and will be tracked and reported by the program."

Thus, an extension allows vehicle owners to register their vehicle even though it has not passed the I/M test and has not had the full $450 in repair attempts. The EPA believes it is reasonable to allow extensions because of the accumulation of defects in older vehicles during the period before enhanced I/M. There may be only one extension in the life of a vehicle. States must modify their registration system to ensure that vehicles which have received an extension are not given a second one in a subsequent year.

Warranty

The Clean Air Act establishes the emissions design and defect warranty. Manufacturers must warrant to the purchaser that the vehicle is designed, built and equipped to meet emissions standards for the useful life of the vehicle and that it will be free of certain defects that could cause a failure to meet emissions standards. The CAAA requires manufacturers to pay for certain repairs necessary to pass an EPA-approved state or local emissions test.

The warranty period for light-duty vehicles and trucks had been 5 years or 50,000 miles (whichever occurs first). The 1990 Amendments extended the warranty for major emissions control components (catalytic converters, electronic emissions control units, etc.) to 8 years or 80,000 miles (whichever occurs first) for 1995 and later model year vehicles. The warranty period for other components was reduced to 2 years or 24,000 miles (whichever occurs first).

An extension of the warranty period for specified major emissions control components is likely to lead to more durable designs for these components. Reducing the warranty period for other emissions control components will pass more repair costs along to consumers and may affect component durability design.

Certification Short Test

Ideally, vehicles which pass the federal Test Procedure for certification should also pass the inspection procedures set forth by the emissions performance warranty provision. To ensure that they do, the CAAA required the EPA to develop new test procedures to be incorporated into the certification process. These test procedures will be capable of determining whether 1994 and later model year light-duty vehicles and trucks, when properly maintained and used, will pass the inspection procedures established for that model year, under conditions reasonably likely to be encountered in I/M programs.

Thus, the EPA finalized a rule on November 1, 1993 requiring a Certification Short Test (CST). This rule has two major components. First, CST replaces the six current, steady-state performance warranty procedures available for use in basic I/M programs. These procedures include three idle tests, a pair of unloaded two-speed tests, and a steady-state loaded test. Second, it also incorporates the CST into the current procedures for obtaining a certificate of conformity for light-duty vehicles and light-duty trucks.

The CST is designed to prevent the occurrence of what are known as *pattern failures*. Pattern failures are groups of vehicles that pass the FTP but show consistent patterns of I/M failure. By ensuring that vehicle designs are compatible with performance warranty procedures through the CST, the occurrence of these pattern failures should be avoided, resulting in a reduction of I/M-related repair attempts and I/M retests.

EPA's Recall Program

Once the EPA determines that a substantial number of a manufacturer's vehicles do not conform to the federal emissions standards, the manufacturer must submit a plan to correct the problem. Light-duty vehicle life is defined as five years or 50,000 miles (whichever occurs first) through the 1993 model year. A longer useful life will be phased in beginning with the 1994 model year, and increasing in the 1996 model year to 10 years or 100,000 miles (whichever occurs first).

The EPA's recall program has been active since the early 1970's. Since the beginning of the program, over 45 million vehicles have been recalled. Currently, the recall program tests about 35 light-duty engine families annually, representing approximately 35 percent of the total light-duty vehicle and light-duty truck production.

There are three basic types of recalls: voluntary, influenced, and ordered. When a manufacturer decides without EPA involvement to correct an emissions problem it has discovered on its own, it is called a voluntary recall. An influenced recall occurs when an EPA investigation identifies a noncomplying family and the manufacturer is convinced to recall without an order being required. If a manufacturer declines to take appropriate action on a noncomplying family and EPA orders the company to correct the vehicles, an ordered recall results. Most vehicle recalls are voluntary, although ordered or influenced recalls affect more total vehicles.

There are two different phases of recall testing: surveillance and confirmatory programs. The less costly surveillance program is used to identify engine families showing a high probability of failure. The confirmatory program is designed to provide a larger sample of vehicles using more rigorous selection and screening procedures. These procedures may be used to support the EPA's decision in the event of a manufacturer challenge.

In selecting engine families for recall testing, data from a variety of sources are considered. First, surveillance or confirmatory test data from a similar engine family for the previous year or an earlier model year are considered. Emissions results from certification testing and manufacturer-provided information on the emissions performance of the vehicle are also considered. Also, I/M data from several states is reviewed, including both idle-only and simulated driving type test results (IM240). Information on any pattern failures documented in these records is incorporated in the selection process.

When an engine family is recalled, the manufacturer is responsible for developing a repair which:

1. Brings the mean emissions, as measured by EPA, into compliance.
2. Is adequately durable.
3. Can be expected to be successfully implemented by the manufacturer's dealership network.

Adjustment fixes, such as an ignition timing change made by rotating the distributor housing, would not be acceptable because of their unsatisfactory durability. The emissions produced after the fix must be equal to or below the federal standards.

The manufacturer must order, acquire, and distribute any parts and labels to be used by the technician performing the recall. The manufacturer must also draft a letter to be sent to every owner explaining the steps to have the recall completed and how the recall will be performed.

Unfortunately, independent service facilities do not typically have access to Original Equipment Manufacturers (OEM) service information, and may not be aware that a vehicle coming into their shop has been subject to a recall. In addition, they may not be able to accurately determine whether a recalled vehicle has had the necessary corrections. Thus, the EPA will soon propose a rule requiring motor vehicle manufacturers to provide and update electronic lists identifying vehicles subject to an emissions recall that have not been repaired. State agencies will be able to use this information to enforce compliance with emissions recall campaigns in areas required to implement enhanced vehicle I/M programs. Such a program will alleviate the problem of having technicians unknowingly attempting to repair an emissions problem for a vehicle under a recall repairs done. This rule is expected to go into effect in July, 1995 and will start affecting motorists in January, 1996.

In addition, several efforts are underway which would provide independent service facilities with greater access to OEM service information. The CAAA requires manufacturers to promptly provide to any person engaged in the repair or service of motor vehicles or motor vehicle engines any and all information needed for making emissions-related diagnosis and repairs. Included in the Supplemental Guidance rule is a requirement that manufacturers make emissions-related repair and service information (including recall information) available to all independent technicians. Beginning in 1996, this information must be provided in a standardized electronic format currently being developed by SAE. Final rules are not yet available.

EFFECTS OF REQUIREMENTS ON VEHICLE REPAIR INDUSTRY

In order to fully assess training needs, it is important to understand how the requirements under the I/M rule affect the vehicle repair industry, and what the technician's task really is with respect to correcting emissions failures.

The EPA's research on the causes of high emissions among in-use vehicles has shown that a wide variety of malfunctions can and do occur. There are a few that are particularly common, especially oxygen sensor deterioration. However, diagnosis of an emissions failure is not a simple matter, since there are frequently multiple causes, and the repair of one component may bring a vehicle into compliance temporarily, but may not solve the underlying cause.

Repair technicians in current I/M programs are encountering a variety of underlying malfunctions as they repair vehicles to pass the idle and 2500/idle tests that are now dominant. The EPA expects that the IM240 tailpipe emissions test will find more of these diverse malfunctions than simpler testing did, especially NO_x-related problems. Evaporative system problems will also be detected for the first time in most I/M programs.

However, the EPA expects that the range of problems will be similar. On one car model, a certain broken part may cause an IM240 failure and an idle test failure. On another car model, the same broken part may not affect idle emissions much at all. There are few, if any, unique problems which appear only with the IM240. The basic challenge of finding the reason for high emissions will be much the same as at present. Technicians will continue to use a combination of routine tune-up checks and parts replacements, visual examinations, on-board diagnostic indicators, special tools, service manual procedures, and the process of elimination.

More vehicles will be identified as having problems, and problems will have to be isolated and corrected more thoroughly for the car to pass the IM240 than for the idle test. This is because the idle test frequently passes cars that still have problems that the IM240 can detect. The IM240 test is less likely to pass vehicles that are out of adjustment or tampered with as a ploy to get a car to pass without performing the necessary repairs. Testing the vehicle under load will tend to cause a vehicle to fail if a ploy has been used to pass HC and CO, and vice versa.

Consequently, technicians will have to find and fix the true causes of a test failure more often. Also, the statutory $450 cost waiver limit is higher than the cost waiver limits typical of basic programs in the past. This higher limit will close off one of the alternatives to repair now available in current I/M programs. Finally, because most enhanced I/M programs will have official inspections performed in test-only networks, there will be no opportunity for an improperly-performed test to be a substitute for a full repair.

Due to the reasons discussed above, it will be necessary for technicians to become even more efficient in diagnosing problems. Education is an important component, although it is not the complete solution. Many factors may affect performance, e.g., lack of technical information, specialized tools, a good working environment, or the time allowed to do a complete job.

The two areas of education that the EPA perceives to be of primary importance are the education of existing technicians in the diagnosis and repair methods required to correct those vehicles that fail the transient test, and bringing the automotive education offered by secondary and post-secondary schools, which educate new technicians, up to speed on the new diagnostic and repair methods required to repair vehicles failing the transient test.

The focus for the education of existing technicians should be on the technicians that already possess a certain level of expertise in automotive repairs. Likewise, the main emphasis of diagnostic and repair education for these experienced technicians should relate primarily to the rapidly changing technology, particularly electronics and computer-controlled emission control systems. The practical effects that these systems, in conjunction with the basic engine mechanical systems, have on transient emissions levels should also be covered. Therefore, such education should assume a basic knowledge of these systems, and should not focus on general repair issues. Technicians not fully understanding the basics should be given the opportunity to take remedial and/or refresher courses. These courses should be coordinated with the advanced level course schedule. Further, the focus should also be on generic training and diagnostic strategies rather than emphasizing instructions for performing specific repairs. The variety of vehicles that may fail the IM240 transient test will require technicians to focus on their problem solving skills.

ROLE OF THE LOCAL OR STATE AGENCY

Natural market forces caused by vehicle owners' interest in a convenient and economical repair can create an incentive for technicians and repair facilities to become more proficient, and for more persons and firms to enter the market. However, considerable stresses in the repair market could occur during a transition phase, to the detriment of the public. Therefore, it is desirable that local agencies take an active role in providing and promoting education for in-service technicians. In addition, the optional certification of technicians and the performance monitoring program, which is required by the I/M Final Rule, will enhance natural market forces.

Technician Education and Certification

While the start of enhanced I/M programs creates a short-term need to educate practicing technicians, in the long term it is preferable to impart emissions repair knowledge during vocational education. The long-term task of upgrading vocational educational programs is one which requires the attention of I/M agencies, but not at the expense of educating in-service technicians in the short term. Considerable coordination between short-term programs to educate practicing technicians and long-term programs to upgrade vocational educational programs might be possible.

The certification of repair technicians has two benefits. First, it allows vehicle owners who need a more qualified technician to locate those who exist and second, it encourages other technicians to get qualified. Certification does present challenges in terms of the development of an appropriate test instrument, administration of the test, motivation of technicians to take the test, and education of the public about the benefit of patronizing certified technicians.

Automotive Service Excellence, Inc. (ASE) has developed a new test, the Advanced Engine Performance Specialist (L1) Test, which is an advanced level exam requiring successful certification in ASE Test A8, Engine Performance. This test is designed to measure the technicians' knowledge of the diagnostic skills necessary for sophisticated emission control systems and engine performance problems.

It should be noted that no certification test is a perfect predictor of job performance under constantly evolving technology and other changing conditions. Thus, certification is a supplement rather than a substitute for a good performance monitoring program.

Performance Monitoring

The concept behind performance monitoring is to let vehicle owners know how well competing repair facilities are doing in terms of getting cars to pass reinspection on their first trip back to the inspection station. Performance monitoring will also let repair facilities know how they are doing so they can improve performance. Finally, performance monitoring should let I/M program officials know each repair facility's success rate so they can actively counsel the technicians who are most frequently causing the public the unnecessary inconvenience and expense of multiple repair trips. Performance monitoring is a new concept in most I/M areas, but has been done in some areas for many years with general acceptance by the repair industry and with public appreciation for the service.

Tools, Methods, and Service Information

A more satisfactory repair process will also be facilitated by technical supports including improved tools, methods, and service information. Perhaps the most important is the development of advanced diagnostic procedures. In the future, the EPA hopes to describe, in detail, some advanced diagnostic procedures that can be used to more accurately and efficiently diagnose emissions failures. Also, the EPA is collecting data to investigate the possibility of using the second-by-second emissions history during the IM240 to provide clues as to the malfunctions responsible for the failing emissions level on the overall test. Furthermore, the equipment manufacturers are starting to undertake the development and distribution of advanced technology repair equipment.

Because the official IM240 transient emissions test is performed with specialized testing equipment, diagnosing failed vehicles and verifying that subsequent repairs were sufficient to adequately reduce emissions to passing levels may be considerably easier. Thus, the EPA has conducted a preliminary study of a relatively low cost repair grade system (RG240). The system should be affordable enough for a repair facility to purchase (in the neighborhood of $25,000 - $40,000, or even less if a BAR '90 analyzer is already owned). It should be accurate enough to indicate when a vehicle has had a substantial emissions reduction and is highly likely to pass the official retest.

Also included in the technical support category is the need for priority to be placed on developing and disseminating generic methods for verifying basic closed-loop operation, since many other malfunctions cannot be isolated if closed-loop operation is not present. (This is covered later on in Chapter 8.) The repair grade RG240 testing package would be able to generate this type of emissions history, allowing the technician to get the information at each step of the repair without returning to a test-only station.

DIAGNOSTIC EQUIPMENT AND TOOLS

Basic Diagnostic Tools

Repair facilities should have a number of basic diagnostic tools, as described below.

1. *Digital Multimeter (DMM)* — A DMM is capable of measuring several properties of electricity (Figure 2-1). These properties are:
   a. *Voltage* — The electrical force that pushes the electrons through the electrical circuit. This force is called Electromotive Force (EMF) and is measured in volts.
   b. *Current* — The movement of electrons in a circuit. Current is measured in terms of amperage, which represents the amount of electrons that pass through a circuit in one second.
   c. *Resistance* — The electrical force which opposes the movement of electrons through the circuit. Electrical resistance is measured in ohms.

   A digital multimeter measures direct current (electrical current flowing in one direction) and alternating current (electrical current which alternates direction). If a repair facility does not have a DMM, they should have an AC/DC voltmeter, an ammeter to measure current, and an ohmmeter to measure resistance.
2. *Scan Tool* — A scan tool is also known as a diagnostic readout tool. It converts computer pulses or signals directly into a digital or numerical display. This device makes it easier to read electronic trouble codes and other important data. (See Chapter 7.)

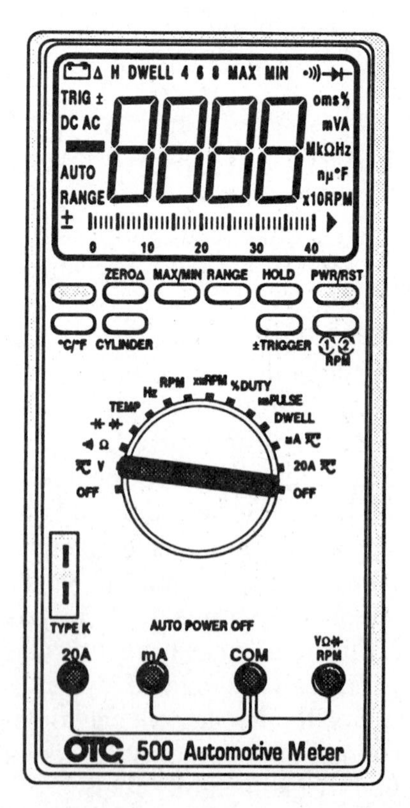

Figure 2-1 OTC 500 Series multimeter (courtesy OTC, a division of SPX Corporation).

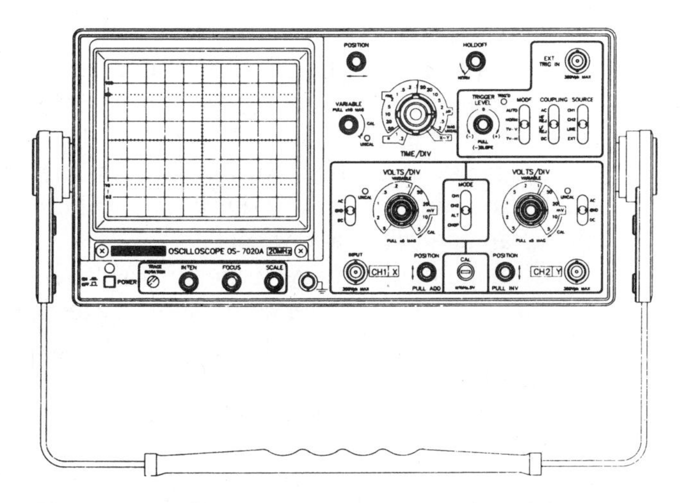

Figure 2-2 Typical dual-trace analog lab scope (courtesy Goldstar Precision Co., Ltd.).

3. *Laboratory Oscilloscope* — An oscilloscope is a cathode ray tube that displays a line pattern representing voltage in relation to time. (See Chapter 9.) Voltage is shown along the vertical axis and time is shown along the horizontal axis. An oscilloscope must have a time base and amplitude resolution adequate for troubleshooting vehicle sensors. (Figure 2-2.)

   Repair facilities may have a laboratory oscilloscope as part of an engine performance analyzer. An engine performance analyzer essentially consists of several pieces of test equipment mounted in one cabinet. Besides an oscilloscope, the analyzer may include a DMM, tach-dwell, exhaust gas analyzer, pressure-vacuum gauge, cylinder balance tester, compression tester, cranking balance tester, vacuum pump, and timing light.
4. *Temperature Gauge (Pyrometer)* — A temperature gauge is frequently used to measure the temperature of various components, such as the radiator temperature. The temperature obtained with the gauge can be compared to specifications. An electronic digital pyrometer is often used to make very precise measurements, such as the temperature of the exhaust manifold at each exhaust port. Lower temperatures at any exhaust port would indicate a misfiring cylinder.
5. *Pressure Gauge* — A pressure gauge is frequently used to measure air and fluid pressure in various systems and components, such as fuel pump pressure. Some pressure gauges can also be used to measure vacuum.
6. *Vacuum Gauge* — A vacuum gauge is used to measure vacuum, such as the vacuum in an engine's intake manifold, vacuum diaphragms, vacuum solenoid switches, and the carburetor mixture solenoid.
7. *Vacuum Pump (Hand-Held)* — A hand-held vacuum pump provides a source of suction and is used in conjunction with a vacuum gauge to test various vacuum devices.
8. *Gas Flowmeter* — This tool is used in the detection and diagnosis of canister purge failures.
9. *Compression Tester or Cylinder Leak-Down (Leakage) Tester* — A compression tester is used to measure the amount of pressure formed during the engine compression stroke. A cylinder leak-down tester performs a similar function, measuring the amount of air leakage out of the combustion chambers. If readings from these devices are out of specifications, such problems as bad intake valves, burned exhaust valves, bad rings, pistons, or cylinders, or a blown head gasket may exist.
10. *Timing Light with Advance Capability* — A timing light with advance capability not only measures engine timing, but also measures exact distributor advance with the engine running at different speeds. This type of timing light has a degree meter built into the back of its case, which shows the exact amount of advance. Most large engine analyzers also have this feature.
11. *Four-Gas or Five-Gas Exhaust Emissions Analyzer* — An exhaust gas analyzer draws a sample of exhaust gas from the car's tailpipe. A four-gas analyzer measures the amount of CO, HC, CO_2, and O_2 in the exhaust. A five-gas analyzer, which also measures NO_x, may be needed in enhanced I/M areas where there is an NO_x performance standard (Figure 2-3). The information provided by a gas analyzer indicates the air fuel ratio entering the engine. Repair facilities may have a gas exhaust emission analyzer as part of an engine performance analyzer. (See Chapter 10.)
12. *Tachometer* — A tachometer measures engine speed in revolutions per minute. It is used to adjust engine speed settings and perform other tests.
13. *Dwellmeter or Duty-Cycle Meter* — A dwellmeter measures the amount of time that voltage is

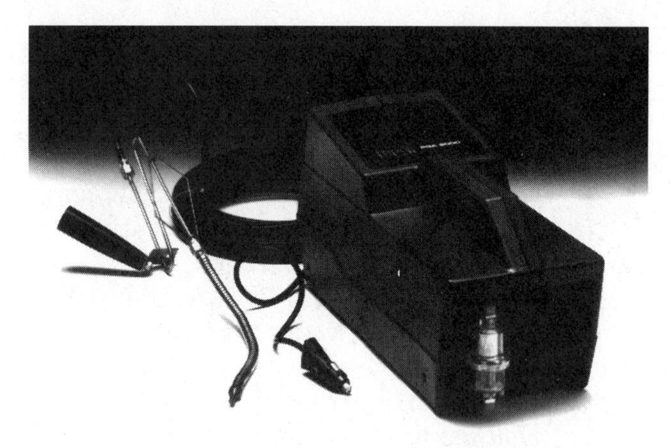

Figure 2-3 MPSI portable five-gas analyzer (courtesy Micro Processor Systems, Inc. – MPSI).

applied to the ignition coil to energize it (duration of distributor point contact) and is measured in degrees of distributor rotation. A dwellmeter is used primarily with older ignition systems. In cars without electronic ignitions, variation in the dwellmeter reading indicates distributor wear problems. With electronic ignitions, dwell changes with engine speed can be normal.

A duty-cycle meter measures the duration of an electronic pulse. The duration is expressed as a percentage representing the width of the pulse relative to the width of the cycle from pulse to pulse. In a vehicle with an electronic ignition, the duty-cycle will increase at high speeds and decrease at low speeds.

A dwellmeter or duty-cycle meter can also be used to measure computer commands to a feedback carburetor solenoid. The dwell indicates the time that the solenoid is on. (See Chapter 8.)

14. *Breakout Box* — A breakout box is essentially a box with a series of test terminals, or lugs, which can be connected to the vehicle's computer wiring harness (Figure 2-4). A DMM can be used to probe specific box terminals which correspond to various systems and test values can be compared to specifications.

Advanced Diagnostic Procedures

In order to diagnose causes of emissions failures and ensure that a repaired vehicle will pass a basic I/M or IM240 retest, advanced diagnostic procedures are

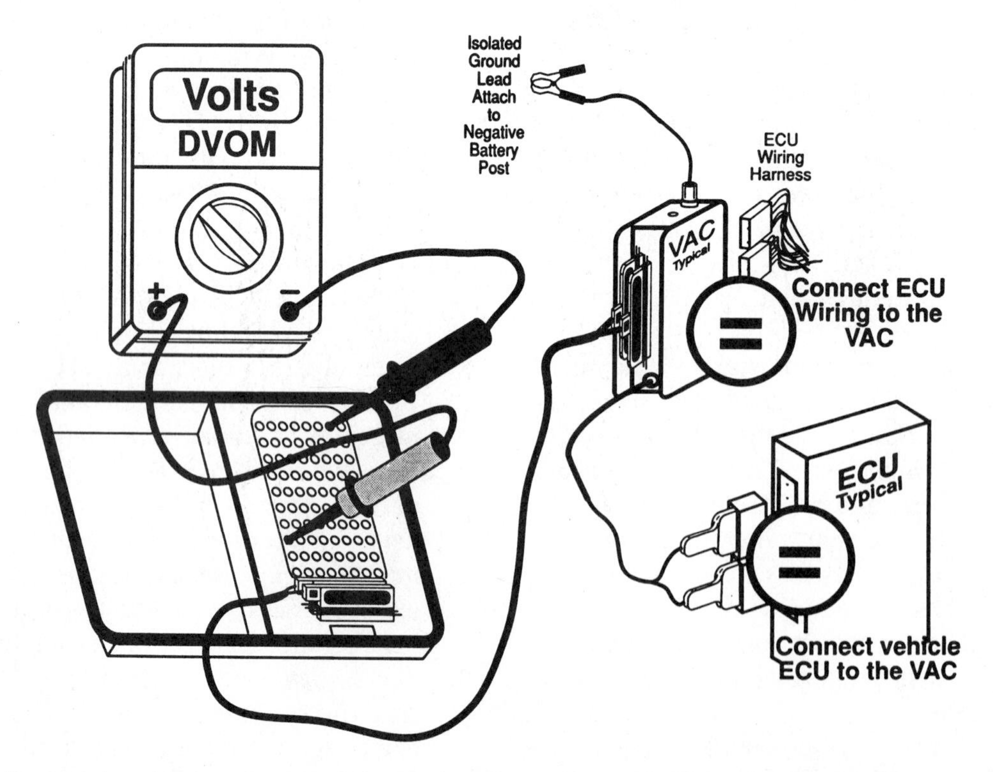

Figure 2-4 Typical breakout setup used in conjunction with a DVOM (courtesy OTC, a division of SPX Corporation).

necessary to pinpoint the causes of many emissions failures.

The following sections briefly discuss how advanced procedures such as waveform diagnostics using an oscilloscope, and using a repair grade IM240 can be used to diagnose emissions failures. The use of new technology to diagnose catalyst failure is also briefly discussed.

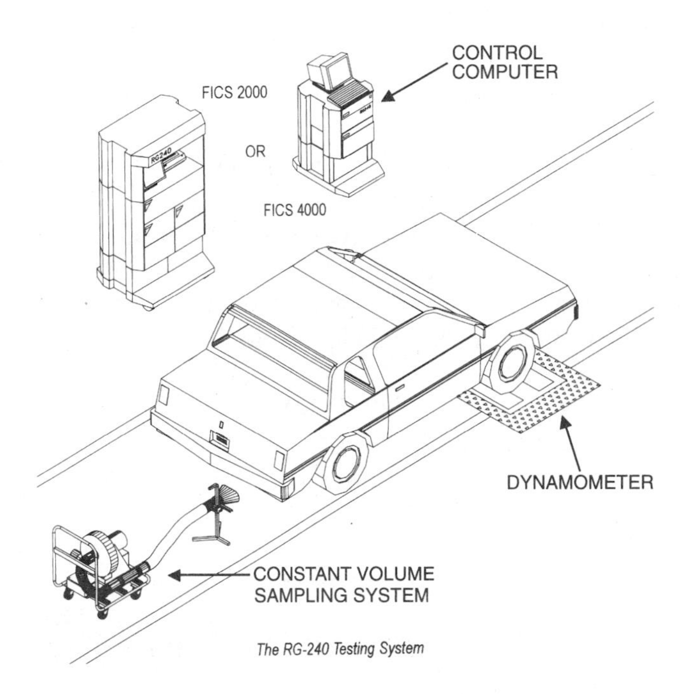

Figure 2-5 RG 240 Lane Setup (courtesy Environmental Systems Products, Inc.).

Advanced Diagnostic Strategy Involving Oscilloscope

An oscilloscope can be used to diagnose problems in any vehicle system that has a voltage output which changes over time. The pattern of voltage values, over time, is known as a waveform. A properly functioning system typically has a characteristic waveform that a repair technician can easily recognize. Deviations from this characteristic waveform represent some system malfunction and specific types of malfunctions have characteristic deviations in some part of the waveform. Waveform diagnostics may be particularly useful in diagnosing problems with fuel injectors and oxygen sensors.

Advanced Diagnostic Strategy Using Repair Grade IM240

The RG240 system (Figure 2-5) consists of a dynamometer with uncoupled twin rollers, a Constant Volume Sampler (CVS) with a flow rate of 100 Standard Cubic Feet per Minute (SCFM), and a BAR '90 emissions analyzer with an additional NO_x analyzer.

Advanced Diagnostic Strategy for Catalyst Failure

It can often be difficult to determine if catalytic converter problems are really the root cause of emissions failures. Only about 30% of vehicles with emissions failures studied by the EPA require catalyst repairs. Sometimes, damage to a catalyst can be diagnosed by tapping on the shell. If the shell sounds hollow, the substrate may be missing. If it rattles, the substrate may be broken up or the inner baffles and shell may be deteriorated. At other times, repair technicians may suspect that high emissions from a vehicle are the result of a damaged catalytic converter, but their suspicions may only be confirmed or contradicted by removing the catalyst and visually inspecting it and/or measuring the pressure drop across the catalyst. However, these tests may not detect catalyst contamination. Also, replacing the catalyst may reduce emissions enough to bring a vehicle into compliance but still not correct some other malfunction, resulting in high emissions shortly after the replacement of the catalyst.

The EPA is investigating additional tools for visual diagnosis of catalyst problems. A boroscope, which is a fiberoptic tool that extends down through the O_2 sensor mounting hole to the catalyst, may be developed so that it is practical for inexpensive and routine inspection of catalysts.

Readings from a gas analyzer may also lead a technician to suspect contamination of a catalyst, even when visual inspection or pressure drop measurements do not indicate a problem. If O_2 readings are above 5%, there is enough oxygen for the catalyst to burn the emissions. However, if the CO readings are still above 0.5% (and other systems are operating properly), the catalytic converter is not oxidizing emissions from the engine and may need to be replaced.

At present, the EPA's advice is as follows:

1. Always check for engine problems before replacing the catalyst.
2. If significant engine problems are found and then fixed, perform another test before replacing the catalyst unless it is obvious from an external inspection that the catalyst content is missing or damaged.
3. Observe EPA rules for the selection of a replacement catalyst. More information on these rules can be found in the following sections.

PARTS AND CATALYST ISSUES

Aftermarket Parts and EPA Tampering Policy

The EPA does not consider it to be tampering for automotive dealers (or any person) to use a non-OEM aftermarket part as a replacement part for the purpose of maintenance or replacement of a defective or worn out part, if the dealer has a reasonable basis for knowing that such use will not adversely affect emissions performance. The same applies to the use of aftermarket parts used as part of an add-on, auxiliary, or secondary part or system. Adjustments or alterations of a particular part or system, if done for purposes of repair or maintenance according to the vehicle or engine manufacturer's instructions, are also acceptable.

Aftermarket Parts Certification Program

The EPA has a Voluntary Aftermarket Parts Certification Program. Essentially, this rule states that any aftermarket part manufacturer that wishes to certify its emissions-related part must demonstrate that the use of its part will not cause a vehicle to fail federal emissions standards during the vehicle's useful life. Furthermore, the rule specifies that vehicle manufacturers cannot deny a performance warranty claim on the basis that the use of the aftermarket part is improper maintenance or repair if the part is certified under the voluntary aftermarket parts certification regulations.

EPA Aftermarket Parts Warranty Requirements

An August 8, 1989 ruling requires all certified aftermarket parts to be warranted by the part manufacturer. They must guarantee the part will not cause an emissions noncompliance of the vehicle on which the part is installed. Certified parts must be warranted for the remaining warranty period of the vehicle, as required under the Clean Air Act. For instance, the warranty period for emissions-related devices is 50,000 miles for pre-1995 light-duty vehicles. If the device is replaced with an aftermarket replacement part at 25,000 miles, the replacement part is warranted for the next 25,000 miles. The vehicle manufacturer is required to repair or replace, without charge, those certified emissions-related components necessary to remedy that emissions failure if it occurs within the prescribed warranty period. The vehicle manufacturer may then obtain reimbursement from the certified part manufacturer for the warranty claim.

ALTERNATIVES TO REPAIR

The Clean Air Act Amendments of 1990 require programs to encourage the voluntary removal from use of pre-1980 model year light-duty vehicles and pre-1980 model year light-duty trucks. A vehicle scrappage program can be used to accomplish this goal. The EPA indicates that scrappage programs for pre-1980 model year vehicles can exhibit a wide range of effectiveness, depending on the program design.

A state or local government can design a scrappage program as a SIP measure or, in conjunction with a private company, as a program to generate emissions credits to satisfy existing or new source-specific requirements. Programs would basically work in the following way. A state or local government or company would advertise for the purchase of certain vehicles. Owners would then voluntarily sell their vehicles to the sponsor of the program and the vehicles would be removed from the fleet. The sponsor would receive an emissions credit for each car removed from operation equivalent to the difference between the emissions from the retired vehicle and the emissions from the replacement vehicle.

Adding a vehicle scrappage option to an I/M program is another way to enhance program benefits and/or reduce costs. Vehicles that fail an I/M test, that have not been successfully repaired, or are known to need repairs costing greater than a predetermined amount, would become eligible for a scrappage program. Scrappage program designs that incorporate an I/M element in this way will not only have greater assurance that they are retiring high-emitting vehicles, but could possibly offer incentives since the vehicle owner is faced with immediate repair costs if the vehicle is not scrapped.

While no owner would be pleased to have to get rid of a difficult-to-repair car and purchase a new or used vehicle as a replacement, the reality is that it might be the logical thing to do for some cars. An option to scrapping the old car is to sell it to an owner living outside the boundary of the I/M program. The free market will surely see some of this happen. I/M agencies may wish to discourage this.

TECHNICAL ASSISTANCE FOR REPAIR INDUSTRY

Providing Information to Repair Facilities

The I/M Final Rule addresses the issue of providing technical assistance to the repair industry:

"The oversight agency shall provide the repair industry with information and assistance related to vehicle inspection, diagnosis, and repair."

This technical assistance requirement, which applies to both basic and enhanced I/M areas, also states:

"The [oversight] agency shall regularly inform repair facilities of changes in the inspection program, training course schedules, common problems being found with particular engine families, diagnostic tips, and the like."

Newsletters

The most obvious approach states can use to meet this requirement is through the distribution of a newsletter. Several I/M areas and states already distribute their own newsletter, developed specifically for that non-attainment area. Examples are the California Bureau of Automotive Repair's *Repair Reporter* (Figure 2-6) and the State of Wisconsin Department of Transportation's *The VIP Analyzer.* Also, some states in the process of developing enhanced I/M programs are requiring the I/M contractor to develop a system to meet the technical assistance requirement.

If a state does not wish to develop its own newsletter, or lacks the resources to do so, it can use I/M newsletters reproduced by other organizations (Figure 2-7).

Service Information

In order to effectively correct emissions problems on current vehicles, it is critical that repair facilities have access to model-specific repair manuals. Of course, a collection of model-specific repair manuals can occupy substantial space in a repair facility, quickly wear out, or be misplaced. These problems can be eliminated by using computer-based (CD-ROM) repair information systems. A number of publishers already have repair manuals available in electronic form.

Also, in order to effectively repair emissions problems, independent repair facilities must have access to the service bulletins that OEMs provide to dealerships (Figure 2-8). As mentioned previously, the EPA's forthcoming final service information availability rule will require OEMs to provide this information to independent facilities in a standardized electronic format currently being developed by SAE.

Electronic Bulletin Boards

The I/M rule does not suggest the use of electronic bulletin boards as an approach for addressing the technical assistance requirement; however, states should consider establishing electronic bulletin boards as a way to communicate with the repair community. Bulletin boards could, in fact, be used in a multitude of different ways to share I/M-related information. For instance, electronic bulletin boards could be accessible to the public as well as repair technicians, and can include information on the status of regulations, down-loadable standardized forms, pattern failure data, and IM240 test data for cars coming in for repairs. Furthermore, bulletin boards can provide an opportunity for technicians to describe hard-to-solve emissions problems and solicit solutions

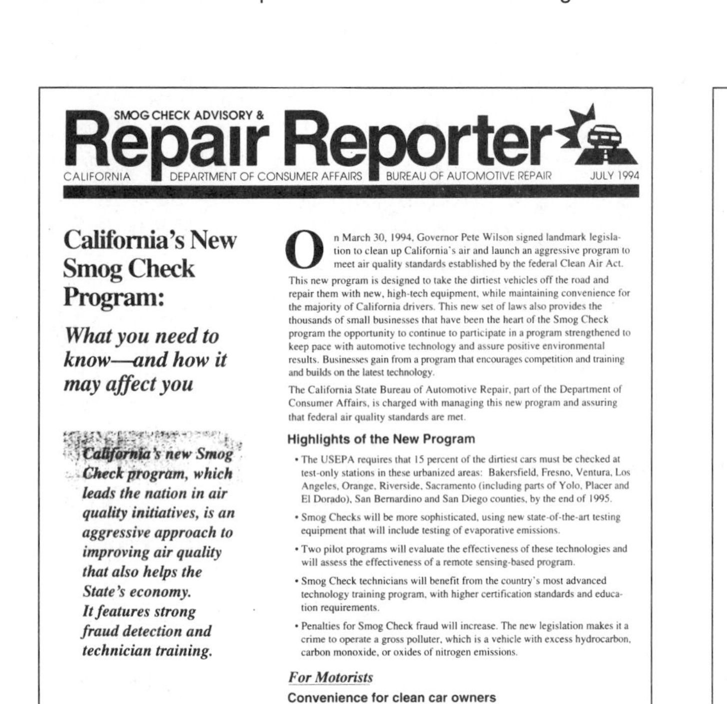

SMOG CHECK ADVISORY &

Repair Reporter

CALIFORNIA ■ DEPARTMENT OF CONSUMER AFFAIRS ■ BUREAU OF AUTOMOTIVE REPAIR JULY 1994

California's New Smog Check Program:

What you need to know—and how it may affect you

California's new Smog Check program, which leads the nation in air quality initiatives, is an aggressive approach to improving air quality that also helps the State's economy. It features strong fraud detection and technician training.

On March 30, 1994, Governor Pete Wilson signed landmark legislation to clean up California's air and launch an aggressive program to meet air quality standards established by the federal Clean Air Act. This new program is designed to take the dirtiest vehicles off the road and repair them with new, high-tech equipment, while maintaining convenience for the majority of California drivers. This new set of laws also provides the thousands of small businesses that have been the heart of the Smog Check program the opportunity to continue to participate in a program strengthened to keep pace with automotive technology and assure positive environmental results. Businesses gain from a program that encourages competition and training and builds on the latest technology.

The California State Bureau of Automotive Repair, part of the Department of Consumer Affairs, is charged with managing this new program and assuring that federal air quality standards are met.

Highlights of the New Program

- The USEPA requires that 15 percent of the dirtiest cars must be checked at test-only stations in these urbanized areas: Bakersfield, Fresno, Ventura, Los Angeles, Orange, Riverside, Sacramento (including parts of Yolo, Placer and El Dorado), San Bernardino and San Diego counties, by the end of 1995.
- Smog Checks will be more sophisticated, using new state-of-the-art testing equipment that will include testing of evaporative emissions.
- Two pilot programs will evaluate the effectiveness of these technologies and will assess the effectiveness of a remote sensing-based program.
- Smog Check technicians will benefit from the country's most advanced technology training program, with higher certification standards and education requirements.
- Penalties for Smog Check fraud will increase. The new legislation makes it a crime to operate a gross polluter, which is a vehicle with excess hydrocarbon, carbon monoxide, or oxides of nitrogen emissions.

For Motorists

Convenience for clean car owners

- Most Californians will continue to have their vehicles tested at neighborhood Smog Check stations.
- Smog Check certificates will be transmitted electronically from the test site to the Department of Motor Vehicles, increasing consumer convenience (effective by January 1, 1996).
- New car buyers have the option of paying a fee of no more than $50 at the time of purchase, which exempts the car from its first biennial Smog Check. The money received through this program goes to help fund the repair and replacement of high-polluting vehicles.
- To facilitate speedy Smog Checks and eliminate errors, vehicle information will be bar coded on the registration form.

(Continued on page 2)

Figure 2-6 California smog check advisory newsletter (courtesy California Department of Consumer Affairs Bureau of Automotive Repair).

CSCV

I&M QUARTERLY UPDATE

A periodic information service for all those involved in achieving cleaner air through Inspection & Maintenance of automotive exhaust & evaporative emission control systems

Published for the Coalition for Safer Cleaner Vehicles by Aftermarket Research Institute, Inc.

I&M Rule Review: examining the regulatory impact

BY NO LATER than January 1, 1995, a majority of states will have to have taken serious steps toward cleaning up their air quality from excessively-polluting automobiles. Some 26 metro areas of the nation in 10 states which now have no I&M programs, must implement basic I&M (exhaust emission Inspection & Maintenance) programs by Jan. 1, 1994. And, by Jan. 1, 1995, 83 other areas of the nation in 20 states plus DC, which have either basic or no I&M programs, but which still fail to meet the national ambient air quality standard—we call these non-attainment areas—must have in operation a more precise Enhanced I&M program designed around the brand-new and more comprehensive IM-240 inspection procedure.

It goes without saying that the new Enhanced I&M inspection procedure will be tighter and much more detailed than the earlier BAR 84 and BAR 90 standards used in almost all I&M programs now in place across the nation.

Despite better, cleaner-performing OE automotive exhaust emission systems on cars of the last five model years, continued and excessive levels of exhaust pollution from older cars are what is causing this new focus on tougher emission system diagnosis and adjustments. The goal is to identify thru inspection the gross polluters, have them brought back into compliance thru maintenance and repair, and then be re-tested to prove their clean-running performance.

With an eye on reducing vehicle-owner inconvenience, EPA will call for a maximum 4-minute test-cycle emission inspection. This has been named the IM-240 test since the equipment which will perform it has been designed to execute the emission check in no more than 240 seconds.

These new developments in the continuing battle to achieve cleaner air for all of us to breathe, will have major impacts on many shops, as well as on many thousands of professional automotive service technicians. States which must implement Enhanced I&M programs by 1/95, will call for IM-240 tests to be performed in test-only (no-repair) facilities. These test-only centers will feature loaded-mode (dynamometer) diagnosis. The results of such a test will be printed out and show in detail where in the 240-second time cycle a vehicle emits excessive levels of HC (hydrocarbons), CO (carbon monoxide) and NOx (oxides of nitrogen).

In addition, an evaporative purge check of each vehicle's fuel-vapor charcoal trap canister will be performed, a step which is not now part of any existing I&M check.

Repairs will not be available at test-only facilities. It will be up to the vehicle owner to choose the repair facility where emission repairs will be made. And, of course, a re-test of every vehicle which fails an initial

continued on page 2

In this issue . . .

This is a sample issue of what CSCV plans will be a regular 4-times-a-year newsletter for all those involved in achieving cleaner air through I&M of automotive exhaust & evaporative emission control systems.

Figure 2-7 Typical industry newsletter (courtesy Coalition for Safer, Cleaner Vehicles).

from other technicians. Electronic bulletin boards could also include performance monitoring statistics and information provided by repair facilities on their skill in repairing specific makes of vehicles or older versus newer vehicles.

Technical Hotline Services

Today's technicians are faced with literally dozens of vehicle makes and with hundreds of car and truck models. The repair of these different makes and models carries a price. The technical reference material necessary to repair all the different vehicles would easily fill a small library. Additionally, every year a whole new set of books and manuals would be needed. It is difficult, if not impossible, for the average shop to maintain 100% coverage of all makes and models.

How does a shop go about acquiring the information it needs to tackle a new vehicle or type of problem? A technical hotline service can be the answer. It is an example of how powerful large numbers of technicians can be. When technicians use a hotline service, they are not just limited to their own training and experiences. A technical hotline is built on the knowledge and experiences of thousands of technicians. Each time a call is handled, a hotline becomes stronger, smarter, and better able to answer future technical questions.

Rather than send the customer back to the dealer for the problem, the independent repair technician can now get help. Technical hotline services are designed to cover a small percentage of a shop's repair problems. Some shops use this service only for the "stumpers." Other shops follow a practice of calling the hotline if they have spent more than 20 to 30 minutes on their initial diagnosis. In other instances, they call for a second opinion, a factory wiring diagram, or to obtain information from a large database of previous repairs.

Hotline services are staffed by experienced technicians. Like the repair technician, the hotline technician cannot be expected to be knowledgeable

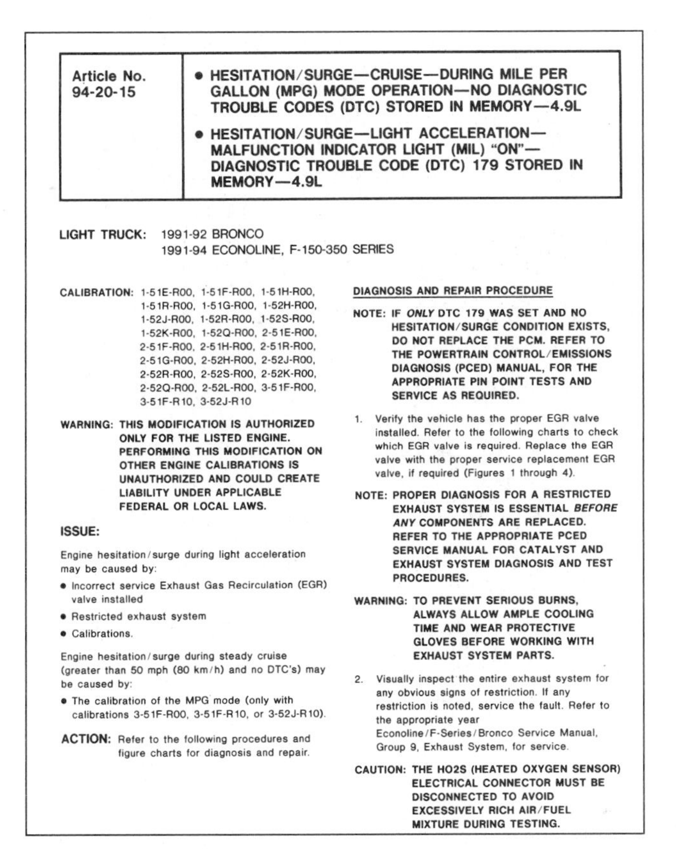

Article No. 94-20-15	• HESITATION/SURGE—CRUISE—DURING MILE PER GALLON (MPG) MODE OPERATION—NO DIAGNOSTIC TROUBLE CODES (DTC) STORED IN MEMORY—4.9L • HESITATION/SURGE—LIGHT ACCELERATION—MALFUNCTION INDICATOR LIGHT (MIL) "ON"—DIAGNOSTIC TROUBLE CODE (DTC) 179 STORED IN MEMORY—4.9L

LIGHT TRUCK: 1991-92 BRONCO
1991-94 ECONOLINE, F-150-350 SERIES

CALIBRATION: 1-51E-R00, 1-51F-R00, 1-51H-R00, 1-51R-R00, 1-51G-R00, 1-52H-R00, 1-52J-R00, 1-52R-R00, 1-52S-R00, 1-52K-R00, 1-52Q-R00, 2-51E-R00, 2-51F-R00, 2-51H-R00, 2-51R-R00, 2-51G-R00, 2-52H-R00, 2-52J-R00, 2-52R-R00, 2-52S-R00, 2-52K-R00, 2-52Q-R00, 2-52L-R00, 3-51F-R00, 3-51F-R10, 3-52J-R10

WARNING: THIS MODIFICATION IS AUTHORIZED ONLY FOR THE LISTED ENGINE. PERFORMING THIS MODIFICATION ON OTHER ENGINE CALIBRATIONS IS UNAUTHORIZED AND COULD CREATE LIABILITY UNDER APPLICABLE FEDERAL OR LOCAL LAWS.

ISSUE:

Engine hesitation/surge during light acceleration may be caused by:

- Incorrect service Exhaust Gas Recirculation (EGR) valve installed
- Restricted exhaust system
- Calibrations.

Engine hesitation/surge during steady cruise (greater than 50 mph (80 km/h) and no DTC's) may be caused by:

- The calibration of the MPG mode (only with calibrations 3-51F-R00, 3-51F-R10, or 3-52J-R10).

ACTION: Refer to the following procedures and figure charts for diagnosis and repair.

DIAGNOSIS AND REPAIR PROCEDURE

NOTE: IF *ONLY* DTC 179 WAS SET AND NO HESITATION/SURGE CONDITION EXISTS, DO NOT REPLACE THE PCM. REFER TO THE POWERTRAIN CONTROL/EMISSIONS DIAGNOSIS (PCED) MANUAL, FOR THE APPROPRIATE PIN POINT TESTS AND SERVICE AS REQUIRED.

1. Verify the vehicle has the proper EGR valve installed. Refer to the following charts to check which EGR valve is required. Replace the EGR valve with the proper service replacement EGR valve, if required (Figures 1 through 4).

NOTE: PROPER DIAGNOSIS FOR A RESTRICTED EXHAUST SYSTEM IS ESSENTIAL *BEFORE ANY* COMPONENTS ARE REPLACED. REFER TO THE APPROPRIATE PCED SERVICE MANUAL FOR CATALYST AND EXHAUST SYSTEM DIAGNOSIS AND TEST PROCEDURES.

WARNING: TO PREVENT SERIOUS BURNS, ALWAYS ALLOW AMPLE COOLING TIME AND WEAR PROTECTIVE GLOVES BEFORE WORKING WITH EXHAUST SYSTEM PARTS.

2. Visually inspect the entire exhaust system for any obvious signs of restriction. If any restriction is noted, service the fault. Refer to the appropriate year Econoline/F-Series/Bronco Service Manual, Group 9, Exhaust System, for service.

CAUTION: THE HO2S (HEATED OXYGEN SENSOR) ELECTRICAL CONNECTOR MUST BE DISCONNECTED TO AVOID EXCESSIVELY RICH AIR/FUEL MIXTURE DURING TESTING.

Figure 2-8 Typical Factory Technical Service Bulletin (courtesy Ford).

and experienced with all makes and models of vehicles. For this reason, hotline technicians usually have extensive experience with a particular make of vehicle. Hotline technicians receive continuous hands-on and classroom training in their area of expertise.

Frequently, calls are solved through the experiences of the hotline technician. These experiences are stored in a database of thousands of previous solutions. The hotline technician is supported by a wide variety of manuals, third party CD-ROM information, and a symptom/repair database. Most technical hotline services provide the technician with the information in one of two ways:

1. A telephone is the most common method of communicating. A fax is not necessary, but it can be very helpful. Rather than trying to explain a wiring diagram, scope pattern, or technical bulletin, a fax can be used to transmit this data. The technician then has a tangible hard copy in hand to work with and review. A modem can also be helpful, especially when transmitting live scan tool data while the engine is running. Not all commercial hotline services are set up to analyze live scan tool data.
2. A Personal Computer (PC) can be used to access electronic bulletin board services. This service allows two or more technicians to simultaneously exchange their experiences by teleconferencing. An E-Mail feature allows technicians to access other technicians, through a "Registry of Users" to leave messages or obtain information about diagnostic and repair solutions. Technicians can also access electronic service information such as repair procedures, technical service bulletins, safety recalls, etc.

The only required equipment to access a hotline service is a phone. Standard repair tools, scan tools, and emissions-related gas analysis equipment should be available to the technician performing the actual repairs.

The technician must perform the normal basic tests plus specific tests pertaining to the symptom experienced before calling the hotline. All data collected from the preliminary diagnostic inspection should be documented. It is difficult to diagnose and repair any vehicle over the phone without preliminary information such as scan tool data, voltage readings, scope patterns, and vehicle make and model.

The biggest problem a hotline service must confront is the inaccurate description of the problem and the work that has already been done. For example, don't say the fuel pressure is okay, or the sensor voltage is 4.5 volts, if it wasn't really checked.

The hotline technician is trained to walk the repair technician through a diagnostic process that should leave nothing to chance. Keep an open mind, accept the help asked for, and do not question it. If the hotline service recommends a test, do it! Do not question it. The hotline technician cannot wave a magic wand over the phone to fix the problem; there must be a two-way effort with both parties working together.

Hotline services provide recommendations for repairs based on the information provided over the phone by the repair technician. The repair technician is ultimately responsible for the final decision to repair the vehicle. Once the repair is successfully completed, report this information back to the hotline technician. The hotline service is on the same learning curve as the repair technician, so it needs this information to build an effective database.

Hotline Services for I/M Programs — The EPA has realized how important a technical hotline service is for supporting emissions repairs. Part of the 1990 Clean Air Act Amendments, subsection 51.369(a)(2), pertaining to Inspection and Maintenance Program Requirements states:

"The (oversight) agency shall provide a hotline service to assist repair technicians with specific repair problems, answer technical questions that arise in the repair process, and answer questions related to the legal requirements of state and federal law with regard to emissions control device tampering, engine switching, or similar issues."

In essence, this regulatory language contains three distinct service areas that a program agency is required to provide through a hotline:

1. Specific repair advice.
2. Technical information during the repair process.
3. Answers to legal and agency policy questions on specific vehicle conditions and/or repair strategies needed for compliance.

The state agency (or the I/M contractor) is to develop a hotline service that provides a basic level of technical service in order to assist the technician (or anyone seeking technical advice, e.g., do-it-yourselfers) during the repair process. With regard to providing repair advice and technical information during the repair process, the hotline technician must have an understanding of the I/M test procedures used in each state, and an understanding of basic vehicle systems and components, as well as a working knowledge of how these relate to each other in order to answer generic repair questions.

It is not necessary that such a service offer repair advice on an in-depth, vehicle-specific level which is available from many commercial hotline services (e.g., "What is the voltage on pin 9 of model xx?" as opposed to a generic question, "How could a purge failure affect IM240 emissions?"). Therefore, the staff of any technical hotline service must at least have a general understanding of emissions repair diagnostic procedures.

With regard to responding to legal and agency policy questions, the hotline staff must be familiar with state and federal specific I/M rules, regulations, and policies in regard to all aspects of the I/M program (tampering, engine switching policy, catalyst replacement policy, consumer protection policy, location of I/M test stations, I/M cutpoints for different model years, waivers, reinspection, etc.).

For minimum access requirements, it is preferable, but not required, that the technical hotline be separate from the public awareness information number which may be established by a state. Technicians (or others) in need of advice during the repair process should not be made to wait while general program information is disseminated. If a state chooses to operate only one hotline, it must have enough capacity or special routing features to ensure that the technicians will not have to wait for repair information.

If the repair questions are more vehicle-specific than generic, hotline personnel must be able to refer the technician to additional sources of information that could aid in the repair process. At a minimum, hotline personnel should be prepared to provide the technician with information on how to contact commercial hotline services that support the specific area of the question, as well as their general capabilities and costs.

Options for Hotline Management — Some of the options available to the states include the following:

1. The state can operate the entire repair information function itself.
2. The state can have a contractor handle all the repair information hotline functions, including detailed repair questions.
3. The state can provide a repair hotline that provides a basic level of technical service and refers detailed repair questions to commercial repair services. With this approach, the referral can be to:
   a. A specific hotline service under contract.
   b. One of several hotline services under contract.
   c. One or many hotline services that meet state requirements.
   d. Any service that the state has identified as being able to provide support for the specific repair questions asked by the technicians.
4. The state can turn over the operation and management of a basic repair hotline to the state I/M

contractor. The state would subsequently refer the more detailed questions to commercial hotline services.

5. The state can work in partnership with an I/M contractor in establishing and operating a hotline service (i.e., a combination of options 3 and 4 above).

Guidelines for Enhanced I/M Areas — A hotline service should provide auto repair facilities with comprehensive technical information. It should be capable of providing assistance in diagnosis and repair of malfunctions in computer-controlled, closed-loop vehicles (e.g., 1981 and later) as well as earlier vehicles (e.g., those with oxidation catalysts or non-catalyst). It should be able to apply emissions control theory and diagnostic data to the diagnosis and repair of problems found during the transient emissions and evaporative system functional checks. It should also be able to use diagnostic information on systematic or repeated failures observed in the transient emissions and evaporative system functional checks.

The capacity of the hotline should be sized to minimize access time during periods of high demand. The information should be provided in a timely manner after receipt of a call. The hotline should be convenient and cost effective (local, 800, or 900 number) with minimum operating hours that cover the hours of normal repair shop activity in the I/M area.

The hotline should provide a mechanism to ensure that the necessary preliminary systems checks have been completed prior to making the initial repair call. This will help to prevent the hotline service from being flooded with inquiries relating to basic service questions. This may take the form of a standard checklist to be provided by the hotline service or initial questioning by the hotline technician to determine if these checks have been made. The existence and necessity of conducting such preliminary checks should be widely distributed to the repair industry.

The service should have a complete collection of factory service manuals, wiring diagrams, factory service bulletins, PROM update information, and a demonstrated ability to acquire and incorporate the most recent information available. This collection should generally be for all model years covered by the I/M program.

Several efforts are underway which would provide independent repair facilities and hotline services with greater access to OEM service information. The EPA's September 24, 1991 proposed rule on on-board diagnostics requires manufacturers to make emissions-related repair and service information (including recall information) available to all independent technicians and services. Also, beginning in 1998, the proposed rules would require this information to be provided in a standardized electronic format currently being developed by the Society of Automotive Engineers (SAE) under SAE J2008.

J2008 will provide a format for setting up an electronic service information system. The hotline should have available current access to various service information in electronic form. The service should be upgradeable to SAE J2008 format when it is available.

The service should remain current with the local I/M fleet as the model year mix changes over time. The hotline service should create a database with the repair knowledge (not IM240 data) gained through the assistance provided by the service. This database will give historical perspective on a particular vehicle and/or vehicle type. The database should be accessible such that it could be easily downloaded to local or county air pollution program databases.

Hotline Companies and Services — The following is a partial bibliography provided for reference only. It is expected that the state and local agencies will determine which, if any, of these hotlines can meet the needs of an I/M program in their state. The information below was provided by some of the companies operating hotline services:

1. ASPIRE
   U.S. Hwy 1
   Morrisville, PA 19067
   Phone: 1-800-435-1050

   The ASPIRE hotline was established in 1980, with subscribers from state and industry. This hotline logs about 20,000 to 25,000 minutes of call time per month. Service charges for actual time used and the price charged depends on the user. Calls are primarily related to driveability, performance, and emissions. There is a separate hotline to support educators.

2. Autoline Telediagnosis
   2714 Patton Rd.
   Roseville, MN 55113
   Phone: 1-800-288-6220

   The Autoline Telediagnosis service was founded in 1987 and handles over 12,000 calls per month. It has 5000 factory manuals and all factory bulletins. The company also has three CD-ROM systems: Expertec, a service from General Motors with PROM updates; ALLDATA; and Mitchell On-Demand. The company has an extensive database with 550,000 fixes logged into the system. Customers include independent shops, service stations, fleets, and car dealers. Autoline provides hotlines for NAPA, Parts Plus, and Mighty Auto Parts. They also provide service for Ammoco, Exxon, Chevron, and British Petroleum oil companies.

   The company can also receive live data over the phone from a modem. Hours of operation are 7:00 a.m. to 7:00 p.m. CST, Monday through Friday. A quarterly newsletter giving technical information is also published and a sample copy can be obtained from the contact listed above. Charges for this hotline service are by the minute, with no monthly fees and no sign-up fees.
3. G.E. Capital Fleet Services
   Three Capital Drive
   Eden Prairie, MN 55244
   Phone: 1-612-828-2799

   G.E. Capital Fleet Services' past experience does not have service center subscribers; its clients are owners of fleet vehicles being serviced at shops not owned by the fleets. The company provides guidance to mechanics, and also issues purchase orders to repair or rental facilities, so drivers do not incur out of pocket expenses.

   At present, G.E. Capital Fleet Services is completing the development of an Enhanced Inspection/Maintenance Technical Hotline. Their proposed system includes online access to emissions results by VIN, the ability to access a database of historical data, the ability to view state parameters regarding emissions tests, and the ability to create statistical reports from the database of calls.
4. Technician ONLINE
   8949 Bluewater Hwy
   Saranac, MI 48881
   Phone: 1-616-642-9271

   The Technician ONLINE hotline is a real-time computer accessed hotline that has been in operation since August, 1992. It is a 24-hour service that enables the technician to access technical service bulletins and service information compiled directly from field experiences and OEM scientific and factual information in the SAE J2008 format. The technician may access whole service information libraries through ONLINE "gateways" to third-party information providers. File libraries can be accessed by the technician to retrieve IM240 emissions traces, OEM emissions/safety product recalls, and OEM emissions recall notices in the SAE J2008 format. ONLINE training in the use of the library system is provided ONLINE when the technician "pages sysop" for assistance. A diskette containing modem communications and graphical user interface (GUI) features is provided free of charge (freeware). This allows technicians to use mouse-driven or keyboard selection methods to view graphical images ONLINE. The cost for this service is flexible, and can be monthly, pre-paid ONLINE time (the technician purchases "credits"), or billable monthly credit usage. A free diskette can be obtained from the contact listed above.

EDUCATION AND CERTIFICATION

The Final Rule for Inspection/Maintenance Program Requirements mandates that all necessary training be made available to all. It further describes what should be included in the training.

Since the I/M Final Rule was published, it was suggested that *education* would be a more appropriate term to use than *training*. As a result, *education* is used in this document in place of *training* whenever possible.

Subject Criteria for an Effective Education Program

The EPA strongly urges states to use their own criteria to evaluate education programs. A state is free to propose other subject matter. For example, some states have in place, or are developing, their own technician certification testing programs which include similar advanced-level tasks. The state may choose to use NATEF Certification, offered by ASE, or its own certification testing program to determine if adequate education is available to meet each state's particular education and certification needs, as long as they meet the requirements of the Final Rule.

Existing and Proposed Certification Programs

The certification of technicians and the licensing or certification of repair facilities is not mandated by the I/M rule. States will be able to use their own discretion when considering these issues; many states have licensing and certification policies already in effect that could be built upon.

ASE Technician Certification Program

The technician certification system of the National Institute for Automotive Service Excellence (ASE) includes eight separate exams with each focusing on one specific area of automobile repair. Upon passing at least one exam and after providing proof of two years of appropriate hands-on work experience, the technician becomes ASE certified in that particular area. When all eight exams have been successfully completed, the technician is then certified as an ASE Master Technician. Recertification tests are given every five years and cover the same content areas as the original exams. However, the number of questions in the recertification test is reduced by about half.

In addition to the eight automobile tests, a new test, the Advanced Engine Performance Specialist (L1) test has been added. It is an advanced-level exam, requiring successful ASE certification in Engine Performance. This test is designed to measure a technician's knowledge of the diagnostic skills necessary for sophisticated emissions and engine performance problems. The L1 test was first administered in May, 1994. For more detailed information about the certificate program and the content of the other test areas, contact:

National Institute for Automotive
Service Excellence
13505 Dulles Technology Drive
Herndon, VA 22071-3415
1-703-713-3800

Test Content Areas for ASE Tests A6, A8, and L1 — States may set technician certification standards by using preexisting certification testing. Several states have adopted the ASE Electrical/ Electronic Systems (Test A6), Engine Performance (Test A8), and the new L1 Test as their basis for establishing technician certification. The content of these three tests is as follows:

1. Electrical/Electronic Systems (Test A6)
   a. General Electrical/Electronic System Diagnosis
   b. Battery Diagnosis and Service
   c. Starting System Diagnosis and Repair
   d. Charging System Diagnosis and Repair
   e. Lighting Systems Diagnosis and Repair
   f. Gauges, Warning Devices, and Driver Information Systems Diagnosis and Repair
   g. Horn and Wiper/Washer Diagnosis and Repair
   h. Accessories Diagnosis and Repair
2. Engine Performance (Test A8)
   a. General Engine Diagnosis
   b. Computerized Engine Controls Diagnosis and Repair
   c. Ignition System Diagnosis and Repair
   d. Fuel, Air Induction, and Exhaust Systems Diagnosis and Repair
   e. Emissions Control Systems Diagnosis and Repair
   f. Engine Related Service
   g. Engine Electrical Systems Diagnosis and Repair
3. Advanced Engine Performance Specialist (Test L1)
   a. General Power Train Diagnosis
   b. Computerized Engine Controls Diagnosis
   c. Ignition System Diagnosis
   d. Fuel Systems and Air Induction Systems Diagnosis
   e. Emissions Control Systems Diagnosis
   f. I/M Failure Diagnosis and Repair

NATEF Education Facility Certification

To help recruit, mentor, educate future technicians, and improve technical instruction, the National Institute for Automotive Service Excellence (ASE) offers certification for education programs. Automotive training programs can earn ASE certification upon the recommendation of the National Automotive Technicians Education Foundation (NATEF). NATEF is a sister organization of ASE.

Through its certification process, NATEF examines the structure and resources of the education programs and evaluates them against nationally accepted standards of quality. For detailed information, contact NATEF directly:

NATEF
13505 Dulles Technology Drive
Herndon, VA 22071-3415
1-703-713-0100

Any secondary, post-secondary, technical institute, or community college program may apply for certification. The program is evaluated to determine if the equipment, curriculum, instructional system, job placement program, staff credentials, safety features, cleanliness, and professionalism meet industry standards.

If the requirements are met, the program will become certified by ASE for five years.

Certification of an automobile training program requires that the facility be certified in at least three of the eight ASE test areas: Brakes, Electrical/Electronic Systems, and Engine Performance. As of January 1, 1996, Suspension and Steering will become the fourth required area. There has also been discussion of NATEF certifying in-service technician education programs. This certification would focus on the criteria in the Engine Performance test and the Advanced Engine Performance test.

Technicians and employers should be aware of these programs. They are an excellent source for technician training and prospective employees. Technicians being trained at these institutions must perform a number of competency-based tasks to successfully complete the program.

CHAPTER 3

Technical Guidance for I/M Programs

UNDERSTANDING IM240 TEST PROCEDURES

The Clean Air Act Amendment of 1990 has had an extensive and significant impact on the automotive service industry. Besides requiring lower tailpipe emissions and standardized service information, this regulation also changes the periodic vehicle emissions inspection requirements.

The EPA has developed three new, high-tech short tests for use in Inspection and Maintenance (I/M) programs across the country. These tests do a more thorough job of identifying which vehicles need emissions repair. Revised regulations make it hard to avoid having the polluting vehicles repaired.

The new high-tech short tests consist of three distinct parts, but have become collectively known as the Inspection and Maintenance 240-second test (IM240). The three short tests that will be used are the IM240 transient (loaded-mode) test, pressure testing the fuel tank and evaporative emissions system, and measuring charcoal canister purge flow volume.

TRANSIENT (LOADED-MODE) TESTING

The first test subjects the vehicle to a 240-second mass (grams per mile) emissions tailpipe test. The test incorporates a Constant Volume Sampling (CVS) machine that collects and measures exhaust emissions while driving the vehicle on a chassis dynamometer (Figure 3-1).

The goal of the transient (loaded-mode) IM240 test is to measure the precise amount of pollutants a

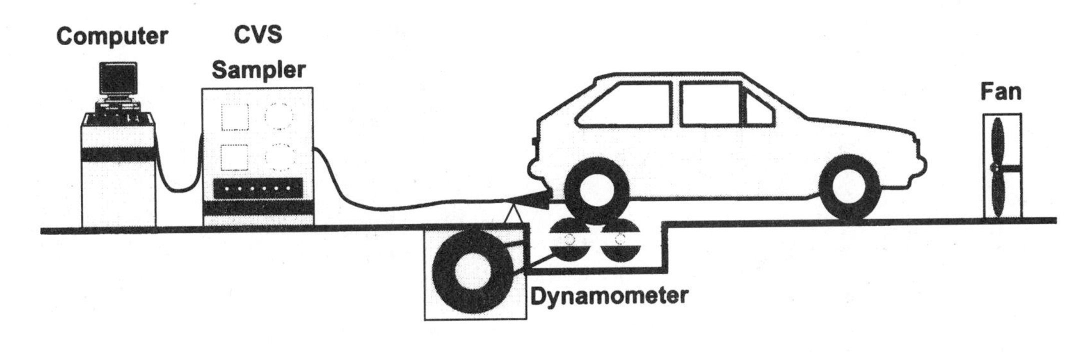

Figure 3-1 Typical IM240 lane.

vehicle produces during normal, on-road driving. The switch to loaded-mode testing makes this precise measurement possible. Unless the analyzing system collects all of the exhaust gases and measures them accurately, the test loses its effectiveness.

This is why the method of gas collection is so important. Standalone and portable in-flight gas analyzers measure only a portion of the exhaust stream. *Partial stream sampling* draws in a low volume of exhaust gases and measures them in concentration of parts per million (ppm) and percentages (%). Concentration readings, by themselves, really do not tell the story. Without controlled measurement of the volume of gases, the readings are not reflective of actual emissions emitted into the atmosphere (Figure 3-2).

A concentration measurement of anything is very limited because it does not tell *how much volume,* it only tells how much there is in relationship to something else. When the total volume is unknown, the concentrations mean very little because partial stream sampling is different for each vehicle. A concentration reading of 2% CO for a 4-cylinder engine is different than a concentration of 2% CO for a big V-8 because the volume of exhaust gases emitted from the V-8 may be twice as much as the four cylinder.

The Constant Volume Sampling (CVS) system guarantees that a constant volume and rate of ambient air and exhaust passes through the gas analyzer

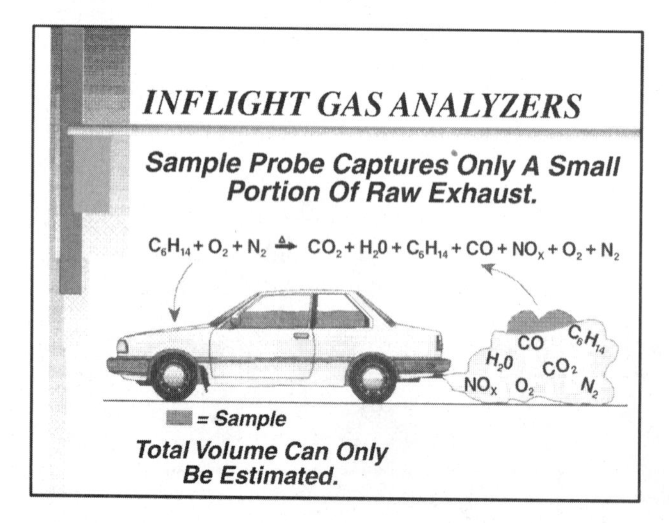

Figure 3-2 Gas concentrations do not effectively reflect the actual exhaust emissions (courtesy Sensors, Inc.).

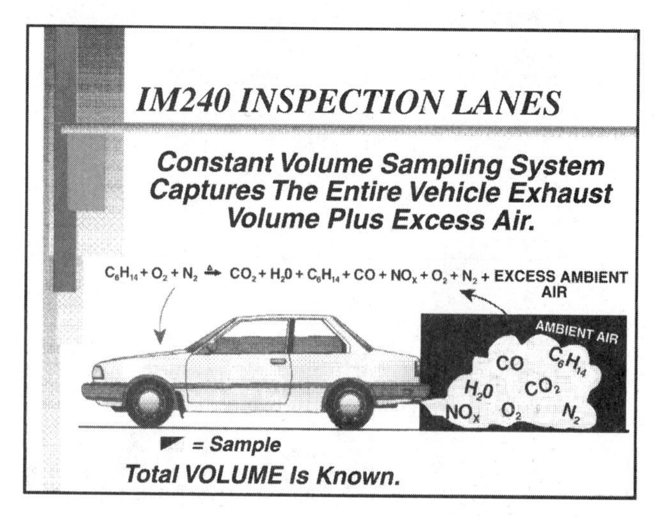

Figure 3-3 Constant Volume Sampling systems capture the entire exhaust emissions (courtesy Sensors, Inc.).

(Figure 3-3). The recommended flow rate for a CVS system is 700 standard cubic feet per minute (SCFM). The main components (Figure 3-4) of a CVS system are the turbine and the venturi. The blower collects exhaust gases and the venturi maintains a constant flow rate by regulating the gas flow. The venturi can be seen to the left of the CVS unit

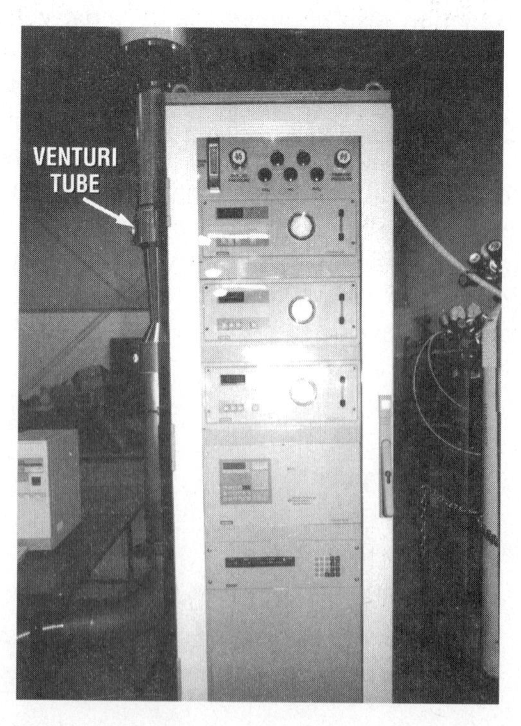

Figure 3-4 A venturi regulates air flow through the Constant Volume Sampling system shown here (courtesy Envirotest Systems, Corp.).

shown in Figure 3-4. It is a fixed opening that allows a fixed amount of gas through per minute. The CVS blower collects the mix at a faster rate than the venturi will let through, keeping the flow through the venturi at the highest level possible throughout the entire test.

The sampling hose(s) from the CVS system slips over the vehicle's tailpipe(s) (Figure 3-5). The sampling hose has a mixing-tee that draws in ambient (makeup) air. If a car emits 100 SCFM of exhaust, the mixing-tee pulls in the makeup air at 600 SCFM. If the exhaust volume increases to 300 SCFM, the mixing-tee only pulls in 400 SCFM of ambient air. This makeup air guarantees that the CVS system receives the constant volume it needs to calculate mass exhaust emissions.

As part of the normal inspection process, the ambient background (diluted air) is sampled and analyzed by the CVS system within seconds of beginning each emissions test. The test computer controls this process to ensure that the sample is taken within 120 seconds of the test and that the sample period is 15 seconds. The average readings for each of the three compounds of interest are recorded in the vehicle test record. Additionally, the ambient air is sampled between tests. If the testing lane's ambient air background check reveals concentrations of 20 ppm HC, 30 ppm CO, 2 ppm NO_x, or greater than the corresponding outside air concentrations, further testing will be prohibited until the ambient background readings are within acceptable limits.

Testing a vehicle while it is being driven on a chassis dynamometer is called transient (loaded-mode) testing (Figure 3-6). A chassis dynamometer is a device that applies a controlled load to the drive wheels of the vehicle. The load is placed at a right angle to the tire across the tread. To correctly simulate the road load for a given vehicle, the dynamometer must be programmed for the correct inertia weight.

To understand the concept of inertia simulation, imagine two vehicles on a highway; one is relatively light-weight (2500 lbs.), and the other is heavy (5000 lbs.). The two are accelerating from 0 to 50 mph over a given time interval. Ignoring the frictional effects of the movement, the energy expended in accelerating the heavy vehicle will be twice that of the lighter vehicle. Also, all other things being equal, the amount of exhaust emissions produced by the heavy vehicle will be twice that of the lighter vehicle.

When these vehicles are placed on a dynamometer and accelerated in a similar manner, the dynamometer must ensure that the energy expended and the exhaust released from both vehicles is the same as on the highway. Wind and tire friction loading are simulated with the dynamometer through the use of a Power Absorption Unit (PAU).

Figure 3-5 Constant Volume Sampling hose covers the entire tailpipe (courtesy Envirotest Systems, Corp.).

Figure 3-6 Chassis dynamometers are used to simulate actual road conditions (courtesy Envirotest Systems, Corp.).

The PAU is an electromagnetic device that absorbs the power produced by the test vehicle.

In order for the dynamometer to do this, it must be told how much inertia to simulate for each vehicle tested. For example, a dynamometer with a fixed 2500-lb. rotational inertia (base inertia) is used to test the two vehicles discussed previously. The 2500-lb. vehicle would experience the same acceleration rate as it would on the highway, not including frictional resistances. The dynamometer is successfully *simulating* the vehicle's inertia. If the 5000-lb. vehicle is driven on this dynamometer, it will accelerate much faster than it would on the road. This fixed inertia dynamometer cannot simulate the correct inertia for this vehicle, so the exhaust emissions will not be the same as on the highway.

Two types of inertia simulation dynamometers can be used for the IM240 test:

1. *Mechanical Inertia Simulation* (Figure 3-7) — This dynamometer is equipped with flywheels (rotating discs) that can vary test inertia weights between 2000 and 5500 pounds, in increments of no greater than 500 pounds.

Figure 3-7 Inertia simulation can be achieved through mechanical flywheels (courtesy Clayton Industries).

2. *Electric Inertia Simulation* (Figure 3-8) — Electric Inertia Simulation (Eddy current), or a combination of electric and mechanical simulation can be used. Eddy current units are electromagnets that

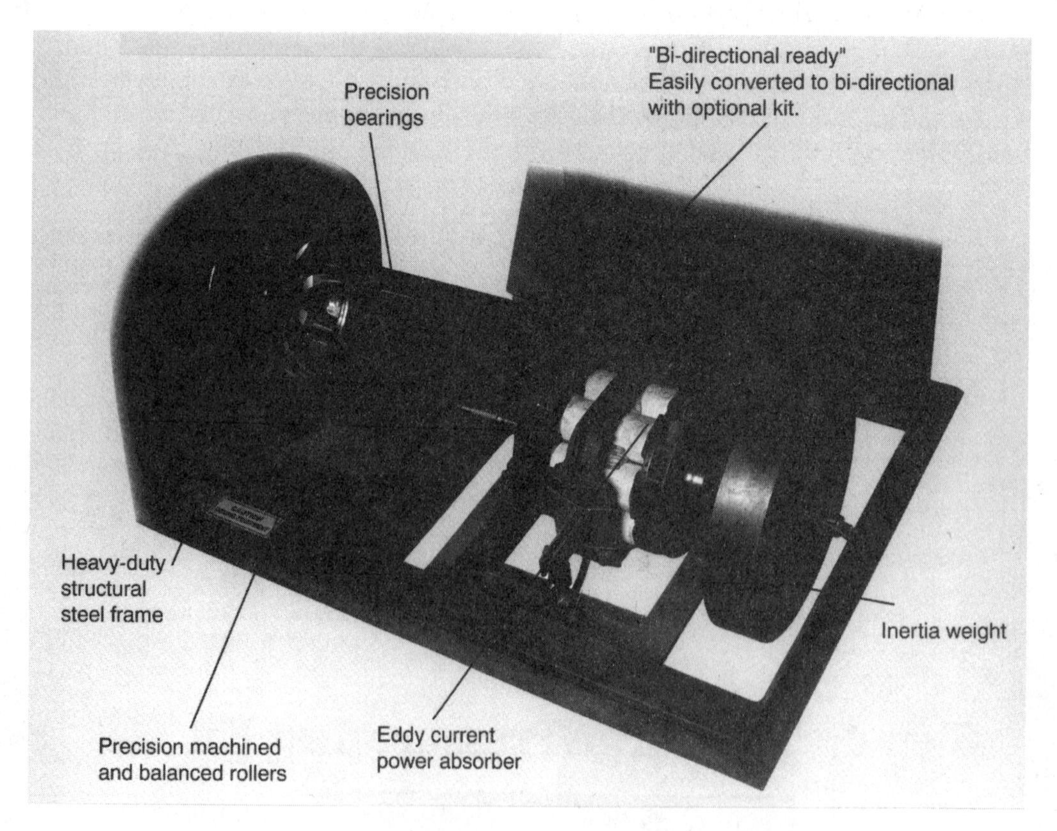

Figure 3-8 Inertia simulation can be achieved through a combination of an Eddy current power absorber and mechanical flywheel (courtesy Mustang Dynamometer).

use magnetic attraction to load the vehicle's drive wheels. An Eddy current PAU can be programmed to duplicate the inertia weight simulation on acceleration through induced current, and deceleration acting like a motor.

For the IM240 test, dynamometer inertia weight settings are automatically chosen from an EPA-supplied look-up table. This information is obtained when the vehicle information is programmed into the computer before the test begins. If the vehicle to be tested is not listed, it can be tested using the default values for inertia setting as shown in Figure 3-9.

A restraint system is required to prevent a vehicle from climbing over the front or rear rolls during acceleration and deceleration. It also prevents excessive sideways movement of the vehicle on the rolls. The restraints are movement-limiting idle rollers rather than a tie-down system; therefore, the restraint mechanism imposes no vertical forces on the vehicle.

During each emissions test, checks are made to determine if the vehicle is being properly loaded. The horsepower is calculated for seconds 55-81 and 189-201 of the driving cycle. The theoretical horsepower for the vehicle under test is calculated using the actual vehicle speeds over the defined time intervals. If the actual horsepower differs from the theoretical horsepower by more than 0.5 HP (2.0 HP for vehicles whose gross vehicle weight exceeds 10,000 pounds), then the test is voided and the lane will be closed until the problem is corrected.

VEHICLE TYPE	NUMBER OF CYLINDERS	ACTUAL ROAD LOAD HORSEPOWER	TEST INERTIA WEIGHT
All	3	8.3	2000
All	4	9.4	2500
All	5	10.3	3000
All	6	10.3	3000
LDGV	8	11.2	3500
LDGT	8	12.0	4000
LDGV	10	11.2	3500
LDGT	10	12.7	4500
LDGV	12	12.0	4000
LDGT	12	13.4	5000

Figure 3-9 Inertia simulation weight can be manually selected when a vehicle is not in the computer (courtesy EPA).

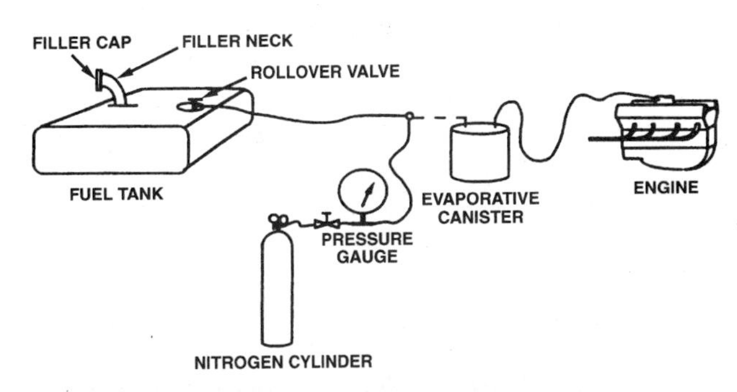

Figure 3-10 Typical fuel and evaporative pressure test setup (courtesy EPA).

EVAPORATIVE SYSTEM TESTING

The other two short tests include a pressure and purge test of the evaporative system. The evaporative system inspections determine whether or not the vehicle's evaporative emission control system is performing correctly. Two different functions of the evaporative emission control system are tested:

1. Pressure test.
2. Purge flow test.

The pressure test (Figure 3-10) determines that the fuel vapors are not leaking into the atmosphere, but are being correctly routed to the charcoal canister where they are stored until being recycled into the engine. This test establishes each system's integrity to hold pressure and not leak. Nitrogen (N_2), or an equivalent non-toxic, non-greenhouse inert gas, can be used to pressurize the system. The evaporative system is tested by pressurizing the system to 0.5 psi or 14" of water. The system must be able to maintain this pressure for 2 minutes without dropping below 8" of water. Components being tested include:

1. Hoses from the canister to the fuel tank.
2. Seams and seals of the fuel tank.
3. Fuel cap/filler seal and the fuel cap pressure relief valve. The fuel cap vacuum relief valve is not tested in this procedure.

The purge test (Figure 3-11) determines that the charcoal canister fuel vapors are being purged back into the intake system. Thus, fuel vapors stored in the charcoal canister are rerouted into the engine, where they are burned.

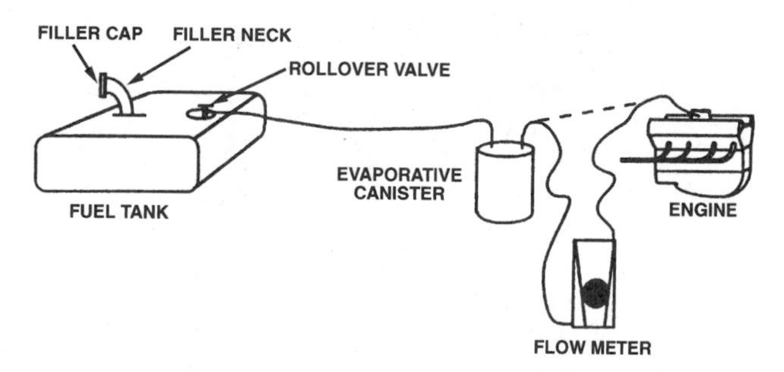

Figure 3-11 Typical evaporative system purge test setup (courtesy EPA).

If the fuel vapor containment system is leaking or the canister purge system is not functioning, then the excess fuel vapors (hydrocarbons) are escaping to the atmosphere. During the IM240 drive cycle, canister flow is measured by a flow transducer and specifications require a minimum purge rate of 1 liter per test.

IM240 TRANSIENT DRIVE CYCLE

The IM240 drive cycle tests the vehicle over a wide range of speeds and operational modes under loaded-mode (transient) conditions. Figure 3-12 shows the entire IM240 cycle in seconds of time and miles per hour.

The inspector driving the trace must follow an electronic, visual depiction (Figure 3-13) of the

Time second	Speed mph	Time second	Speed mph	Time second	Speed mph	Time second	Speed mph	Time second	Speed mph
0	0	48	25.7	96	0	144	24.6	192	54.6
1	0	49	26.1	97	0	145	24.6	193	54.8
2	0	50	26.7	98	3.3	146	25.1	194	55.1
3	0	51	27.5	99	6.6	147	25.6	195	55.5
4	0	52	28.6	100	9.9	148	25.7	196	55.7
5	3	53	29.3	101	13.2	149	25.4	197	56.1
6	5.9	54	29.8	102	16.5	150	24.9	198	56.3
7	8.6	55	30.1	103	19.8	151	25	199	56.6
8	11.5	56	30.4	104	22.2	152	25.4	200	56.7
9	14.3	57	30.7	105	24.3	153	26	201	56.7
10	16.9	58	30.7	106	25.8	154	26	202	56.3
11	17.3	59	30.5	107	26.4	155	25.7	203	56
12	18.1	60	30.4	108	25.7	156	26.1	204	55
13	20.7	61	30.3	109	25.1	157	26.7	205	53.4
14	21.7	62	30.4	110	24.7	158	27.3	206	51.6
15	22.4	63	30.8	111	25.2	159	30.5	207	51.8
16	22.5	64	30.4	112	25.4	160	33.5	208	52.1
17	22.1	65	29.9	113	27.2	161	36.2	209	52.5
18	21.5	66	29.5	114	26.5	162	37.3	210	53
19	20.9	67	29.8	115	24	163	39.3	211	53.5
20	20.4	68	30.3	116	22.7	164	40.5	212	54
21	19.8	69	30.7	117	19.4	165	42.1	213	54.9
22	17	70	30.9	118	17.7	166	43.5	214	55.4
23	14.9	71	31	119	17.2	167	45.1	215	55.6
24	14.9	72	30.9	120	18.1	168	46	216	56
25	15.2	73	30.4	121	18.6	169	46.8	217	56
26	15.5	74	29.8	122	20	170	47.5	218	55.8
27	16	75	29.9	123	20.7	171	47.5	219	55.2
28	17.1	76	30.2	124	21.7	172	47.3	220	54.5
29	19.1	77	30.7	125	22.4	173	47.2	221	53.6
30	21.1	78	31.2	126	22.5	174	47.2	222	52.5
31	22.7	79	31.8	127	22.1	175	47.4	223	51.5
32	22.9	80	32.2	128	21.5	176	47.9	224	50.5
33	22.7	81	32.4	129	20.9	177	48.5	225	48
34	22.6	82	32.2	130	20.4	178	49.1	226	44.5
35	21.3	83	31.7	131	19.8	179	49.5	227	41
36	19	84	28.6	132	17	180	50	228	37.5
37	17.1	85	25.1	133	17.1	181	50.6	229	34
38	15.8	86	21.6	134	15.8	182	51	230	30.5
39	15.8	87	18.1	135	15.8	183	51.5	231	27
40	17.7	88	14.6	136	17.7	184	52.2	232	23.5
41	19.8	89	11.1	137	19.8	185	53.2	233	20
42	21.6	90	7.6	138	21.6	186	54.1	234	16.5
43	23.2	91	4.1	139	22.2	187	54.6	235	13
44	24.2	92	0.6	140	24.5	188	54.9	236	9.5
45	24.6	93	0	141	24.7	189	55	237	6
46	24.9	94	0	142	24.8	190	54.9	238	2.5
47	25	95	0	143	24.7	191	54.6	239	0

Figure 3-12 Second-by-second vehicle speeds during an IM240 test (courtesy EPA).

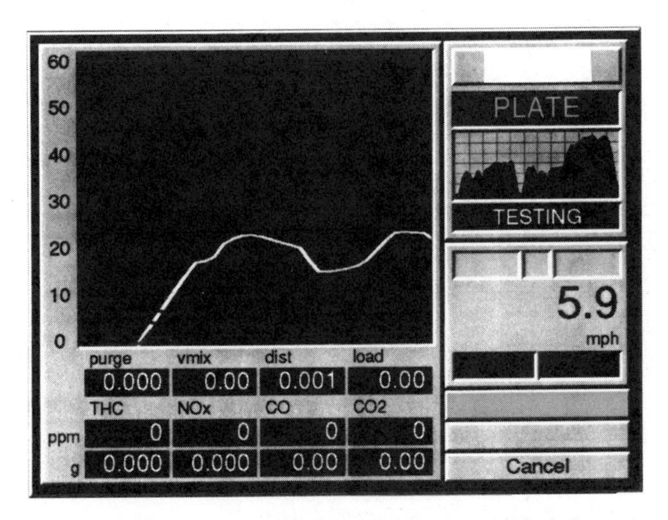

Figure 3-13 Typical IM240 driving trace that the Lane Inspector follows. The trace includes idle, acceleration, and deceleration modes (courtesy Environmental Systems Products, Inc.).

Shift Sequence *gear*	Speed *miles per hour*	Nominal Cycle Time *seconds*
1 - 2	15	9.3
2 - 3	25	47.0
De-clutch	15	87.9
1 - 2	15	101.6
2 - 3	25	105.5
3 - 2	17	119.0
2 - 3	25	145.8
3 - 4	40	163.6
4 - 5	45	167.0
5 - 6	50	180.0
De-clutch	15	234.5

Figure 3-14 Manual transmissions must be shifted during the IM240 test (courtesy EPA).

speed, time, acceleration, and load relationship of the transient driving cycle. For vehicles equipped with a manual transmission, the inspector is prompted by the computer to shift gears following the schedule shown in Figure 3-14.

Figure 3-15 shows the IM240 drive cycle converted to a trace. The trace is based on road speed versus time. The test incorporates two phases (Phase 1 and Phase 2) that when combined are called the *Test Composite.*

Phase 1 is the first 95 seconds of the drive cycle. The vehicle travels 0.560 miles and is equivalent to a flat highway driven at a maximum speed of

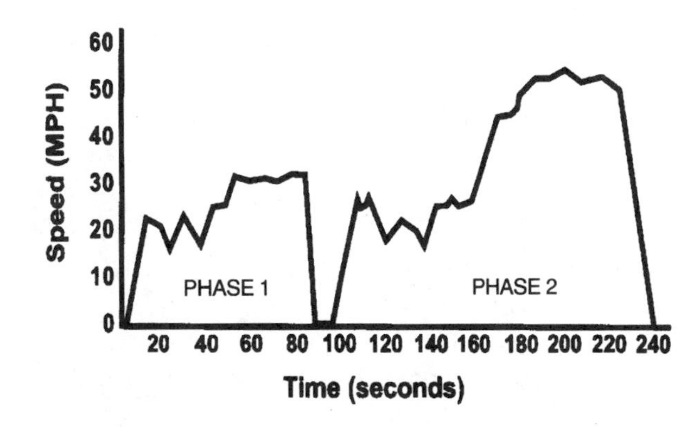

Figure 3-15 Phase 1 + Phase 2 = Composite IM240 Test.

32.4 miles per hour (mph) under light to moderate loads. Phase 2 is from 96 seconds to 240 seconds of the drive cycle. The vehicle travels 1.397 miles on a flat highway at a maximum speed of 56.7 mph, including a hard acceleration to highway speeds.

The length of the IM240 test varies depending on the emissions levels from the vehicle. As soon as the emissions levels indicate the vehicle is exceptionally clean or dirty, and that the purge test has exceeded the minimum, the computer notifies the inspector to stop the test. This is called a fast pass/fast fail.

IM240 EMISSIONS STANDARDS

Two sets of standards are used for each separate emission as shown in Figure 3-16. The *Test Composite* standards are based on the cumulative emissions levels during the full driving cycle. Phase 2 NO_x standards (Reserved) have yet to be established.

The start-up standards for exhaust emissions are in grams per mile (Figure 3-16). They are considered phase-in standards and are recommended by the EPA for use during calendar years 1995 and 1996. Tier 1 standards are recommended for 1996 and later vehicles and may also be used for 1994 and later vehicles certified to Tier 1 standards.

The final standards for vehicles tested in calendar years 1997 and later are shown in Figure 3-17. Tier 1 standards are required for all 1996 and newer vehicles, but may also be used for 1984 and newer vehicles.

There are two ways to pass the standard for each emission. A vehicle will pass the IM240 test for a

(i) Light Duty Vehicles.

Model Years	Hydrocarbons		Carbon Monoxide		Oxides of Nitrogen	
	Composite	Phase 2	Composite	Phase 2	Composite	Phase 2
1994+ Tier 1	0.80	0.50	15.0	12.0	2.0	*(Reserved)*
1991-1995	1.20	0.75	20.0	16.0	2.5	*(Reserved)*
1983-1990	2.00	1.25	30.0	24.0	3.0	*(Reserved)*
1981-1982	2.00	1.25	60.0	48.0	3.0	*(Reserved)*
1980	2.00	1.25	60.0	48.0	6.0	*(Reserved)*
1977-1979	7.50	5.00	90.0	72.0	6.0	*(Reserved)*
1975-1976	7.50	5.00	90.0	72.0	9.0	*(Reserved)*
1973-1974	10.0	6.00	150	120	9.0	*(Reserved)*
1968-1972	10.0	6.00	150	120	10.0	*(Reserved)*

Figure 3-16 EPA-recommended start-up standards for IM240 programs (courtesy EPA).

(i) Light Duty Vehicles.

Model Years	Hydrocarbons		Carbon Monoxide		Oxides of Nitrogen	
	Composite	Phase 2	Composite	Phase 2	Composite	Phase 2
1994+ Tier 1	0.60	0.40	10.0	8.0	1.5	*(Reserved)*
1983-1995	0.80	0.50	15.0	12.0	2.0	*(Reserved)*
1981-1982	0.80	0.50	30.0	24.0	2.0	*(Reserved)*
1980	0.80	0.50	30.0	24.0	4.0	*(Reserved)*
1977-1979	3.00	2.00	65.0	52.0	4.0	*(Reserved)*
1975-1976	3.00	2.00	65.0	52.0	6.0	*(Reserved)*
1973-1974	7.00	4.50	120	96.0	6.0	*(Reserved)*
1968-1972	7.00	4.50	120	96.0	7.0	*(Reserved)*

Figure 3-17 EPA-recommended final standards for IM240 programs (courtesy EPA).

given exhaust emission if either of the following conditions occur:

1. During the full driving cycle the cumulative exhaust readings for all emissions are below the composite standards for each emission.
2. If the composite emissions rates calculated exceed the standard for any exhaust emission, additional analysis of the results will look at Phase 2 of the driving cycle separately. If Phase 2 grams per mile emissions levels are equal to or below the applicable Phase 2 standard, then the emission in question passes.

If the vehicle continues to fail, then second-by-second grams and composite exhaust emissions rates in grams per mile will be calculated for the entire test. If the second-by-second and/or composite emissions levels are equal to or below the composite standard, then the emission in question will pass.

IM240 INSPECTION AND TEST PROCEDURES

This section is a walk-through of an IM240 inspection from beginning to end. It is based on a centralized testing station with multiple positions and lanes. Lane Inspectors are the people responsible for conducting the inspection.

The test is conducted in an inspection lane that is comprised of three work stations. Each station is responsible for different aspects of the inspection and testing. During the course of the emissions test, the vehicle stops at each of the three stations, where Lane Inspectors conduct different portions of the test:

Station #1:
1. Pre-inspection checks.
2. Vehicle Identification and verification.
3. Database search for manufacturer recalls.
4. Preparation for emissions and evaporative tests.

Station #2:
1. Emissions testing.
2. Purge flow volume testing.
3. Removal of some test equipment.

Station #3:
1. Evaporative pressure test.
2. Removal of all remaining test equipment.
3. Report test results to the motorist.

Pre-Inspection Qualifications and Safety Checks (Station 1)

As the vehicle enters the inspection lane, the motorist is greeted by a Lane Inspector. The Lane Inspector initiates the fee collection process, if required, and the appropriate fee is collected. The Lane Inspector completes the entry of vehicle data. The motorist is directed to the waiting area for their safety and comfort. The vehicle is moved to Station 1, where the Lane Inspector begins the safety check process.

Before the vehicle is allowed to begin the transient test, it must first pass the pre-inspection checks. The vehicle's engine is shut off while this portion of the inspection is being conducted. It is safer for the Lane Inspector installing the purge and pressure test harnesses to have the engine off while working under the hood. The Lane Inspector examines each vehicle for several items (Figure 3-18).

Each vehicle should be inspected at Station 1 for the following items before driving the vehicle onto the dynamometer:

1. The vehicle should be at operating temperature, and all accessories should be off. If the vehicle is not at operating temperature, it will be rejected until it reaches operating temperature.

Figure 3-18 A safety inspection is required of all vehicles undergoing an IM240 test (courtesy Envirotest Systems Corp.).

2. If the vehicle has an engine knocking noise indicative of internal engine damage, it will be rejected. Also, if the vehicle exhibits a condition that could pose a hazard to the Lane Inspectors, it will be rejected.
3. Exhaust pipes must be accessible to allow for the collection of the exhaust gases (short pipes ending near the rear differential will be rejected).
4. A visual check of the catalytic converter is done to assure its presence and the appearance of correct connections.
5. Tires with less than 1/16 inch of tread depth in two adjacent grooves at three locations on drive wheels will cause the vehicle to be rejected. Measurements are taken when the tire condition appears to be borderline. With the motorist's permission, grossly underinflated tires are inflated to the correct pressure. Other drive-wheel tire conditions that will cause rejection for safety reasons are:
   a. Wear, cuts, snags, or knots that expose ply or cord.
   b. Temporary tires (emergency spares).
   c. Unmatched tire sizes on the same drive axle.
6. Drive wheels that are mounted without all of the required lug nuts, based on a visual inspection conducted without removing wheel covers, will be rejected.
7. Other checks involve identification of full-time four-wheel drive vehicles and vehicles with

automatic traction control systems. These systems are tested on a dynamometer that is specifically designed for them.

8. An under-hood inspection of the air-injection system is done to check for tampering.
9. If the evaporative canister is determined to be accessible, it will be inspected for obvious damage and the evaporative hoses will be visually checked for proper routing and connection. If the canister is missing, obviously damaged, or if the hoses are improperly routed or disconnected, the vehicle will fail the purge test.
10. The vehicle should be checked for a missing non-vented fuel cap. The tightness of the cap should also be checked. If loose, the cap should be tightened, and the owner informed of the air quality impact of a loose fuel cap.
11. Purge test and pressure test adapters are attached to the vehicle (Figure 3-19). The purge test unit is normally connected in series with the canister purge line at the canister end. If the canister is inaccessible but the purge line can be identified, the test can be conducted with the purge test unit connected where the line is accessible.

If a defect is observed, the vehicle is rejected. If the vehicle is rejected, the customer is given a Vehicle Inspection Report (VIR) detailing the reason(s) for rejection, including a description of the defect and what corrective actions are necessary before the vehicle can return for inspection.

Vehicle Identification/Verification

As the first step in the vehicle identification process, the Lane Inspector enters the vehicle's license plate number into the vehicle ID workstation computer (Figure 3-20). The plate number will be used as a key into the host vehicle identification database. The vehicle information needed to perform the inspection will be transmitted back to the lane.

If the vehicle is not found in the host database, the Vehicle Identification Number (VIN) will be decoded to obtain most of the necessary vehicle parameters. Any remaining parameters can be manually entered by the Lane Inspector. In addition to identifying the vehicle, the vehicle ID data is used to automatically select inspection parameters such as test type (IM240 or two-speed idle), inertia weight setting, purge test requirements, and so on.

The vehicle data retrieved from the host system database contains a test number indicating whether the current inspection request is an initial inspection or a reinspection. If the test is to be a reinspection, the previous test VIR and repair reports are collected.

Manufacturer's Recall

An emissions recall information system can provide a complete corrective action process for handling all vehicles targeted for emissions recall repairs. The system is a closed loop to ensure that all vehicles

Figure 3-19 Evaporative emissions equipment setup (courtesy Envirotest Systems Corp.).

Figure 3-20 All data is entered into the inspection lane computer (courtesy Envirotest Systems Corp.).

are identified and entered into the system. Action is initiated on them, a resolution is achieved, and a detailed status is maintained and made available on a monthly basis.

Queuing (In-Line) Time

It is necessary to measure or establish the length of time a vehicle is waiting in queue for two reasons. First, if the queuing time for a vehicle is greater than 20 minutes, the vehicle may be eligible for a second-chance emissions test. Second, the vehicle queuing times are used to compute the wait times that are displayed on the station's projected wait time sign. The wait time sign is an informational sign that gives motorists approaching the station an indication of the approximate time it will take before their vehicles can be inspected.

Queuing time is automatically determined by the facility's computer. When the vehicle enters the station's driveway, its presence is detected by a magnetic loop or a light beam device. The signal from the detection device triggers a video camera system to capture a video image of the vehicle's license plate. Character recognition software on the station manager computer captures the plate number and records the time when that vehicle arrived.

When the vehicle arrives at Station 1, it is identified by plate number entry and its time of entry is recorded. The time the vehicle arrived at the station is retrieved from the station computer and the wait time is computed from the difference between the time the vehicle entered the driveway and the time it entered Station 1.

Queuing time is the time difference between the vehicle's arrival at the facility and its arrival at the emissions test position. The vehicle qualifies for a second-chance test if all emissions readings are less than or equal to 1.5 times the vehicle's standards and the vehicle has been in queue for more than 20 minutes before being tested. Otherwise, the vehicle fails the emissions test. The pass/fail determination during a second-chance test is as if it were an initial test.

Discretionary Preconditioning

At the discretion of the Lane Inspector, a vehicle may be preconditioned using any of the following methods:

1. *Non-Loaded* — Engine speed should be increased to 2500 RPM, for up to four minutes, with or without a tachometer.
2. *Loaded Mode* — Drive the vehicle on the dynamometer at 30 mph for up to 240 seconds at road load.
3. *Transient* — Drive the transient cycle consisting of speed, time, acceleration, and load similar to that of the transient driving cycle.

IM240 Emissions Test (Station 2)

When the Lane Inspector enters the vehicle in preparation for proceeding, the engine is restarted and the vehicle is moved to Station 2. The Lane Inspector monitors the time from engine restart to the beginning of the driving cycle to ensure that a minimum of 30 seconds has elapsed.

This portion of the test consists of operating the vehicle under test on an inertia simulation dynamometer. The dynamometer portion of the test includes the following procedures:

1. Maneuver the vehicle onto the dynamometer. The vehicle lift feature of the dynamometer should be raised and locked. The hood of the vehicle must be raised for the duration of the inspection.
2. The Lane Inspector releases the dynamometer roll brakes, engages the vehicle's transmission, and allows the tires and rolls to turn slowly until the vehicle sits squarely on the dynamometer. This self-centering process works very well for rear-wheel drive vehicles and for most front-wheel drives. For some front-wheel drive vehicles, the Lane Inspector will steer the vehicle into a square position prior to activating the restraints (Figure 3-21).
3. A remote vehicle cooling system must be positioned and activated as shown in Figure 3-22. The cooling capacity of this fan must be 5400 ±300 SCFM. It should be placed within 12 inches of the intake air to the vehicle's cooling system. The system is designed to avoid improper cooling of the catalytic converter.
4. Wet tires should be dried by running them on the rollers with the preselected load and inertia weight set.
5. The vehicle should be allowed to idle at zero speed for a minimum of 10 seconds.

Figure 3-21 Automatic front-wheel drive restraint system senses the tire position and reverses to set correct running clearance (courtesy Clayton Industries).

Figure 3-22 A large blower prevents the vehicle from overheating during the IM240 transient drive cycle (courtesy Envirotest Systems Corp.).

6. The transient emissions test driving cycle should now begin. The transient driving cycle is 240 seconds and includes periods of idle, cruise, and varying accelerations and decelerations.
7. During the execution of the driving cycle, the vehicle's exhaust is collected by the Constant Volume Sampling (CVS) system. The exhaust gas is analyzed for *total hydrocarbons* (THC), carbon monoxide (CO), carbon dioxide (CO_2), and oxides of nitrogen (NO and NO_2). By having measured the equivalent distance traveled on the dynamometer, the vehicle's mass emissions in grams per mile for each exhaust emission can be calculated. These mass emissions results are compared to the applicable standards to determine the vehicle's pass or fail status. Fuel economy is also calculated as a confirmation of test validity.
8. Upon completion of the transient emissions test, the Lane Inspector disconnects purge hoses, stows the cooling fan and exhaust collector, and lowers the hood. The vehicle then advances to Station 3.

Evaporative Pressure Test (Station 3)

The Lane Inspector drives the vehicle to Station 3. The evaporative pressure test unit is connected, and the computer-controlled pressure test begins. While the pressure test is in progress, the Lane Inspector removes the purge test adapter and reconnects the vehicle's purge hose.

At the completion of the pressure test, the computer makes the final pass/fail decision. The Lane Inspector disconnects the evaporative pressure test unit, removes the adapter, reconnects hoses, and lowers and latches the hood.

Vehicle Inspection Report

The Vehicle Inspection Report (VIR) is not printed until the test data is stored on both hard disk and floppy disk in the lane. This ensures that the vehicle does not leave the lane with a valid VIR until the data is securely stored in at least two locations. Once the vehicle inspection test data is transmitted to the facility's computer, the system prints the vehicle test results and any diagnostic information or reports that may be applicable.

Each VIR contains a variety of data, including the following minimum information specified by the EPA:

1. Facility number, lane number, and Lane Inspector number.
2. Type of test(s) performed.
3. Test date and time.
4. Test serial number.
5. Test certificate number.
6. Test system number.
7. Vehicle make and model year.
8. Vehicle type.
9. License plate number.
10. Transmission type.
11. Number of cylinders.

12. Weight (GVWR).
13. Vehicle Identification Number (VIN).
14. Vehicle odometer reading.
15. Pass/fail result of applicable visual tampering inspections.
16. Test results, including exhaust concentrations and pass/fail results for each mode measured, and pass/fail results for evaporative system checks.
17. The applicable standards for each test.
18. Type of vehicle preconditioning performed.
19. Exhaust emissions test start time and end time.
20. The pass or fail status for the complete emissions inspection.
21. Warranty message.
22. Recall information and recall indicator if appropriate.

While the results are being printed, the motorist is directed back to the vehicle. The Lane Inspector retrieves the VIR and diagnostic reports from the printer and presents them to the motorist with a brief explanation.

The motorist is given a printed copy of the computer-generated test results, whether the vehicle passed, failed, or was rejected. If the vehicle passes the test, the motorist is given the emissions test results and a Certificate of Compliance.

If the vehicle fails the test, the motorist is given a VIR and a certificate of denial, plus supplementary diagnostic information. If the vehicle is a reinspection failure, the Lane Inspector explains the waiver eligibility, answers questions, and directs the motorist to speak to the Inspection Facility Manager, if needed.

Vehicles rejected during the test for safety or performance-related problems will be tagged as an *Abort*. For example, if a vehicle cannot complete the full IM240 trace due to engine stalling, it can be failed after the third occurrence. The VIR results will show an Abort and state the reason(s) why the test was discontinued.

Retest Repair Form

Instructions about how to fill out the retest and repair form are given to motorists of vehicles that do not pass the test. The person performing the repairs is responsible for filling in the pertinent information on the retest repair form.

This data is entered by the Lane Inspector during the reinspection procedure and subsequently transmitted and stored on the facility's computer system. Repair data is used to develop a diagnostic repair database that can provide failing motorists with diagnostic messages based on historic repair scenarios. These reports will feature the categorization of vehicles by make, model, engine type, year, and the itemization of successful repairs completed for the same types of vehicles with the same types of failures. The repair industry should be able to improve their diagnostic and repair techniques appreciably by reviewing these statistics on successful vehicle repairs.

Repair Shop Listing/Report Card

A special repair shop listing/report card will be available at each facility to all motorists, and will be of particular interest to a motorist whose vehicle fails the inspection. It will provide a summary of the performance of repair facilities within the program area that have repaired vehicles presented for retest. The following information will also be provided:

1. Total number of repaired vehicles presented for retest.
2. Number and percentage of repaired vehicles passing the first retest.
3. Number and percentage of repaired vehicles passing subsequent retests.
4. Number and percentage of repaired vehicles receiving waivers.
5. Cost effectiveness of repairs for each facility will be indicated by listing the repair facilities in order from the most to the least cost effective.

CHAPTER 4

Alternative Emissions Testing

INSPECTION NETWORK TYPES

Two basic types of inspection networks have existed since the inception of I/M programs. Concerns exist with both types because each has advantages and disadvantages. Figures 4-1 and 4-2 illustrate the differences between decentralized and centralized programs.

A *centralized* network consists of inspection and retest at high volume, multi-lane, usually highly automated, test-only stations, run by either the state or a single contractor (franchised) within a defined area. Typically, these test facilities are strategically placed and designed to handle the high volumes of vehicles seeking inspection during peak times of the test cycle without long lines. Vehicle repairs and/or other business is not performed or permitted in centralized facilities. In existing centralized networks performing steady-state emissions tests and tampering checks, the ratio of test lanes to annual vehicles tested is about 1 to 35,000.

Figure 4-1 illustrates the operation of a centralized I/M program at which all inspections are done by a special purpose, *test-only* facility. As illustrated in the figure, the testing facility is a large drive-through facility with several different positions at which different elements of the inspection are performed. Vehicles failing an inspection are required to be repaired at a facility of the vehicle owner's choice. After repair, the vehicle must return to the test-only facility for a retest. Under EPA requirements, failing vehicles must continue going back and forth between the inspection facility and a repair facility until they pass the test or the motorist spends a predetermined repair amount.

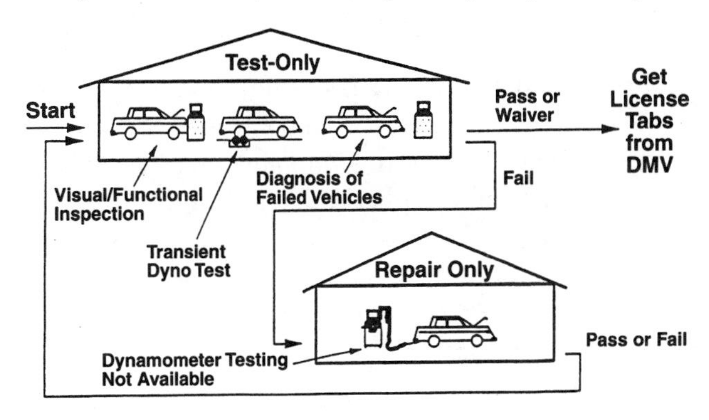

Figure 4-1 Centralized test-only program (courtesy California I/M Review Committee Report).

A *decentralized* network consists of inspection and retesting at privately owned, licensed repair facilities. Typically, there are thousands of stations, depending on the number of vehicles subject to the I/M requirement and the size of the program. The station-to-vehicle ratio in decentralized networks is typically on the order of 1 to 1000, e.g., in the New

York City metropolitan area, 4300 stations test approximately 4,600,000 vehicles annually.

As shown in Figure 4-2, test and repair functions are combined under the decentralized program. Vehicles are inspected at licensed stations, most of which are also in the business of performing emissions-related maintenance and repairs. If the vehicle passes, it receives a *Certificate of Compliance,* which is required by Department of Motor Vehicles (DMV) prior to registration renewal. If the vehicle fails, necessary repairs can often be performed at the same facility. Sometimes the vehicle is taken to another facility for repair and retest, either out of necessity or by owner preference.

If a vehicle cannot be repaired under the applicable repair cost waiver limits, it may be taken to a *referee* facility operated by an independent contractor. The owner must present the referee with a work order from a licensed repair facility showing that the repair cost ceiling has been exceeded, or would be exceeded with further repair. The referee determines if the vehicle is:

1. Free of any tampering (missing, modified, or disconnected emissions-related components).
2. Set to factory specifications and that the repair technician did not overlook anything.

If the referee is satisfied that everything has been done correctly, then a Certificate of Compliance may be issued by the referee.

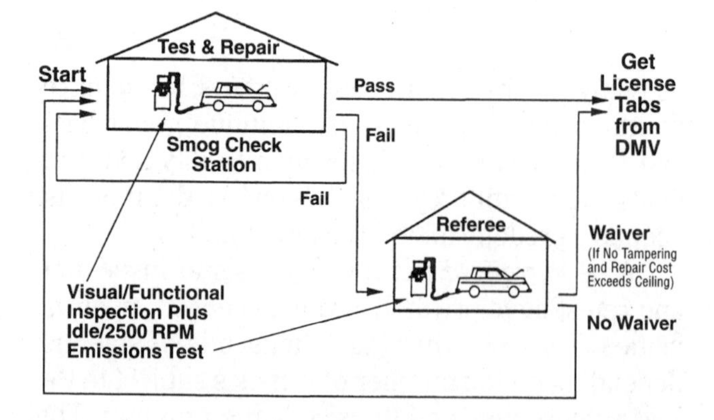

Figure 4-2 Decentralized test and repair program (courtesy California I/M Review Committee Report).

Advantages of Centralized and Decentralized Networks

The advantages offered by centralized programs are lower test fees, reduced oversight and enforcement costs, and an objective test for the consumer. This is because the centralized station performs the inspection and test and does not need to be concerned about the repairs that might be required.

One advantage offered for decentralized test and repair programs by station owners and The Society of Automotive Vehicle Emission Reductions, Inc. (SAVER) was that decentralized programs are more convenient for the public. The consumer can choose where to have a vehicle inspected and repaired. Because repairs and testing are not separated, it is easier for the technician to verify that repairs were performed effectively.

Disadvantages of Centralized and Decentralized Networks

There are potential problems that arise with convenience in both centralized and decentralized test systems. Long driving distances and high prices are a disadvantage of centralized programs. In nearly all cases, this has been in government operated centralized systems. The problem occurs as the result of a combination of factors:

1. Inadequate numbers of stations or lanes to handle peak volumes.
2. Poor station locations.
3. Under-staffing so that all lanes cannot be opened when needed.
4. Insufficient resources.
5. Inadequate equipment and technical expertise.

For the most part, the facility was not originally designed for such high volume testing. Many facilities were originally safety inspection stations, to which emissions testing was later added.

Another motorist convenience issue is the fact that in test-only networks motorists must go to separate facilities for tests and repairs. Centralized facilities are not equipped or licensed to perform the necessary repairs. The motorist has to go back and forth (called *ping-ponging*) between the repair facility and the inspection center when the vehicle continues to fail a retest after repairs have been performed.

In a decentralized system, convenience problems include having to wait an excessive amount of time for a test (excessive waits also occur in poorly-designed centralized programs), having to leave the vehicle behind because testing on demand is not available, being refused testing, and having to return at another time or go to another station. Decentralized stations are rarely originally designed for the purposes of testing and the manual nature of many operations that go into the process can result in a much longer wait time.

Adequate numbers of licensed test stations have been a problem in some decentralized programs, but this is mainly a function of the limited fee that stations in these programs have been allowed to charge the motorist for doing a test. In some states, the test includes safety as well as emissions-related inspections and, when performed correctly, these tests can take as long as 20-30 minutes. Given the rise in shop labor rates, doing inspections in such state programs became a money loser for good repair shops that could better spend their time on higher value services. Thus, insufficient numbers of stations signed up to do testing. In states where there is no test fee cap, such as California, there is a lower vehicle-to-station ratio, indicating that there are more suppliers willing to enter the market.

Based on past performance, the EPA believes that a decentralized test and repair program will not achieve emissions reductions equal to that of a similarly designed centralized program. The fundamental problems with the test and repair approach, especially those related to conflict of interest, have not been successfully controlled in a test and repair program. EPA has looked for strategies that would be sufficient to equalize test and repair program performance. Some have suggested that better emissions analyzers would solve the problem, but it is clear from the experience in programs that have already adopted such equipment that this is not an adequate solution. Similarly, a few states have also implemented rigorous quality assurance programs, but still suffer from significant levels of improper testing. Clearly, performance can be substantially improved in the poorly-run test and repair programs. Better surveillance and more rigorous enforcement should reduce the high incidence of improper testing found in these programs.

Centralized testing programs using transient (loaded mode) testing will create new challenges for the technician. In a high-tech centralized test system, repair technicians will be faced with a more rigorous exhaust emissions test procedure than the idle, two-speed, or loaded steady-state tests now used in I/M programs. These tests will differ in three respects.

First, the transient tests more accurately and selectively determine which vehicles need repair. The steady-state tests pass more gross emitters and fail more vehicles that are close to or below the standards for which the vehicles were originally designed than the transient test. Second, the transient test cannot be fooled by strategies aimed merely at passing a test, such as doping the gasoline with additives or disconnecting vacuum hoses. Third, typical repairs in responding to steady-state tests may not always sufficiently reduce emissions to allow a vehicle to pass a transient test. For example, vehicles without a catalyst or with an empty shell of a catalyst can pass a steady-state test if they are operating in a lean condition during the particular test mode. In actuality, however, such vehicles are gross emitters and could not pass the transient test. The real defects in the emission control system will need to be repaired in a transient test program.

Repairs to pass the transient test may require greater diagnostic proficiency on the part of technicians than what is generally needed in response to a steady-state test failure. Furthermore, some repair facilities may return a vehicle to its owner without verifying that it actually passes the transient exhaust test, due to lack of test equipment or unwillingness to get the vehicle retested at the state inspection station prior to owner pickup. If the entire repair industry is not prepared or able to respond adequately and in a timely manner to the challenge, motorists may be put in the awkward position of failing the retest at higher than necessary rates.

TEST PROCEDURES

Most I/M programs currently operating in North America use idle emissions measurements to determine whether a vehicle passes or fails the tailpipe emissions standards. Some also include a *preconditioning* mode involving a 2500 RPM *high idle* operation of the engine for 20-30 seconds or dynamometer operation at about 30 mph. Some programs use a measurement of emissions at 2500 RPM as well as idle to determine whether a vehicle passes or fails. A few programs (e.g., British Columbia and Arizona) are making pass/fail decisions based on emissions measured with the vehicle operating under a steady load on a chassis dynamometer, in addition to measurements made at idle. Most of the test procedures involve the measurement of both hydrocarbon and carbon monoxide emissions; however, British Columbia and Arizona are also measuring NO_x emissions.

Some programs also include visual and/or functional inspections of the emission control system to determine whether a vehicle passes or fails. The visual inspection includes applicable emissions systems:

1. Positive crankcase ventilation (PCV).
2. Air injection reaction (AIR).
3. Thermostatically-controlled air cleaner (TAC).
4. Evaporative emissions (EVAP).
5. Exhaust gas recirculation (EGR).
6. Catalytic converter (OC and TWC).
7. Spark controls.
8. Computer controls.
9. Other related emissions components.

The visual inspection (Figure 4-3) is designed to identify components or systems that have been tampered with or have deteriorated with age. Tampered or deteriorated conditions include:

1. *Missing* — All or part of an emission control system has been removed from the vehicle.

```
--------------------- UNDERHOOD INSPECTION RESULTS ---------------------

                         **** passed ****

     PCV system
     Thermostatic air cleaner
     Pulse air reed valves
     Fuel evaporative controls
     Catalyst
     Exhaust gas recirculation
     Ignition spark controls
     Wiring of other sensors/switches/computers
     Vacuum line connections to sensors/switches
     Carburetor
     Other emission related components
-------------------------- GENERAL INFORMATION --------------------------

                    ****  test information ****

     Test date: 11/23/94                Test type: after repairs
     Test start time:  9:17 am

                    **** vehicle information ****

     License number: 2AAE487            VIN: JHMST5439CS035080
     Vehicle make: HONDA                Model year : 1982
     Vehicle type: Passenger car        Vehicle model: CIVIC
     Transmission type: Manual          Fuel type: Gasoline
     Odometer reading: 146024           Number of cylinders:  4
     Exhaust type: single               Engine size:   1.5 Liters
     Vehicle certification type:  California
```

Figure 4-3 Inspection and maintenance visual inspection test results.

Example: The manual shows that a catalytic converter is required. During the visual inspection the technician notices that a pipe has been installed in place of the catalytic converter.

2. *Modified* — An emission control system or component has been modified if:
   a. It has been physically or functionally altered.
   b. It has been replaced with a non-original equipment manufacturer (non-OEM) part that has been identified by the manufacturer as not legal for use.
   c. A replacement part designed for one application is used on a different application for which it was not designed.

   Example: An altered device could be an EGR valve where the diaphragm cover has been crushed to prevent the valve from opening.
3. *Disconnected* — An emission control system has been disconnected if a hose, wire, belt, or component required for the operation of the system is present, but has been disconnected.

   Example: The engine is equipped with a thermal vacuum switch normally located between a ported vacuum source and the distributor vacuum advance mechanism. However, the hose has been routed directly from the vacuum advance to a manifold vacuum source, thereby bypassing the TVS switch.
4. *Defective* — An obvious condition that is noticed and is due to normal wear or deterioration. The emission control system or component is not operational, but is not the result of tampering. The component or system must still be in place.

   Example: The AIR system is connected and working properly, but the air injection tubes have rusted out and have obvious holes in them. The system is not tampered with, just worn-out.

Functional testing (Figure 4-4) is done to assure that an emission control system is working correctly. Systems requiring functional testing include:

1. Emissions control warning indicator lights.
2. Idle ignition timing.
3. Exhaust Gas Recirculation (EGR).
4. Fuel fill pipe restrictor.

```
-------------------------- FUNCTIONAL TEST RESULTS -----------------

                           **** pass check ****

Fuel Fillpipe Lead Restrictor
Exhaust Gas Recirculation System
Ignition Timing
        The timing is 10 degrees BTDC.

     ----------------------- GENERAL INFORMATION ------------------

                      **** test information ****

         Test date: 11/22/94                  Test type: after repairs
         Test start time: 11:01 am

                      **** vehicle information ****

         License number: 2EYG254              VIN: JS4JC5LCLH4L76557
         Vehicle make: SUZUKI                 Model year : 1987
         Vehicle type: Truck                  Vehicle model: SAMURAI
         Transmission type: Manual            Fuel type: Gasoline
         Odometer reading: 28308              Number of cylinders:  4
         Exhaust type: single                 Engine size:    1.3 Liters
         GVWR: 5999
         Vehicle certification type:  California
```

Figure 4-4 Inspection and maintenance functional inspection test results.

Two-Speed Idle Emissions Test – No-Load

This test consists of an idle mode followed by a high-speed mode (2,500 ± 300 RPM). Some programs reverse the test procedure by performing the high-speed mode followed by the idle mode. For most model years, there are pass/fail cutpoints for both modes. If a vehicle fails the initial idle mode but passes the high-speed mode, then the vehicle will be given a second-chance idle mode test. This preconditioning can last up to three minutes to assure that the engine, oxygen sensor, and catalytic converter are at operating temperature. During the test, raw exhaust is sampled and analyzed and engine RPM is measured to ensure that the vehicle is operating in the correct mode and within the mode limits.

During the execution of the test, several parameters are monitored to ensure test validity. These include engine RPM, sample dilution (CO + CO_2), sample flow rate, etc. (Figure 4-5).

If the sum of CO + CO_2 does not reach a minimum percentage, which varies between 7-8% at idle, the vehicle cannot be tested. Figure 4-6 shows the results of a test conducted on a vehicle experiencing high dilution at idle. The CO + CO_2 total less than 7%, which is the minimum sample dilution for this vehicle. This is reflected in the idle emissions results; because the BAR '90 machine would not allow the idle test to be conducted, no emissions were measured. The actual problem will be covered in Chapter 11.

The final pass/fail emissions test results for the two-speed idle test is based on ppm of HC and the % of CO. (See Figure 4-6.) The vehicle must pass both the idle mode (initial or second-chance) and the high-speed mode to pass the emissions test.

The problem with idle mode I/M programs is further aggravated by the fact that modern day, computer-controlled emission control systems may not be fully functional when the transmission is out of gear and the engine is idling. To ensure that false failures do not occur, the level of the idle emissions standards is set at a relatively high concentration that results in many defective vehicles passing the test.

Steady-State – Loaded Mode

Steady-state, loaded-mode testing (testing the vehicle at one or more constant speeds) avoids some of the costs associated with transient (IM240) testing.

```
------------------------- EMISSION TEST RESULTS -------------------------

TRAINING MODE: NOT A VALID REPORT

                          2500 RPM TEST RESULTS
                                MAXIMUM
                           ALLOWED EMISSIONS
                                  220
              MEASURED            1.2
HC(ppm)            140
CO(%)             0.92
CO2(%):           14.2
O2(%):             0.9
Engine RPM:       2507

                           IDLE TEST RESULTS
                                MAXIMUM
              MEASURED     ALLOWED EMISSIONS
HC(ppm)            110            150
CO(%)             0.00            1.2
CO2(%):           13.1
O2(%):             2.8

Engine RPM:        931
```

Figure 4-5 Inspection and maintenance emissions test results.

2500 RPM TEST RESULTS

	MEASURED	MAXIMUM ALLOWED EMISSIONS
HC(ppm)	33	220
CO(%)	0.00	1.2
CO2(%):	6.7	
O2(%):	11.2	
Engine RPM:	636	

IDLE TEST RESULTS

	MEASURED	MAXIMUM ALLOWED EMISSIONS
HC(ppm)	0000	100
CO(%)	00.00	1.0
CO2(%):	00.0	
O2(%):	00.0	
Engine RPM:	0000	

GENERAL INFORMATION

**** test information ****

Test date: 11/21/94 Test type: initial
Test start time: 8:51 am
Reason for test: Biennial inspection

**** vehicle information ****

License number: 2MSX628 VIN: 1G2FS2185JL250410
Vehicle make: PONTIAC Model year : 1988
Vehicle type: Passenger car Vehicle model: FIREBIRD
Transmission type: Automatic Fuel type: Gasoline
Odometer reading: 59785 Number of cylinders: 8
Exhaust type: single Engine size: 5.7 Liters
Vehicle certification type: California

Figure 4-6 Emissions test results for a vehicle with high sample dilution.

Although a number of steady-state test procedures have been developed in the past, most do not correlate well with the FTP; however, a new approach to steady-state testing that uses higher loads has recently been shown to substantially improve the correlation with the FTP, especially for NO_x emissions.

The California Bureau of Automotive Repair (BAR) has developed a *hybrid* test that incorporates both steady-state and transient testing called *Acceleration Simulation Mode* (ASM). The ASM procedure is a high-load, steady-state procedure that allows testing to be performed using a relatively simple chassis dynamometer and an NO_x analyzer.

Figure 4-7 shows how the BAR test, in its current state of development, compares to the IM240 test recommended by the EPA. The transient portion of the BAR test, which is only about 90 seconds in length, was created by removing the redundant portions of the IM240. An important feature of the new BAR test is that it adds a variable length preconditioning mode at the beginning of the transient cycle. The proposed preconditioning mode is a 15 mph test under which the dynamometer load is set to load the engine to the same degree that would occur during a moderate acceleration (this is also called the ASM5015 test). This mode is much better for preconditioning than the 2500 RPM, no-load

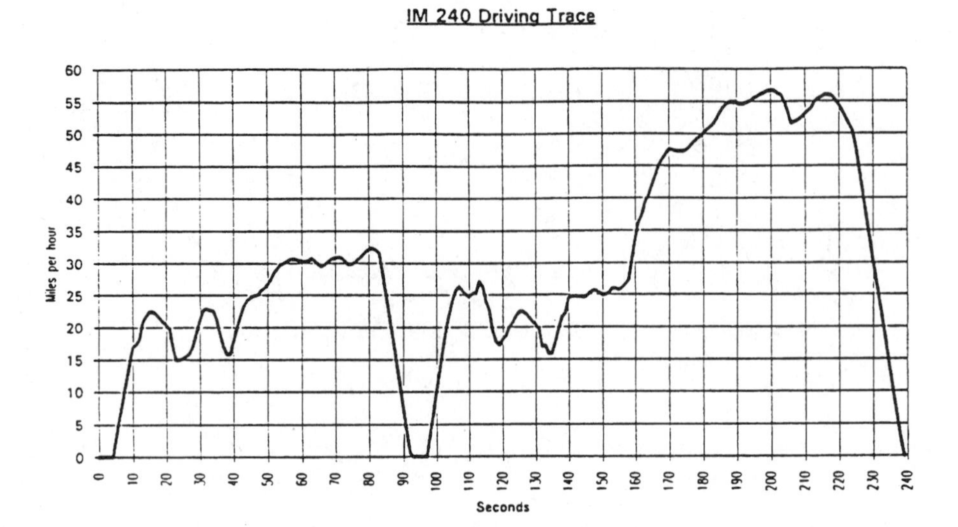

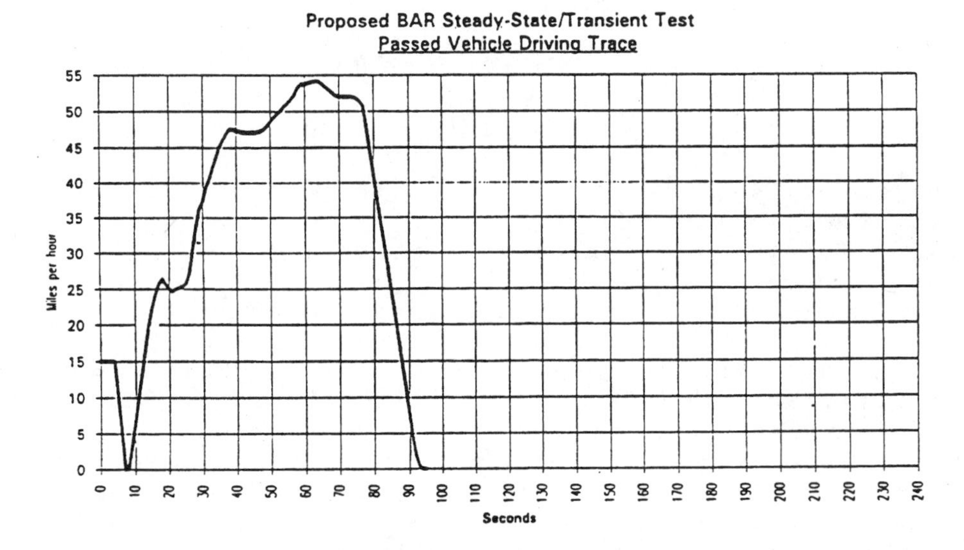

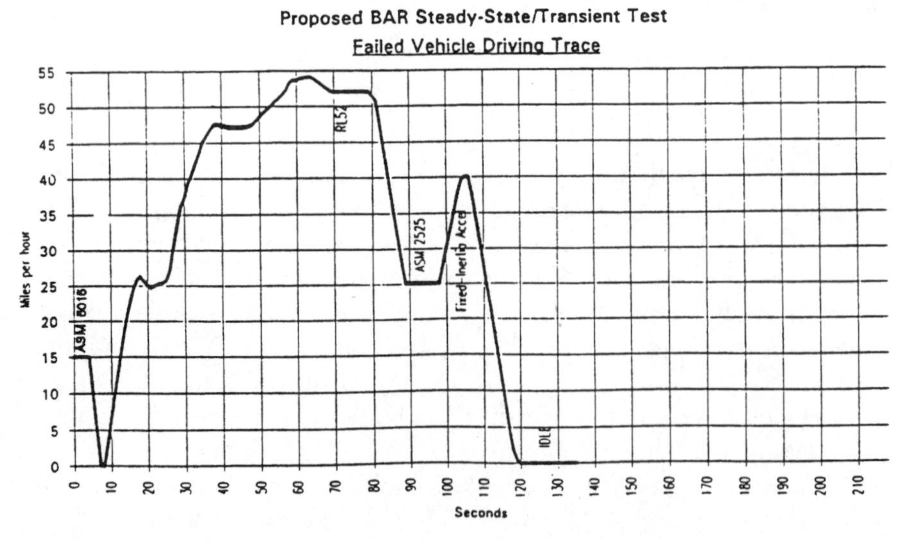

Figure 4-7 IM240 driving trace versus steady-state/transient testing (courtesy California I/M Review Committee Report).

operation used by current programs. Depending on the emissions measured during this mode, the test may be as short as 5 seconds. Longer preconditioning times, perhaps up to 1 minute, would be used for vehicles that appear not to be fully warmed up.

As shown in the bottom portion of Figure 4-7, additional modes are added to the BAR test for vehicles exhibiting high emissions during the first 70 seconds of the transient test. The current proposal is to add a 52 mph road load cruise (RL52), a 25 mph moderate load test (ASM2525), a "fixed-inertia" acceleration test, and an idle test. Using this combination of steady-state and transient testing, the most accurate possible measure of actual emissions for each vehicle could be determined using the transient test, and the steady-state testing could be used on failing vehicles only to provide a benchmark against which repair facilities could determine the effectiveness of repairs using less expensive testing equipment.

Another *hybrid* program concept is being considered. Figure 4-8 illustrates an alternative to a centralized test program that would minimize the disadvantages associated with complete separation of inspection and repair. Under this hybrid concept, all vehicles would be required to be inspected at a test-only facility. Many failed vehicles could avoid returning to the test-only facility for retesting after repair if the repairs are performed at *Gold Shield* Smog Check Stations.

Gold Shield stations would use steady-state dynamometers to determine whether the vehicle has been effectively repaired. Vehicles with tampering or gross emitters would be required to return to the test-only facility for retesting after repairs are made. Using this approach, independent verification of proper repairs would be provided for vehicles with the most serious emissions problems, just as would occur under a centralized program, but other failing

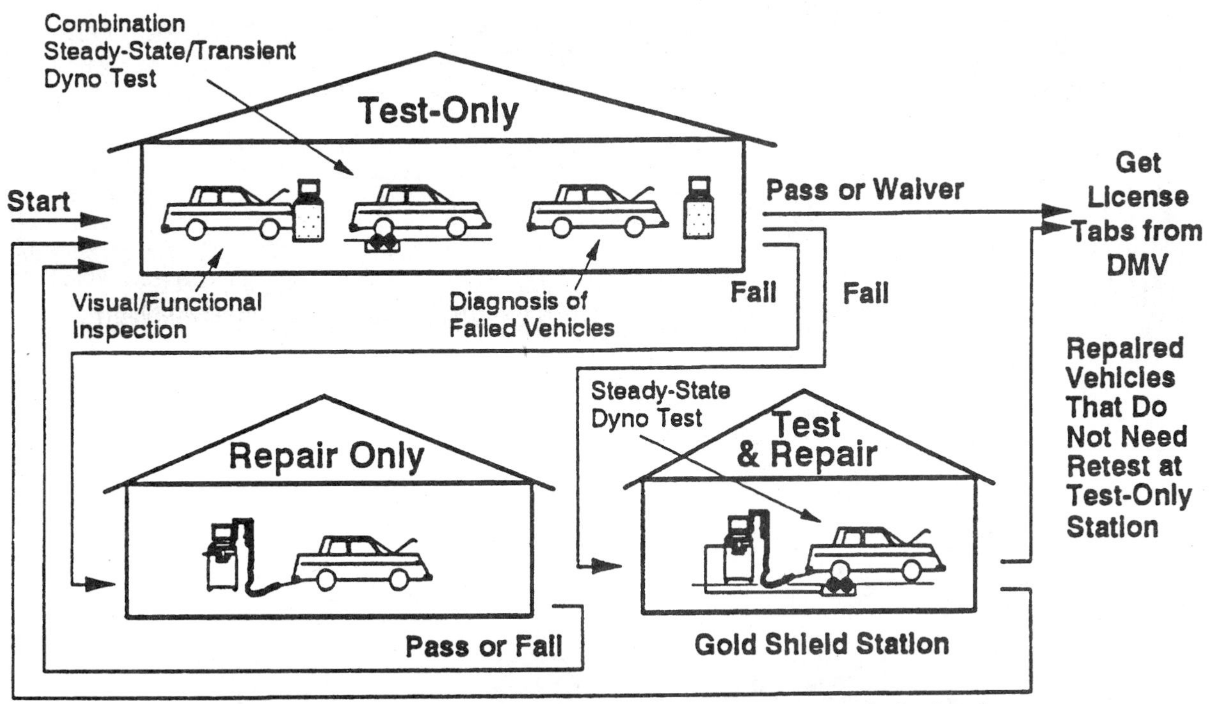

Figure 4-8 Hybrid test and repair program (courtesy California I/M Review Committee Report).

vehicles would be required to make only one stop after the initial inspection.

Repair Grade (RG240) Testing

Hybrid testing programs are now a real possibility because of new equipment called *Repair Grade* or RG240. This equipment includes chassis dynamometers, Constant Volume Sampling (CVS) machines, and portable five-gas analyzers at a much lower cost.

While RG240 equipment may not substitute as a full-fledged IM240 emissions inspection system, it certainly represents an economic way to assist in verifying repairs and/or performing an IM240 test before re-inspection. The projected cost for this type of setup is about $25,000 to $40,000. The entire setup is shown in Figure 4-9.

The setup shown in Figure 4-9, which is designed by Environmental Systems Products (ESP) uses a computer to control the test process, operator prompts, and IM240 driving trace, and interfaces the analyzer to the dynamometer. The computer interface allows the computer to regulate the dynamometer by controlling the driving cycle and vehicle inertia weight. A constant-volume sampling

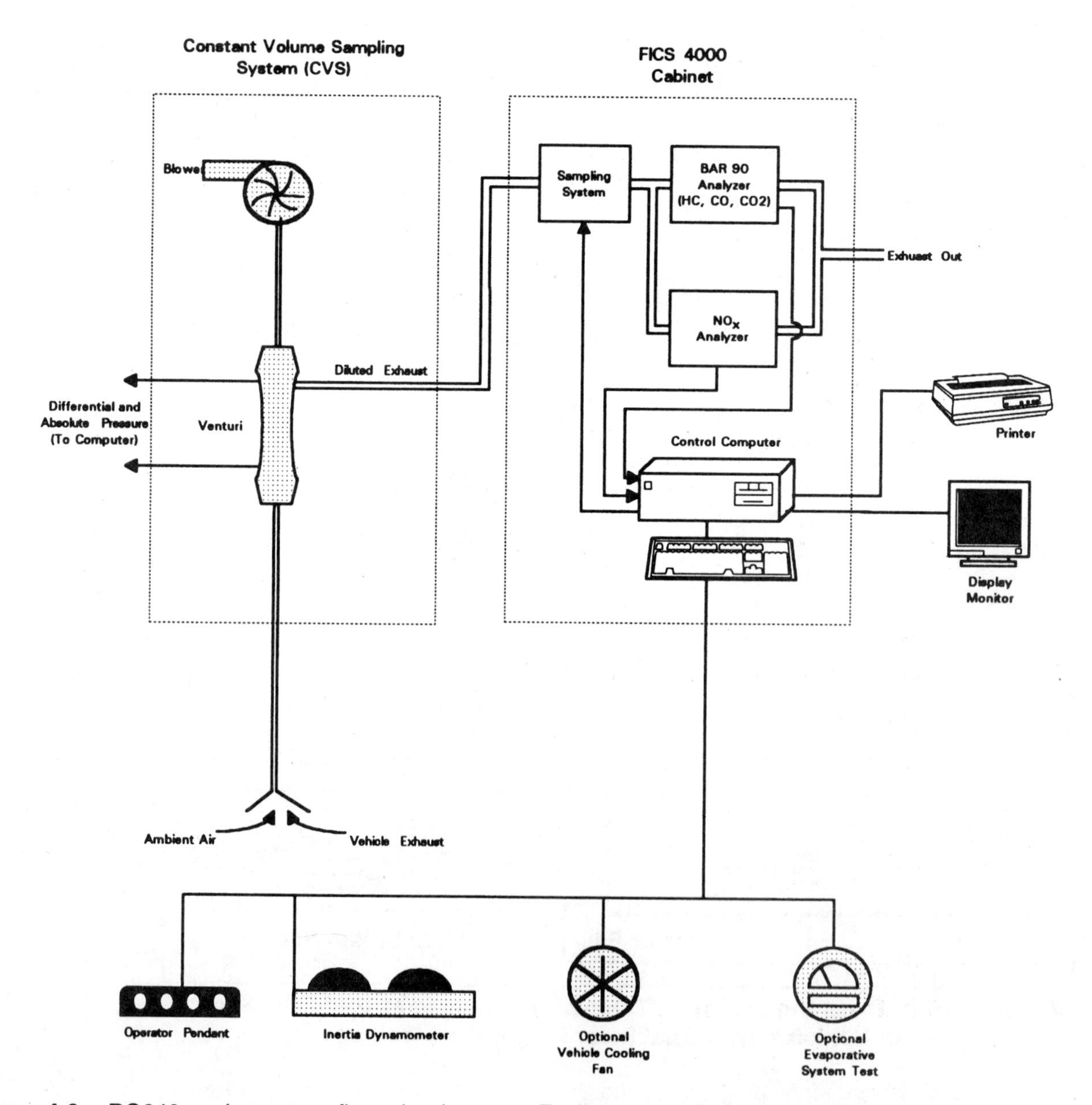

Figure 4-9 RG240 equipment configuration (courtesy Environmental Systems Products, Inc.).

system is used to determine the total exhaust flow used to calculate the grams-per-mile pollutant readings and fuel economy.

During the driving cycle, all vehicle exhaust passes through a constant-volume sampling system, which mixes and dilutes the vehicle exhaust with makeup ambient air drawn in by the blower (Figure 4-10).

The total diluted air passes through a subsonic venturi and the pressure drop across the venturi is sufficient to accurately calculate the total flow in grams-per-mile calculations. A sample extraction line collects a small fraction of the exhaust just after the venturi and passes it to the emissions analyzer.

The diluted exhaust flows through the BAR '90 and NO_X analyzer, which measure the pollutant concentration. These measurements, in conjunction with the corrected total flow, are used in the computer to calculate the grams per mile for each pollutant. The measured levels can be compared to the standards for specific vehicles to determine whether the vehicle will pass or fail the re-inspection. The computer will provide second by second readings of each pollutant during the drive cycle that will further enhance diagnostics.

Why the Delay in Loaded-Mode Testing?

The potential benefits of loaded-mode (dynamometer) testing in I/M programs have been recognized since the 1960s. By more accurately simulating vehicle operation in stop-and-go driving, loaded-mode testing has the potential to improve the correlation between the *short tests* used in an I/M program and the actual emissions of a vehicle in customer service. Improved correlation increases the ability to identify vehicles with emissions in excess of the standards they were certified to meet. The greatest potential benefit from loaded-mode testing is for NO_x emissions. Vehicles do not emit significant NO_x emissions during no-load operation; therefore, it is not possible to accurately identify vehicles with excess NO_x emissions from tailpipe emissions measurements.

Figure 4-10 RG240 Constant Volume Sampling system (courtesy Environmental Systems Products, Inc.).

Despite the potential benefits of loaded-mode testing, there has been extremely little interest in the use of dynamometer testing by states that operate I/M programs until quite recently. Most states have done little more than meet the minimum requirements for I/M programs established by the EPA. The EPA did not encourage the states to do more than simple concentration measurements of HC and CO at idle until the enactment of the Clean Air Act Amendments of 1990.

There are two principal reasons for the lack of focus on loaded-mode testing. First, much of the research on the potential benefits of I/M was done during the 1970s when most vehicles were equipped with easily adjustable carburetors and low energy, breaker point ignition systems. Studies conducted by the EPA showed that most of the excess emissions of HC and CO were due to maladjustment of idle air/fuel ratio and ignition misfire. These defects were easily detected at idle.

Second, there was little attention focused on the problem of excess NO_x emissions because the EPA had concentrated on a hydrocarbon-only control strategy for reducing ozone levels. Few areas outside of California have experienced violations of the ambient air quality standard for oxides of nitrogen (NO_x). The EPA has recently begun to show an interest in California's long-held belief that a strategy of HC and NO_x controls is needed to reduce ozone violations. Since NO_x control had not been considered a priority, the EPA took the position that a simple idle test was sufficient for I/M programs.

Since the flurry of I/M research in the 1970s, there have been substantial changes in the motor vehicle fleet. Air/fuel ratios are no longer easily adjustable and ignition system defects have become less important with the transition to solid-state, high-energy ignition systems. Instead of simple maladjustments and misfires, excess emissions are now related to defective sensors and degraded catalyst efficiency. These kinds of defects are more difficult to detect at idle. Even a degraded catalyst has a relatively high efficiency with the low exhaust flow rate that occurs at idle. In addition, the catalyst can "mask" other causes of excess emissions.

DETECTION AND DETERRENCE OF TAMPERING

Ideally, an I/M program should detect high-emitting vehicles and correct them to their properly functioning state. Current I/M programs, however, are less than effective for several reasons, one of which is the inability of the idle mode tailpipe test procedure to identify all of the vehicles in need of repair. According to a March 1991 publication by the EPA entitled "Inspection and Maintenance Policy Issues," it has been shown that the present idle test procedures have become less effective at identifying high-emitting 1980 and newer model year vehicles that utilize computer controlled closed-loop feedback systems with fuel injection. Another concern with the present I/M program is that all vehicles must be tested in order to identify the relatively small percentage of high-emitting vehicles.

Other causes of reduced I/M effectiveness include poor mechanic performance for inspection and/or repair, readjustment of the vehicle's emission control systems by technicians or motorists after an I/M test, and limitations imposed by low repair cost limits. Although California has the most comprehensive visual and functional tampering checks required in any I/M program, the accuracy of these checks relies on the integrity and competence of technicians working in the decentralized I/M program.

Ideally, any system intended to detect high-emitting vehicles should be cost effective, cause the least amount of consumer inconvenience, be highly efficient in detecting malfunctioning vehicles while not failing correctly functioning vehicles, and provide the repair industry with as much information as possible for correction of the malfunction. Two systems that are currently available are described below.

Remote Sensing Device (RSD)

One method for detecting high-emitting vehicles is the use of remote sensing devices that measure the exhaust emissions of on-road vehicles. Remote sensing involves a computer-controlled device that transmits an infrared beam across a single lane of traffic to a receiving unit. (See Figure 4-11.)

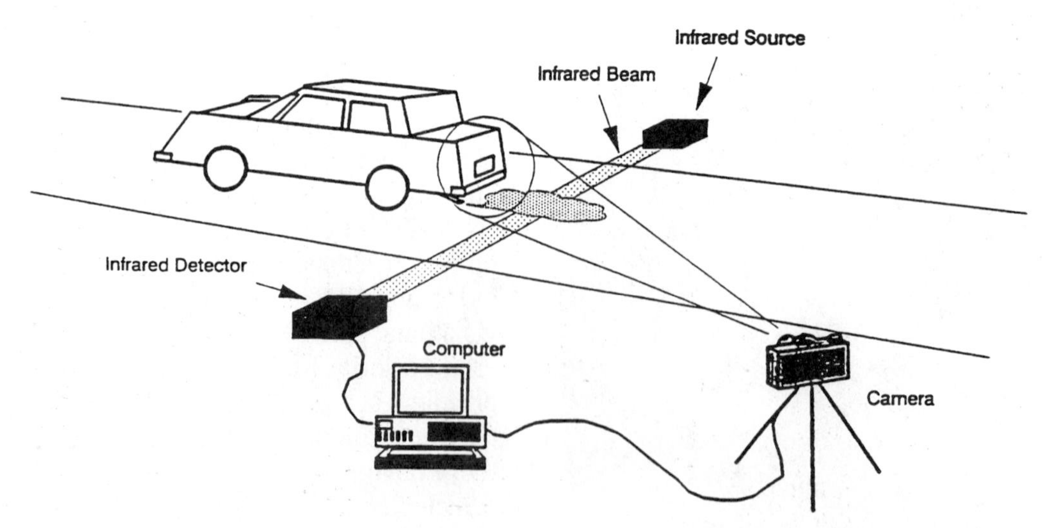

Figure 4-11 Remote sampling equipment positioning for maximum accuracy (courtesy California I/M Review Committee Report).

As a vehicle crosses this beam, the instantaneous concentrations of carbon monoxide (CO), hydrocarbons (HC), and carbon dioxide (CO_2) are determined from the exhaust plume and recorded into computer memory. In addition, there is a video camera that records the vehicle's license plate for later identification.

Application of RSD — Remote sensing systems are ideally suited for detecting high-emitting vehicles with a large throughput of vehicles tested. It is estimated that about 1,000 vehicles per hour and over 8,000 vehicles per day may be monitored by the system. When positioned in selected traffic lanes, the system provides information on the number of high-emitting vehicles that traverse the selected lane after recording their emissions and corresponding license plates (Figure 4-12).

The information gathered by the device could be used to assess the occurrence of high-emitting vehicles in the fleet and also may be used as a means of identifying individual vehicles that could be repaired or corrected to reduce excess emissions. In using remote sensing to identify tampered vehicles, the advantages of a high throughput and a low per-test cost may be compromised by the need to immediately physically inspect the vehicles identified as high emitters.

There are two reasons why immediate physical inspection is probably necessary. First, the identification and subsequent notification of a vehicle owner based on videotaped license plate information may take weeks, and the conditions causing the high emissions, especially in the case of deliberate tampering, may no longer exist when the vehicle is finally inspected. Second, the highly variable nature

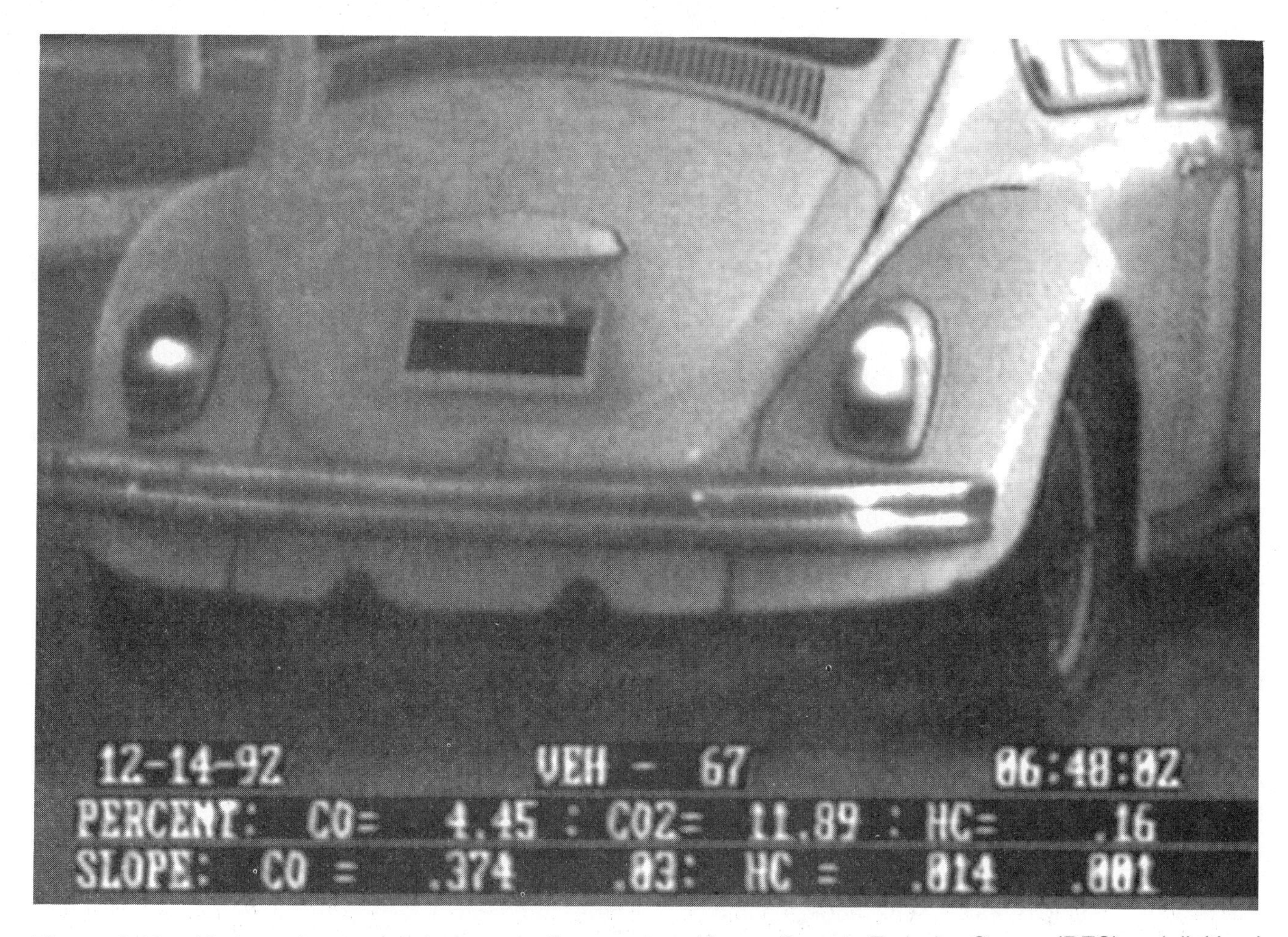

Figure 4-12 License plate and emissions readings captured from a Remote Emission Sensor (RES) and digitized camera (courtesy Santa Barbara Research Center).

of emissions as a function of throttle position and degree of warm-up makes it difficult to justify that a vehicle is tampered with or otherwise defective based on one or two instantaneous measurements taken with a remote sensor.

Advantages and Disadvantages of RSD—There are several advantages to the use of remote sensing relative to the conventional idle test currently used in the inspection and maintenance program. The test is unobtrusive, it is not constrained to idle conditions, throughput is high, and per-test costs are low. The remote sensing device is currently limited to the measurement of volume concentration of CO, CO_2, and HC (in % and ppm), as is the current idle test.

Because the remote sensing equipment is portable, it allows testing to be conducted at various sites. In addition, the test is not limited to the idle test, and since the vehicle to be tested need not stop during the measurement phase, the test would not inconvenience the driver.

Nevertheless, to capitalize on the advantages of remote sensing as a tool for detecting high emitters, the limitations should be identified and understood. Remote sensing is effective only if the testing conditions are well controlled. Motor vehicle emissions levels are sensitive to variations in speed and are greatly influenced by accelerations and decelerations. For example, an inherently low-emitting car could be falsely identified as a high emitter by the remote sensor if the speed of the vehicle is not controlled while crossing the infrared beam or if the driver is accelerating at that instant.

The results of the testing show that the variability for exhaust CO is at its lowest during the cruise speeds of 15-45 miles per hour, and for light acceleration. The greatest variations in exhaust CO occur under hard acceleration. For exhaust HC, the variability is lowest during accelerations, but greatest during deceleration.

In addition, automobile exhaust emissions are normally higher during cold start operation since the catalyst has not reached peak efficiency. This is especially true for emissions of CO. The remote sensing device would not be able to determine whether the vehicle is in cold start mode, leading to possible errors. Falsely failing too many clean vehicles would undermine public confidence and result in low acceptance of programs that utilize the device.

Another concern with using the remote sensing device arises from the limitations imposed by its physical setup. A single driving lane is required, because the remote sensing device is limited to testing traffic flow in single file. Thus, multi-lane highways would need to be channeled down to one lane, resulting in some traffic congestion and limiting the times and location of device operability. This limitation may result in traffic delays and avoidance of the test area. The road grade should also be considered in order to standardize the engine load during testing. In a report entitled "Identifying Excess Emitters with a Remote Sensing Device: A Preliminary Analysis," (Glover and Clemmens, EPA, 1991), the effectiveness of remote sensing was evaluated with respect to uphill and level road testing. It was found that the uphill results showed a fairly high level of repeatability, which may be due to smaller load variations than would be found by driving on a level road.

Testing conducted by EPA during light rain found that water splashed up by the tires interfered with the infrared beam. Measurements performed on wet pavement resulted in an increase in the number of invalid readings. A plume interference analysis was conducted to determine the effects on the measurement of a plume if the remnants of a previous vehicle's exhaust were still present. This analysis indicated that the remote sensing readings will be about 0.5 percent higher for a vehicle following within one or two seconds of another vehicle having emissions greater than 5% CO.

It must also be realized that not all exhaust plumes would pass through the infrared beam at a uniform height. Vehicles with relatively high or vertical tailpipes would pass by without being tested, thus contributing to errors. For the above reasons, careful site selection is critical in obtaining accurate results.

It is important to note that remote sensing, as an independent system, cannot detect tampered vehicles directly. Instead, suspect vehicles are identified by their high emissions, and tampering is then verified through physical inspection. Owners who have intentionally tampered with their vehicles would not have the opportunity to reconnect or repair their emission control systems prior to the test.

Theoretically, remote sensing technology could be used as a screening tool under controlled conditions to detect high-emitting vehicles at a fast throughput for subsequent testing and repairs. In a high volume, test-only program, the device would be used to screen out the majority of clean cars, allowing resources to be focused on the high emitters that may have been tampered with. Here, fewer devices would be needed and the per-test cost would be low. This approach does not appear to be practical, however, because of the poor correlation between the instantaneous emissions of a vehicle as it drives by the remote sensor and the presence of emissions-related defects. Based on the correlation studies discussed above, RSD technology is capable only of identifying vehicles with extremely high emissions. Many vehicles with easily correctable defects would likely be missed. Another concern is that excessive NO_x emissions do not always occur under the same driving modes that cause high HC or CO emissions. Depending on the particular mode of operation when the vehicle passes by a remote sensor, certain problems are unlikely to be detected.

Remote sensing, in conjunction with roadside inspections, could be used to identify some of the vehicles that have become high emitters between biennial smog inspections. This could help deter tampering, especially if a fine were imposed on owners of tampered vehicles, as has been proposed by some groups. Owners of cars with high emissions that have no tampering could be advised or required to obtain needed maintenance or repairs. Overall, the presence of remote sensing could increase the desire of vehicle owners to seek proper inspections, encourage better repairs, and deter tampering.

Remote sensing is not a viable option to completely replacing a conventional inspection program because of the problems discussed above and because it would not evaluate emission control systems that have no effect on tailpipe emissions, such as Positive Crankcase Ventilation (PCV) and evaporative emission control systems. Failures of these systems can be identified only by a visual and/or functional inspection. As previously mentioned, the inspection process is necessary to actually determine whether a vehicle was in fact tampered with, and to provide insight to the cause of high emissions.

On-Board Diagnostic (OBD) Systems

OBD systems have the potential to reduce the occurrence of tampering. The new second generation of OBD systems (OBD-II) are very comprehensive and should detect nearly all the problems that can lead to emissions increases. Also, OBD-II can alert owners of the need to service their engine and emission control systems. Under these circumstances, the principal function of the I/M program will be to enforce timely and proper vehicle response to problems identified by the OBD system. However, the system will also inform the motorist if tampering with the emission control systems has occurred during service.

The third generation of OBD (OBD-III) may provide the capability to further discourage tampering. The California Air Resources Board (CARB) is considering a requirement for radio transponders to be incorporated into the OBD system. With such a system, tampering could be detected when the vehicle drives past receivers installed along heavily traveled roadways. Vehicles that fail a drive-by test could be required to go for an inspection before registration renewal. This could reduce the number of inspection facilities required under the basic I/M program while allowing tampering to be detected more frequently than once every two years.

OBD-II will be covered in detail in Chapter 12.

CHAPTER 5

Computerized Engine Controls

INTRODUCTION TO AUTOMOTIVE COMPUTERS

In the past, automotive engineers found it difficult to design mechanical, vacuum, and hydraulic systems that could deliver the needed results. These systems failed to meet environmental mandates because they could not monitor combustion and accurately control air/fuel ratios. They needed elaborate emission control systems to reduce exhaust emissions. These systems required a great deal of maintenance and were sometimes complicated, making them difficult to repair. If the engineers concentrated on increasing fuel economy and lowering emissions, then performance and driveability suffered. If they concentrated only on lowering emissions, then economy, driveability, and performance suffered, creating many dissatisfied customers. Electronic engine controls allowed the engineers to restore good driveability, increase economy, lower exhaust emissions, and increase performance.

Good driveability is what the driver expects from a properly functioning engine, including easy starting, smooth idle, good acceleration through all speeds, instant response, and full power. Driveability complaints arise when a vehicle stalls, hesitates or stumbles, idles rough, surges, lacks power, is hard to start, or fails to start.

Good engine performance reduces throttle response time, while maximizing horsepower and torque. It can be accomplished by installing:

1. A supercharger (Figure 5-1) or a turbocharger.
2. Variable induction intake manifolds.
3. Multiple valves per cylinder.

These factory-installed components are designed to optimize performance and to maintain a

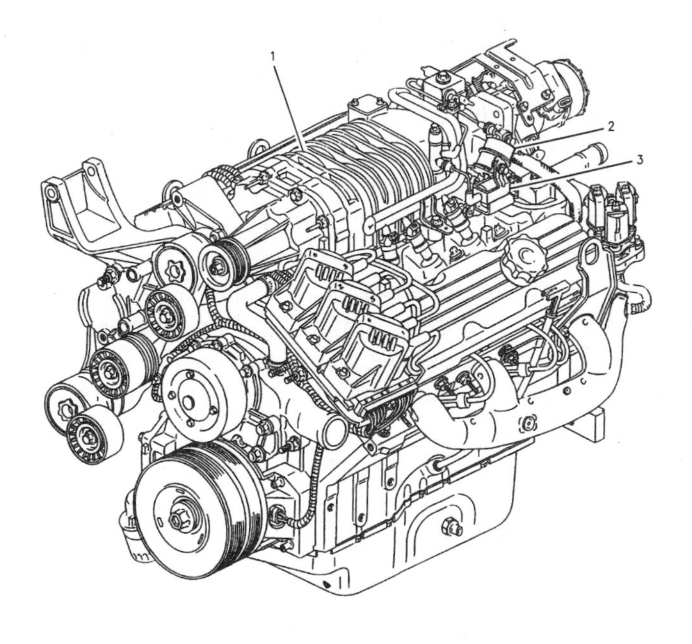

Figure 5-1 Factory-installed superchargers are one way to increase engine performance and still comply with emissions standards (courtesy GM).

balance between good driveability, increased fuel economy, and low exhaust emissions.

Computer Terminology

Prior to OBD-II, each manufacturer referred to the computer by a different name. Since they all function alike and use the same basic inputs and outputs, they were generically called a *computer.* According to OBD-II guidelines, all engine control computers must now be called *Powertrain Control Modules* (PCMs).

Computer Theory

A computer found on a vehicle is really no different from any other computer encountered in our daily lives. Computers rely on data from input devices, then follow an internal program to calculate the required output. An input device may be a keyboard or a coolant temperature sensor, and the output may be a video display monitor or a fuel injector. The program a computer follows may be for word processing or for controlling fuel metering and engine timing. Computers can process a great deal of data quickly and accurately, making them useful for controlling many systems on an automobile.

The computer does not perform complicated operations. It performs thousands of simple operations incredibly fast and accurately. It is a *logical* device that *processes* inputs from sensors and switches, then calculates output commands based on internal instructions called a *program.* The computer does not *think*, it follows instructions *programmed* into memory. This can result in the computer making decisions based on faulty information, resulting in poor or erratic engine performance. As the saying goes: *garbage in, garbage out.*

Problems can be traced to three areas:

1. Quality of incoming information.
2. Processing of information.
3. Execution of output commands.

To solve driveability complaints, the technician needs to:

1. Understand how the computer processes information.
2. Visualize how the system operates.
3. Develop a logical approach to diagnostics.

Computer Operation

Computer operation is divided into three categories (Figure 5-2):

1. Input
2. Processing
3. Output

The computer relies on inputs from sensors and switches to monitor various vehicle conditions. These inputs send voltage signals to the computer that change often depending on the sensor and the conditions it monitors. The computer keeps track of and controls engine operation by learning from the various driving conditions. This is accomplished by monitoring:

1. Engine coolant and manifold air temperature.
2. Intake manifold pressure and barometric pressure.
3. Throttle position opening rate and direction of travel.
4. Exhaust gas oxygen content.
5. Input switches to see if the transmission is in gear, the steering wheel is turned, the brake is applied, or the air conditioning has been turned on.

Signals to the computer are either analog or digital. An analog signal, shown in Figure 5-3, is like a waveform on an oscilloscope. It appears as a flowing line with curved peaks and valleys, indicating rises and drops in voltage.

All analog signals must be converted to digital signals because the computer requires digital inputs. This is done inside the computer through an analog to digital converter (A/D).

A digital signal, shown in Figure 5-3, is one value or another and does not have waves, peaks, or valleys. They create vertical rises and drops in voltage, and a horizontal line with sharp square corners. The top horizontal lines show when the voltage is high and the bottom lines show when the voltage is low.

On/off switches, Hall-Effect switches, and some air flow sensors generate digital signals. The computer can interpret digital signals directly because

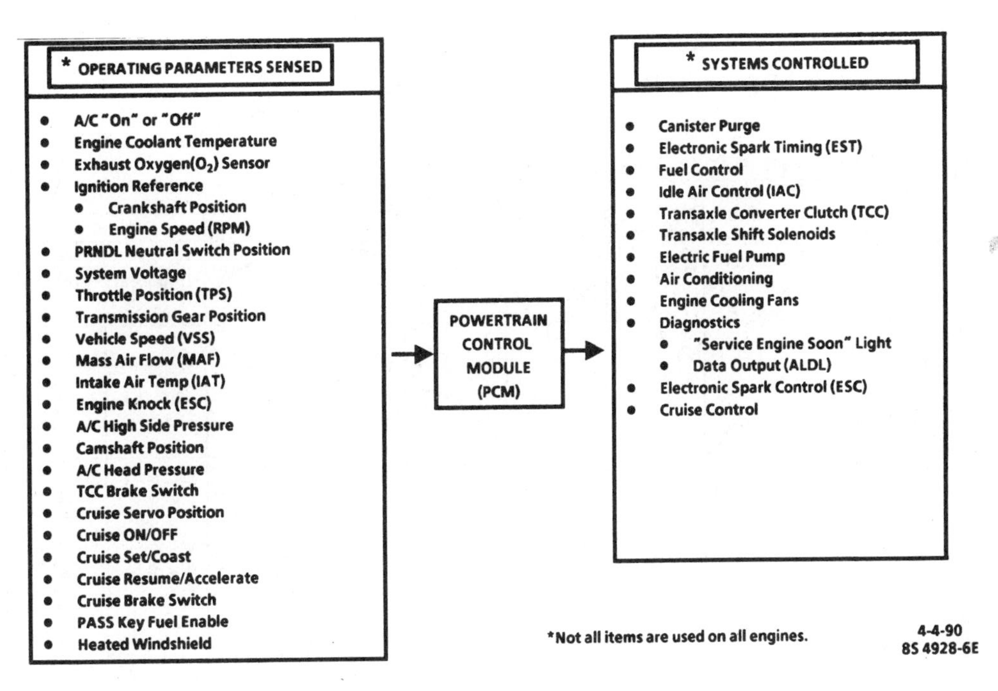

Figure 5-2 The computer (PCM) needs a variety of inputs to calculate each output (courtesy GM).

they are either *on* or *off.* The A/D converter is bypassed and signals are sent directly to an Input/Output (I/O) processor.

Processing information is the next stage of computer operation. The *Central Processing Unit* (CPU), which is the *brain* of the computer, is responsible for executing output commands based on inputs from the sensors and switches. The CPU compares sensor input to programs stored in memory and calculates how to react to changing engine conditions.

A program is a *plan* or set of instructions that guide the CPU. (See Figures 5-4 and 5-5.)

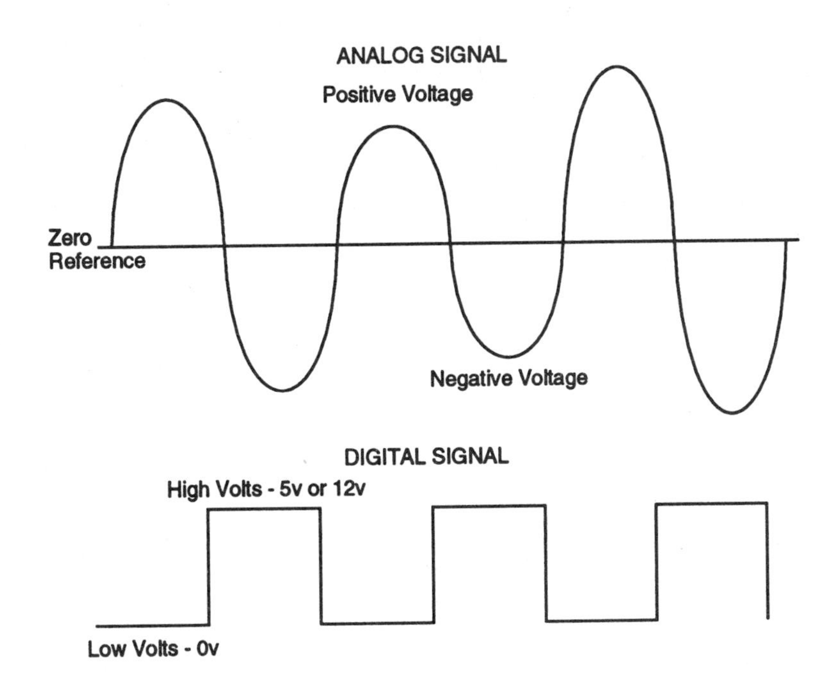

Figure 5-3 Analog and digital voltage signals.

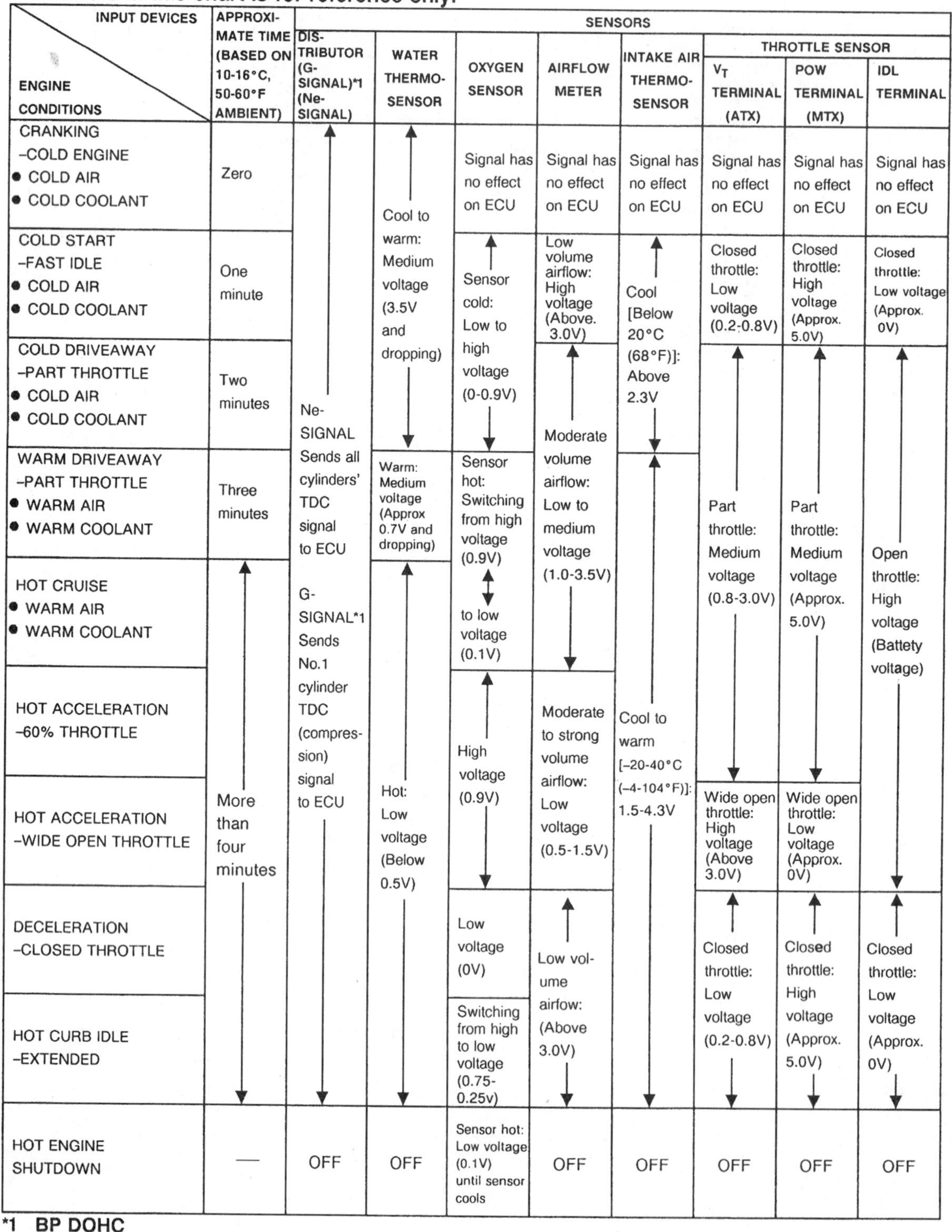

ENGINE CONTROL OPERATION CHART
Input Devices and Engine Conditions

Note
- The data in this chart is for reference only.

INPUT DEVICES / ENGINE CONDITIONS	APPROXIMATE TIME (BASED ON 10-16°C, 50-60°F AMBIENT)	SENSORS: DISTRIBUTOR (G-SIGNAL)*1 (Ne-SIGNAL)	SENSORS: WATER THERMO-SENSOR	SENSORS: OXYGEN SENSOR	SENSORS: AIRFLOW METER	SENSORS: INTAKE AIR THERMO-SENSOR	THROTTLE SENSOR: VT TERMINAL (ATX)	THROTTLE SENSOR: POW TERMINAL (MTX)	THROTTLE SENSOR: IDL TERMINAL
CRANKING –COLD ENGINE • COLD AIR • COLD COOLANT	Zero			Signal has no effect on ECU	Signal has no effect on ECU	Signal has no effect on ECU	Signal has no effect on ECU	Signal has no effect on ECU	Signal has no effect on ECU
COLD START –FAST IDLE • COLD AIR • COLD COOLANT	One minute		Cool to warm: Medium voltage (3.5V and dropping)	Sensor cold: Low to high voltage (0-0.9V)	Low volume airflow: High voltage (Above. 3.0V)	Cool [Below 20°C (68°F)]: Above 2.3V	Closed throttle: Low voltage (0.2-0.8V)	Closed throttle: High voltage (Approx. 5.0V)	Closed throttle: Low voltage (Approx. 0V)
COLD DRIVEAWAY –PART THROTTLE • COLD AIR • COLD COOLANT	Two minutes	Ne-SIGNAL Sends all cylinders' TDC signal to ECU G-SIGNAL*1 Sends No.1 cylinder TDC (compression) signal to ECU			Moderate volume airflow: Low to medium voltage (1.0-3.5V)		Part throttle: Medium voltage (0.8-3.0V)	Part throttle: Medium voltage (Approx. 5.0V)	Open throttle: High voltage (Battety voltage)
WARM DRIVEAWAY –PART THROTTLE • WARM AIR • WARM COOLANT	Three minutes		Warm: Medium voltage (Approx 0.7V and dropping)	Sensor hot: Switching from high voltage (0.9V) to low voltage (0.1V)		Cool to warm [–20-40°C (–4-104°F)]: 1.5-4.3V			
HOT CRUISE • WARM AIR • WARM COOLANT	More than four minutes		Hot: Low voltage (Below 0.5V)						
HOT ACCELERATION –60% THROTTLE				High voltage (0.9V)	Moderate to strong volume airflow: Low voltage (0.5-1.5V)				
HOT ACCELERATION –WIDE OPEN THROTTLE							Wide open throttle: High voltage (Above 3.0V)	Wide open throttle: Low voltage (Approx. 0V)	
DECELERATION –CLOSED THROTTLE				Low voltage (0V)	Low volume airfow: (Above 3.0V)		Closed throttle: Low voltage (0.2-0.8V)	Closed throttle: High voltage (Approx. 5.0V)	Closed throttle: Low voltage (Approx. 0V)
HOT CURB IDLE –EXTENDED				Switching from high to low voltage (0.75-0.25v)					
HOT ENGINE SHUTDOWN	—	OFF	OFF	Sensor hot: Low voltage (0.1V) until sensor cools	OFF	OFF	OFF	OFF	OFF

*1 BP DOHC

Figure 5-4 The computer uses inputs to identify different engine conditions (courtesy Mazda Motor of America).

Output Devices and Engine Conditions

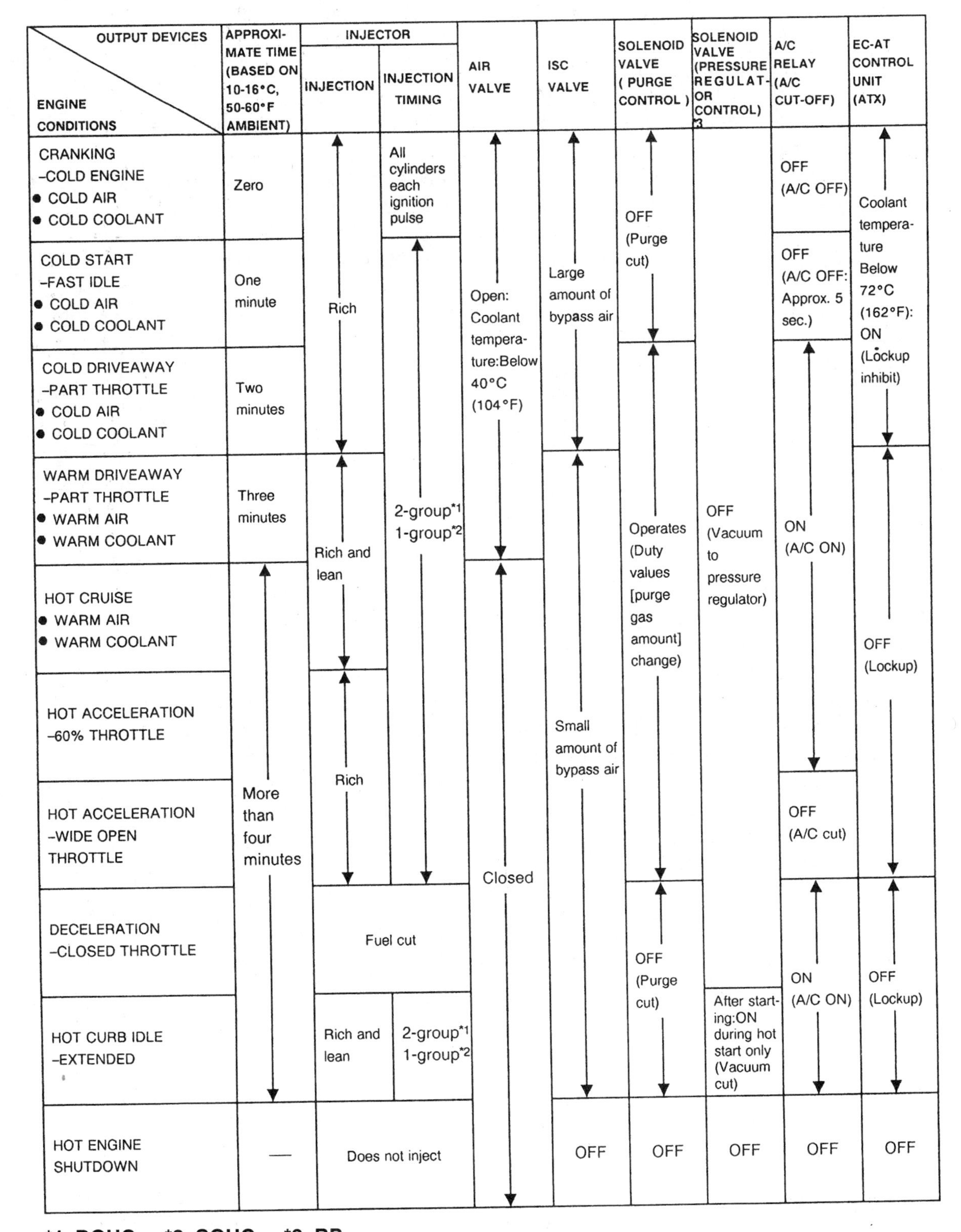

ENGINE CONDITIONS \ OUTPUT DEVICES	APPROXIMATE TIME (BASED ON 10-16°C, 50-60°F AMBIENT)	INJECTOR: INJECTION	INJECTOR: INJECTION TIMING	AIR VALVE	ISC VALVE	SOLENOID VALVE (PURGE CONTROL)	SOLENOID VALVE (PRESSURE REGULATOR CONTROL) *3	A/C RELAY (A/C CUT-OFF)	EC-AT CONTROL UNIT (ATX)
CRANKING –COLD ENGINE • COLD AIR • COLD COOLANT	Zero		All cylinders each ignition pulse					OFF (A/C OFF)	
COLD START –FAST IDLE • COLD AIR • COLD COOLANT	One minute	Rich		Open: Coolant tempera-ture:Below 40°C (104°F)	Large amount of bypass air	OFF (Purge cut)		OFF (A/C OFF: Approx. 5 sec.)	Coolant tempera-ture Below 72°C (162°F): ON (Lockup inhibit)
COLD DRIVEAWAY –PART THROTTLE • COLD AIR • COLD COOLANT	Two minutes								
WARM DRIVEAWAY –PART THROTTLE • WARM AIR • WARM COOLANT	Three minutes	Rich and lean	2-group*1 1-group*2			Operates (Duty values [purge gas amount] change)	OFF (Vacuum to pressure regulator)	ON (A/C ON)	
HOT CRUISE • WARM AIR • WARM COOLANT									OFF (Lockup)
HOT ACCELERATION –60% THROTTLE					Small amount of bypass air				
HOT ACCELERATION –WIDE OPEN THROTTLE	More than four minutes	Rich		Closed				OFF (A/C cut)	
DECELERATION –CLOSED THROTTLE		Fuel cut				OFF (Purge cut)		ON (A/C ON)	OFF (Lockup)
HOT CURB IDLE –EXTENDED		Rich and lean	2-group*1 1-group*2				After starting:ON during hot start only (Vacuum cut)		
HOT ENGINE SHUTDOWN	—	Does not inject			OFF	OFF	OFF	OFF	OFF

***1 DOHC *2 SOHC *3 BP**

Figure 5-5 Once the inputs are processed, the computer can send commands to the various outputs (courtesy Mazda Motor of America).

Figures 5-4 and 5-5 show the plan Mazda uses to operate a 1993 323/Protégé. To carry out this plan, the computer uses mathematical formulas and logic gates to *decide* how an engine should respond to the input information. The CPU reacts to the same input with the same adjustments every time. Once it decides how to react, it sends out a set of commands to the output component.

Outputs, also called actuators, are components that convert electrical signals from the CPU into mechanical action. This can be accomplished by:

1. Controlling a fuel pump relay.
2. Regulating fuel injector pulse-width.
3. Turning on the *check engine* light when a circuit or component has failed.
4. Maintaining the correct idle speed.
5. Switching on or off an emissions-related component.
6. Computer timing through the ignition module.

The computer controls actuators by using transistors to open and close a circuit. This is called *ground side switching* because the transistors are located in the computer in the ground side of the circuit. With this digital control of actuators, the CPU turns the actuator fully on or off.

Like a CPU, the technician should follow a program when diagnosing a problem. For example, a check engine light is on and a *hard fault* code is present (Figure 5-6). The technician identifies the code and the circuit in question. This data is the technician's first *input* to diagnosing the problem.

The next step is for the technician to follow a step-by-step set of instructions called a *diagnostic chart.* A diagnostic chart, shown in Figure 5-7, is a *program* found in a repair manual that guides the technician through the diagnosis. The technician uses the diagnostic chart to collect the data. Collecting this data allows the technician to select the correct repair procedure. Through the entire procedure, the technician is *processing* information.

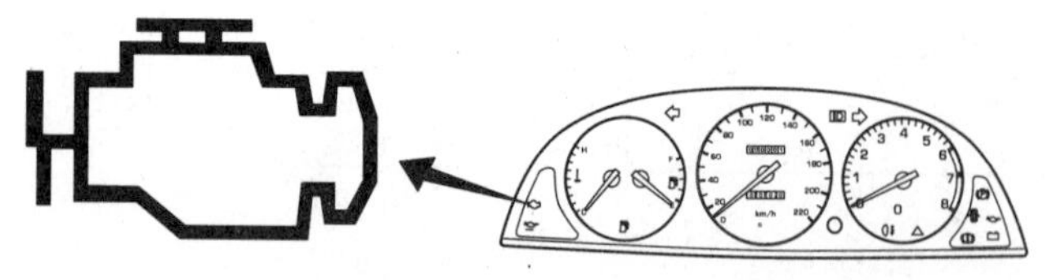

Figure 5-6 The *check engine* light is activated when a *hard fault* has occurred, providing the technician with a *clue.*

Finally, the appropriate repair is made and the repair is verified. This is *output.* The technician learns from this experience and stores the information in memory. This helps the technician quickly diagnose and repair the same type of problem next time.

Computer Memory

Inside the CPU, the computer has a filing system called *memory* (Figure 5-8). There are three types of memory:

1. Random Access Memory (RAM).
2. Read-Only Memory (ROM).
3. Programmable Read-Only Memory (PROM).

Random Access Memory temporarily stores data and instructions. It allows the computer to watch engine performance over time so it can constantly learn and adjust the performance level of the vehicle. Diagnostic trouble codes are stored in the RAM.

Most systems incorporate two types of RAM: *volatile* and *non-volatile.* Think of the RAM as a scratch pad. If written on using a pencil, it can be erased—this is temporary or *volatile* memory. Removing battery power from the computer erases the codes and adaptive values stored in the volatile RAM. *Non-volatile* memory is like writing with an ink pen and is considered permanent. To erase this information requires a specific procedure.

Volatile or *Keep Alive Memory* RAM receives power directly from the battery. Information about codes and adaptive learned values is stored here. Turning off the ignition switch does not clear the memory. Sometimes, disconnecting a fuse or battery cable will clear codes and data stored in memory.

Adaptive learning or strategies are part of a Keep Alive Memory. They allow the computer to learn and remember the most recent calibrations and operating experience. It learns from experience and constantly attempts to adjust factory-best average settings to maintain maximum engine performance, driveability, economy, and emissions. In executing adaptive learning, the computer monitors and detects signs of wear or age in calibrated components and makes the necessary corrections. If

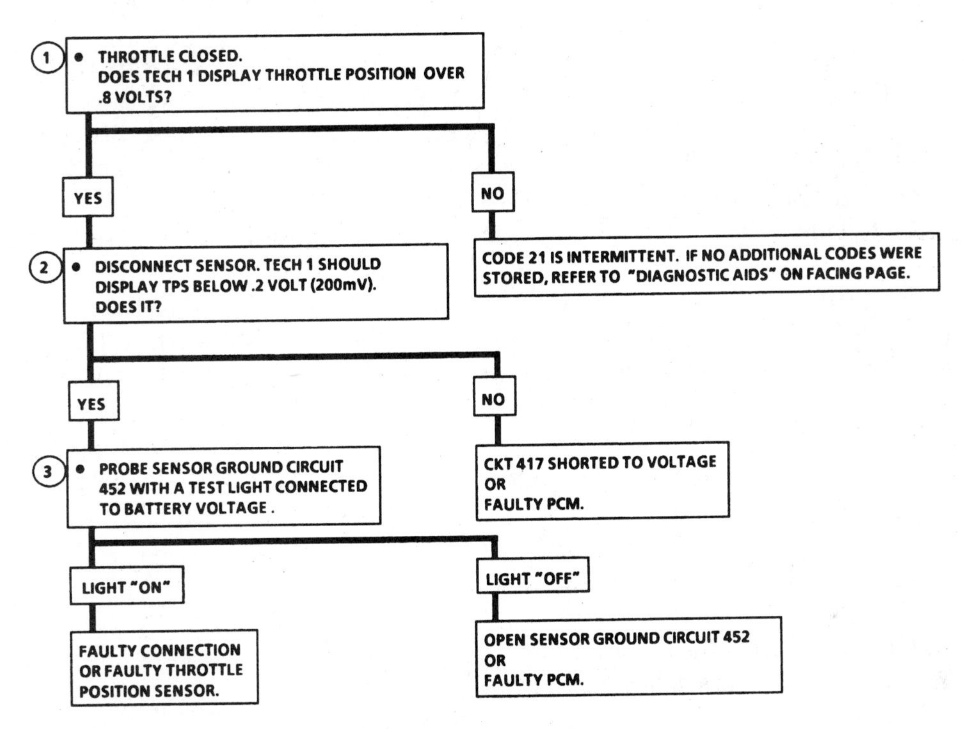

Figure 5-7 A diagnostic chart (trouble tree) should be used when a *hard fault* code has been detected. Follow the plan closely without skipping steps (courtesy GM).

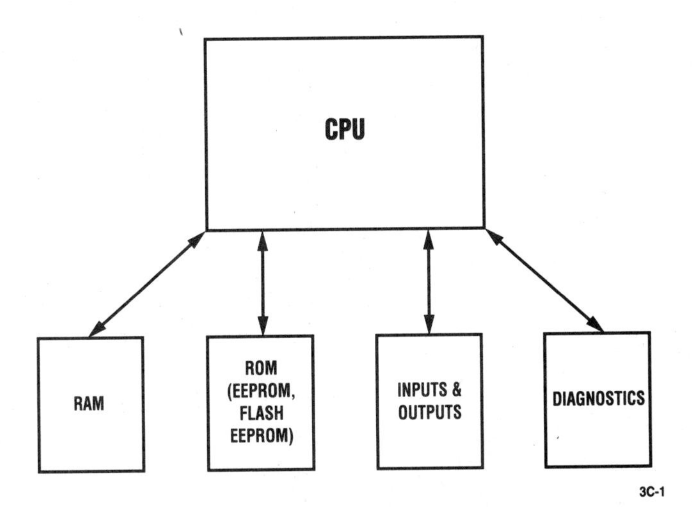

Figure 5-8 The computer uses memory to process inputs and to calculate output commands. Information can also be stored in memory (courtesy Chrysler Corp.).

the signal from a sensor swings erratically, the CPU may even ignore input from the sensor.

Most computers are equipped with adaptive learning capabilities. Those that are require a short adjustment or learning period after:

1. A new vehicle is put into service.
2. The battery has been discharged or disconnected.
3. The computer has been disconnected or replaced.
4. A defective sensor, switch, or actuator has been replaced with a new part.

If any of above should occur, then the adaptive strategies learned will return to basic or neutral settings. This will cause driveability to suffer until the computer has had time to learn and adapt to new strategies.

Read-Only Memory is where basic operating instructions for the computer are stored. As the name implies, while the CPU can read information from ROM, it cannot write new information into it. When the computer is manufactured, instructions are programmed into the ROM chip and cannot be changed. Disconnecting the vehicle's battery does not cause information in ROM to be lost.

Programmable Read-Only Memory contains specific vehicle information about:

1. Engine displacement.
2. Transmission type: automatic or manual.
3. Drive axle ratio and type.
4. Tire size.
5. Accessories such as A/C and power steering.

Included in the PROM are specific instructions about spark advance, fuel delivery, idle speed, etc. First generation PROM chips cannot be reprogrammed. If defective, the technician has to replace the PROM or the computer to repair or update the system.

Always consider the PROM when diagnosing a driveability problem. A PROM may have glitches or imperfections programmed into it. If the PROM is not updated, the problem cannot be successfully repaired.

Information regarding PROM updates can be found in a variety of sources:

1. CD-ROM information systems like Mitchell and AllData.
2. Factory technical bulletins.
3. Aftermarket scan tool manufacturers, Snap-On and OTC, include a PROM booklet with each scan tool or scan tool software update.

Electronically Erasable Computer Memory (EEPROM) — Why the need for *Electronically Erasable Programmable Read-Only Memory*? The development of erasable computer memory allows the technician to service customers more efficiently. The ability to update computers without having to replace them reduces downtime that occurs while waiting for parts. EEPROMs provide greater system functionality and flexibility, and allows the PROM to be serviced by dealership personnel.

Two types of second generation PROM chips are in use today. The first is an *Erasable Programmable Read-Only Memory* (EPROM). To erase and reprogram an EPROM, it has to be removed from the circuit board and placed under UV (Ultra-Violet) light for a specified amount of time. EPROMs that are hard-wired (soldered) into the circuit must be replaced with a new or remanufactured computer.

The second type is a *Flash EEPROM.* It uses an inexpensive bulk erase of the chip, eliminating the time-consuming UV-erase procedure. Without removing it from the circuit board, a Flash EEPROM can be selectively erased (certain data can be erased while other data is left alone) and reprogrammed.

Flashing an EEPROM requires special equipment. Figure 5-9 shows the equipment and setup required to reprogram a Chrysler computer. When certain pins on the EEPROM receive a Flash voltage of 12 or 21 volts, this will "open the door" for the CPU to accept the down-loading of new program instructions for the Flash EEPROM.

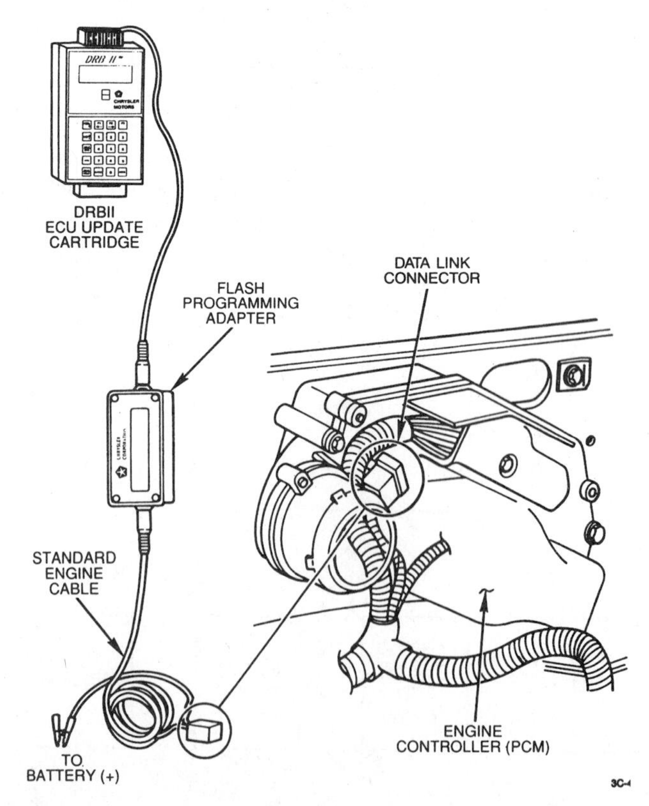

Figure 5-9 Flash EEPROM updating system (courtesy Chrysler Corp.).

After programming is completed, the EEPROM is like new. This means the computer must learn and adapt to its operating environment. In addition, the technician benefits from the Flash EEPROMs enhanced diagnostics and repair capabilities.

In the future, programmable memory will be designed for use in:

1. Body control.
2. Safety and convenience.
3. Keyless entry.
4. Instrumentation clusters.
5. System personalization.

Researchers are projecting 100% of engine and 80% of the transmission control systems will be electronically programmable by the year 2000.

Computer Location

The computer is usually located in the passenger compartment for protection against moisture, extreme temperatures, and excessive vibration, all of which are present in the engine compartment. The use of improved electronic components that can dissipate heat more efficiently has allowed manufacturers to install computers in the engine compartment if they choose to.

Handling the Computer

Computers contain electronic components that are extremely sensitive to static electricity and physical shock. When handling the computer, follow these precautions:

1. Touch the vehicle chassis to discharge any electricity that may be stored in your body.
2. Turn off the ignition switch before disconnecting, removing, or installing the computer.
3. When performing diagnostic tests, use a VOM meter that has a minimum of 10 megohms of resistance in the voltmeter circuit. A digital meter is preferred, but some analog volt meters include this feature.
4. Do not use an ohmmeter to test the computer itself unless instructed to do so by the manufacturer.
5. Be careful not to damage pin connectors when backprobing computer terminals.

INPUT SENSORS AND SWITCHES

The computer obtains information about engine operation from different sensors located in various places on the engine, or in the passenger compartment. There are many types of sensors that serve one purpose, and that is to provide a voltage or frequency signal to the computer.

Sensors convert mechanical or physical engine conditions such as: RPM, air flow, coolant and inlet temperature, and throttle angle into an electrical voltage or frequency signal. These signals allow the computer to detect engine and driving conditions. Once the information is processed, the computer calculates fuel, timing, idle speed, and transmission output commands.

Despite the number of systems in use, and the variety of names given them, there are only five types of sensors used. All inputs are one of the following types:

1. Variable resistors (voltage modifying).
2. Variable DC frequency.
3. Variable DC voltage generators.
4. Variable AC voltage/frequency generators.
5. Switches.

Before testing any sensor, the technician needs to know its basic operation and how the computer uses its signals. Keep in mind that not every one of these sensors and switches are on all engines.

Variable Resistors – Thermistors

A *thermistor* is a device that changes resistance with temperature. The computer uses the thermistor to learn the temperature of the:

1. Engine coolant.
2. Intake manifold air.
3. Inlet air ahead of the throttle plates.
4. Air temperature at the throttle body.
5. Fuel temperature at the pressure regulator.
6. Battery temperature and exhaust gas temperature.
7. Transmission fluid temperature.

Two types of thermistors are used. The first type is an NTC (Negative Temperature Coefficient).

Resistance	Temperature
100,000	-40°F
25,000	32°F
1000	100°F
500	180°F
150	212°F

Figure 5-10 Typical thermistor values.

As *temperature increases, resistance decreases.* The opposite is also true—as temperature decreases, resistance increases. The second type of thermistor is a PTC (Positive Temperature Coefficient). As *temperature increases, resistance increases,* and as temperature decreases, resistance decreases. Figure 5-10 shows the relationship between resistance and temperature for an NTC thermistor.

A thermistor circuit consists of two wires. The computer supplies a 5-volt reference voltage to one side of the sensor (Figure 5-11). Current is limited by a fixed resistor located inside the computer. The ground side of the sensor returns to the computer.

The computer constantly monitors this circuit for a *voltage drop* across the sensor resistance. Voltage drop changes as the temperature increases or decreases. The voltage drop across the thermistor is a variable analog voltage signal, typically 0.02v to 4.8v.

Coolant Temperature Sensors — The coolant sensor is threaded into a coolant passage near the thermostat housing or in the side of the engine block. It works best when immersed in a mixture of antifreeze and water.

The function of this sensor is to convert engine coolant temperature into an electrical voltage signal. The computer uses the input signal to:

1. Calculate *open-loop* air/fuel ratios.
2. Calculate spark timing curves.
3. Turn on/off solenoids used to control EGR flow, evaporative canister purge, and heat riser position.
4. Calculate *closed-loop* air/fuel ratios.
5. Control torque converter clutch operation.
6. Control cooling fan relay operation.

It is important to remember that high resistance readings indicate a cold engine and low resistance readings indicate a warm engine. Specific resistances will vary between manufacturers. Since the computer is monitoring a voltage drop across the sensor, system voltage drops will be high (2.5 to 4.0v) when cold and low (0.5 to 1.5v) as the engine reaches operating temperature.

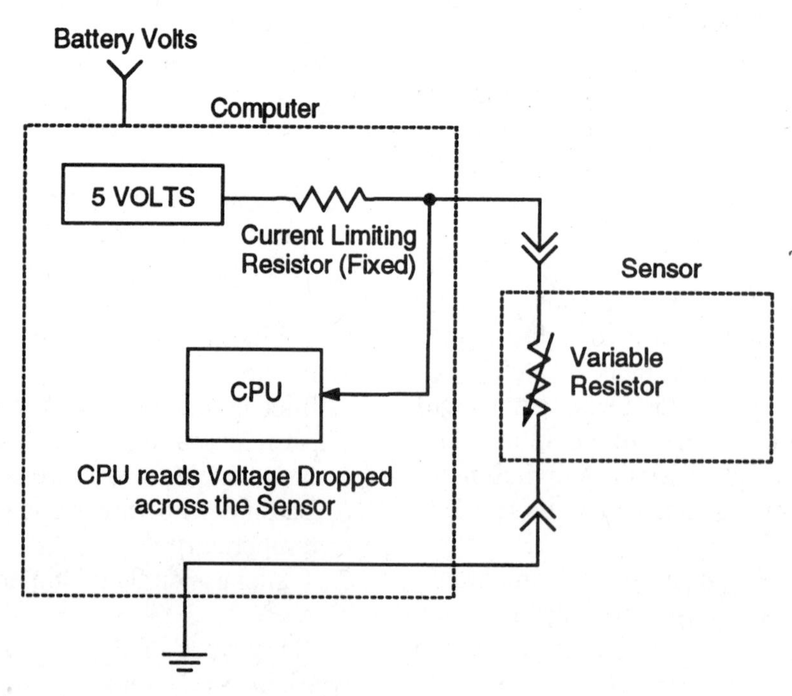

Figure 5-11 Typical thermistor circuit.

Chrysler EFI systems, first used in 1985, incorporated a dual-stage coolant sensor. The computer uses a dual-range circuit to monitor sensor temperature that is converted to a voltage. The operation of this circuit is about the same as other coolant sensors; the difference is inside the computer. This can cause some unusual voltmeter readings if the technician does not understand the internal workings of the circuit. The circuit still uses a 5-volt reference with a current-limiting resistor, but the circuit is shunted. The circuit uses a 10,000-ohm and 909-ohm resistor. For a more accurate reading during open-loop warm-up, 5 volts are fed through the 10,000-ohm resistor. At approximately 120°F the voltage will drop to 1.25v and jump back up to about 4.2v, because the computer internally switches the 5 volts to the 909-ohm resistor. At this temperature, many warm-up functions of the computer have been completed, and the coolant sensor reading is not as critical. It might be helpful to think of an engine analyzer analog tachometer that changes scales as RPM increases.

In 1986, General Motors began using a dual-stage coolant sensor similar to that used by Chrysler. Unlike Chrysler, this type of coolant sensor is not used on all GM products. The basic purpose and operation are the same, but operating voltages and resistances will be different. Figure 5-12 shows the circuit and voltage curve used by GM to monitor coolant temperature.

Inlet or Intake Manifold Air Temperature Sensors — Inlet or intake manifold air temperature sensors convert incoming air temperature to an electrical voltage signal. The computer uses this input to calculate the temperature of the incoming air.

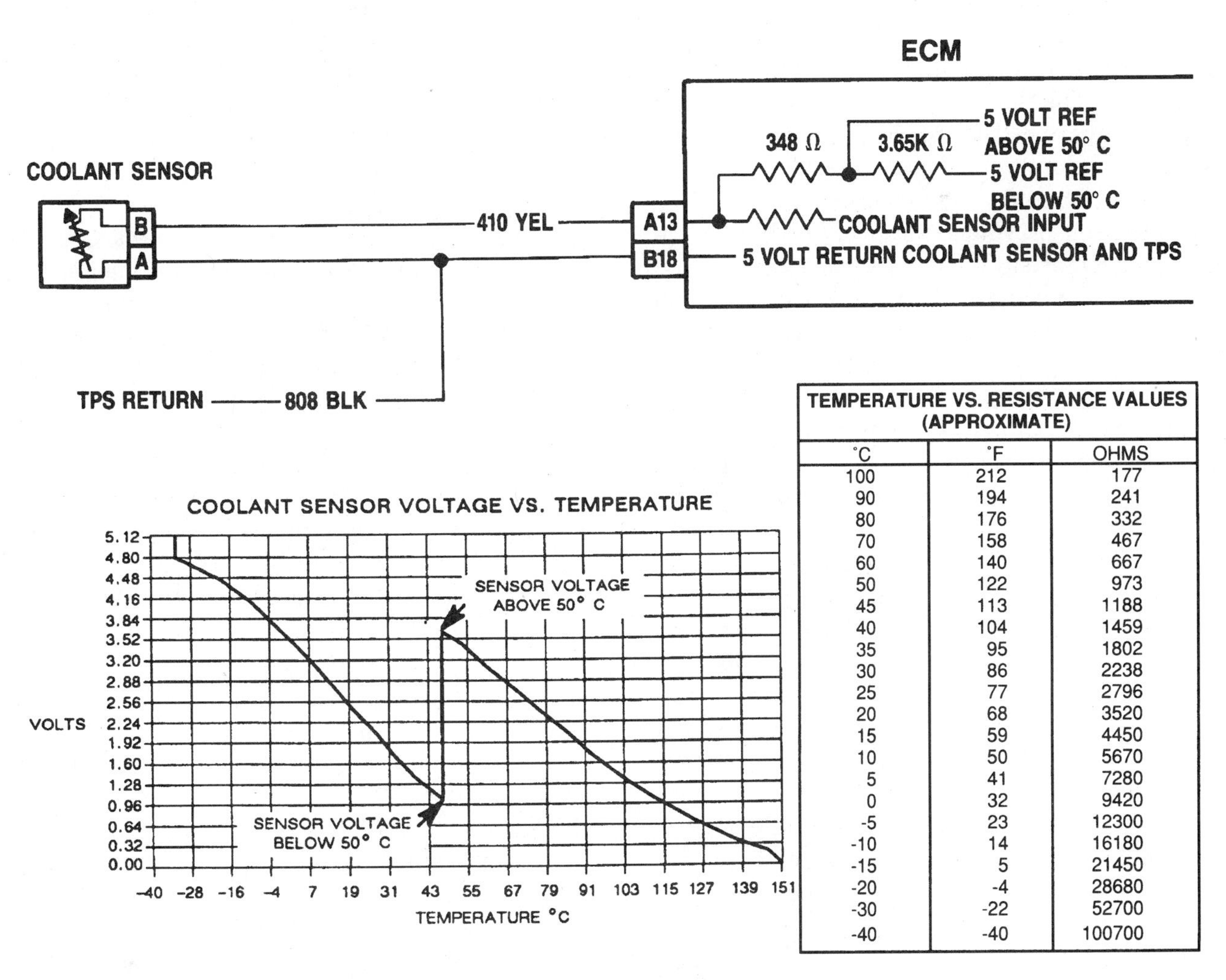

TEMPERATURE VS. RESISTANCE VALUES (APPROXIMATE)		
°C	°F	OHMS
100	212	177
90	194	241
80	176	332
70	158	467
60	140	667
50	122	973
45	113	1188
40	104	1459
35	95	1802
30	86	2238
25	77	2796
20	68	3520
15	59	4450
10	50	5670
5	41	7280
0	32	9420
-5	23	12300
-10	14	16180
-15	5	21450
-20	-4	28680
-30	-22	52700
-40	-40	100700

Figure 5-12 Dual-stage (shunted) coolant sensor (courtesy GM).

Knowing the intake air temperature allows the computer to fine tune the air/fuel mixture to compensate for air density by:

1. Increasing/decreasing injector pulse-width.
2. Increasing or decreasing spark timing to control NO_x emissions.
3. Accurately control charging system voltage based on the air temperature near the battery.

This sensor can be located in an intake manifold runner, air cleaner housing, vane air flow meter, throttle body injection unit, or near the battery inside the computer. This sensor functions like a coolant sensor, except the tip is open and the thermistor is exposed to the air passing over it.

Specific resistances will vary between manufacturers, but the voltage curve will be similar to that of the coolant temperature sensor. It is important to remember that high resistance readings indicate cold, dense incoming air and a need for more fuel, while low resistance readings indicate warmer, less dense incoming air and a need for less fuel. Since the computer is monitoring voltage drop across the sensor, voltage drop will be high when cold and decrease as the engine warms.

This sensor is mostly found on fuel-injected engines, but has been used on carbureted feedback systems used by Ford, Jeep, and Honda.

Throttle Body Air and Fuel Temperature Sensors — The computer can look at the throttle body air or fuel temperature during engine start-up. The sensor threads into the throttle body near the fuel lines or into the pressure regulator on port-injected engines.

Throttle body temperature and fuel temperature sensors can be found on engines with poor engine compartment air circulation or cast iron intake manifolds because of their inability to dissipate heat properly. The computer uses the sensor during hot engine restarts because higher engine temperatures could cause fuel to boil in the throttle body or fuel rail. If the fuel does boil and vaporize, the engine can suffer from a long crank time and rough running after start-up due to lack of fuel.

At higher throttle body temperatures the computer will increase the injector pulse-width. When the fuel temperature sensor is higher than a specified level, the computer will cut off the intake manifold vacuum to the fuel pressure regulator, which will raise the fuel pressure at the injector nozzles.

The result of ***hot soaked*** fuel in the throttle body or fuel rail is a lean mixture, so the increased pulse-width is only for a short time. The engine gets enough fuel to assure a quick start-up and a smooth running engine. This sensor is monitored during start-up, and should not cause a driveability problem once the engine is running.

Battery Temperature Sensor — In all Chrysler EFI systems since 1985, a ***battery temperature sensor*** and battery voltage sensing circuit work together to regulate charging system voltage. Charging system regulation control is built into the computer.

The battery temperature sensor is built into the computer located in the inlet air stream near the battery. Battery temperature and battery voltage are used to calculate charging system ***target voltage.*** Target voltage allows the computer to determine alternator field control if the charging system is operating correctly. Voltage regulation will vary somewhere between 12.9 and 15.0 volts.

Exhaust Gas Flow Sensors — An *Exhaust Gas Recirculation* (EGR) flow sensor is a thermistor that monitors the temperature of the exhaust gases entering the intake manifold. The exhaust gas recirculation temperature sensor is located (Figure 5-13) in the intake (vacuum) side of the EGR system, where it is exposed to the flow of hot exhaust gas whenever the EGR valve opens.

If the EGR valve is closed and no exhaust gas flow is present, then resistance and voltage drop remains high. As exhaust gas flows, it causes a decrease in sensor resistance voltage drop. The computer monitors voltage drop to detect if adequate flow is occurring.

Another type of EGR measures the content of oxygen in the exhaust gases and is a PTC thermistor. This type uses a Titania oxygen sensor. This sensor consists of a semiconductor material that varies resistance with the density of oxygen in the surrounding atmosphere. It does not create a voltage like a zirconium sensor, but its resistance value changes with oxygen content in the exhaust. The sensor is sealed and does not require an ambient oxygen reference.

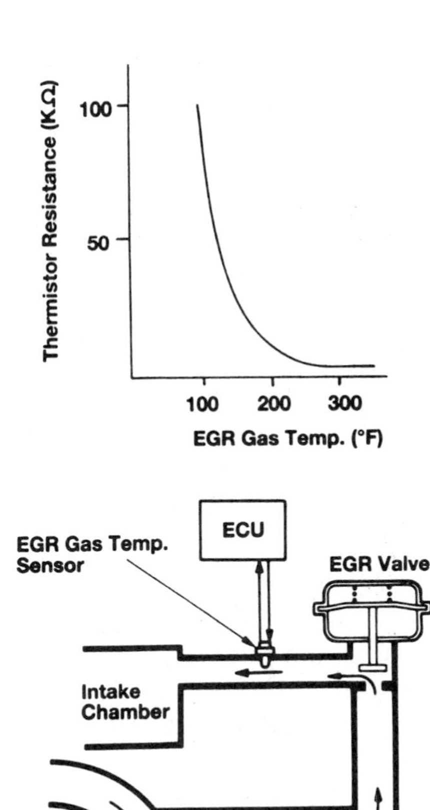

Figure 5-13 EGR temperature sensor informs the computer that exhaust gases are flowing.

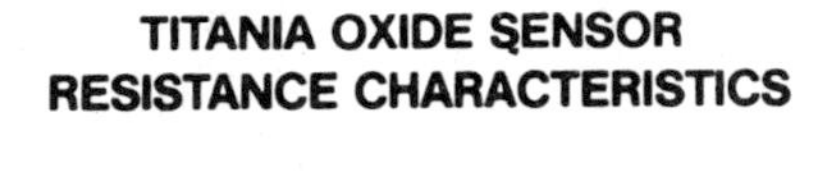

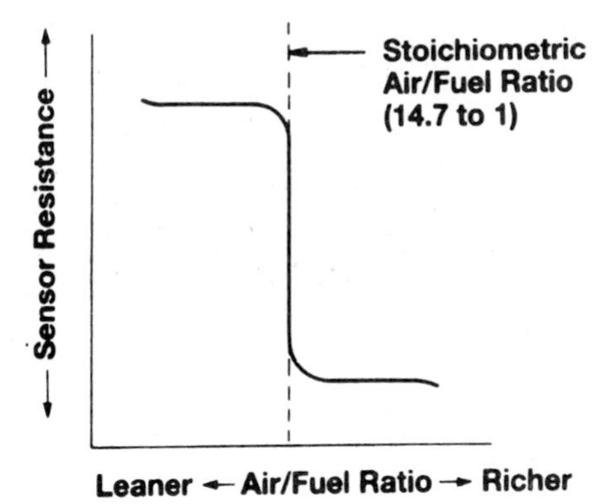

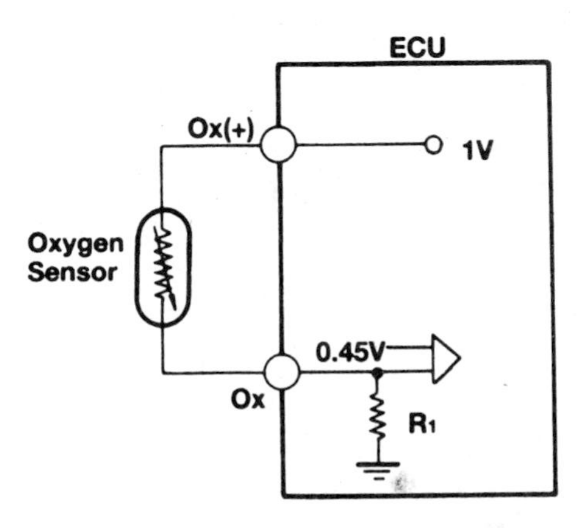

Figure 5-14 Titania oxygen sensor circuit that uses a 1-volt reference.

Reference voltage to the sensor is 1 or 5 volts. Renault and Jeep use a 5-volt reference, while Toyota and Nissan use a 1-volt reference (as shown in Figure 5-14). Depending on circuit design, sensor voltage will be the same as a zirconium sensor or just the opposite.

As the air/fuel ratio changes, so does the resistance of the titania sensor. A lean exhaust will raise the exhaust temperature and produce a high voltage drop because resistance increases as the temperature increases. A rich exhaust will cool exhaust gas temperature and produce a high voltage drop because resistance decreases as temperature decreases.

Voltage Dividers — Voltage dividers are a type of variable, resistor that produce a continuously variable, analog DC voltage input signal, proportional to a mechanical position. A reference voltage from the computer is varied by the voltage divider and then returns a signal voltage to the computer.

Voltage divider circuits all have three wires. The circuit shown in Figure 5-15 consists of a:

1. 5-volt reference.
2. Signal return.
3. Ground return to the computer.

The difference between voltage dividers has to do with their internal design. The first type of variable resistor includes a single fixed resistor with a wiper arm. A voltage divider can measure throttle position, EGR valve position, and air flow. A second type uses a resistor network built into a flexible diagram. Resistive network, or Strain Gauge-type

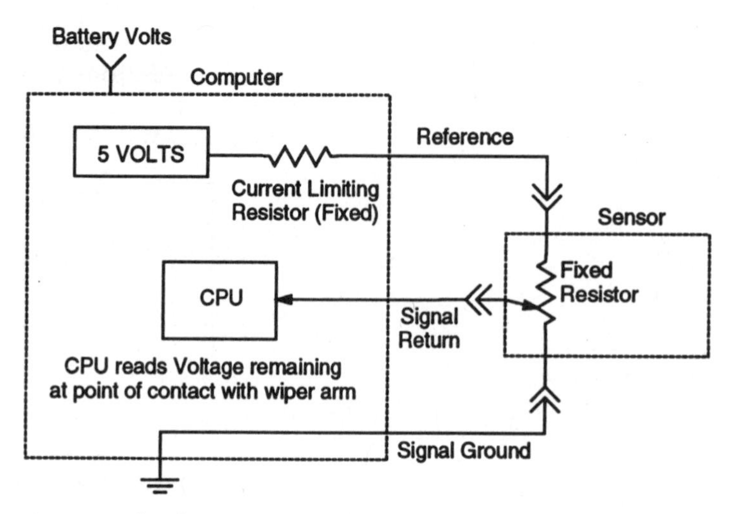

Figure 5-15 Typical potentiometer circuit.

sensors measure engine vacuum, manifold pressure and barometric pressure.

Potentiometer — The computer looks at the *throttle position sensor* to learn the position of the throttle plates and the status of their movement. The angle of the throttle plates determines the amount of air entering the engine to mix with the fuel. Through the throttle position sensor, the computer identifies closed throttle, acceleration, or wide-open-throttle (W.O.T.). By watching the rate of change, the computer can see if the throttle is being opened or closed, and how quickly or slowly this happens. This input can be used for:

1. Fuel enrichment and deceleration lean out.
2. Clearing a flooded engine during cranking.
3. Spark advance calculation.
4. Air injection directional flow.
5. A/C cutoff during wide-open-throttle.
6. Evaporative canister purge flow.

The *EGR valve position sensor* (EGR-EVP) provides the computer with a signal indicating the position of the EGR pintle valve. It is located on top of the EGR valve (Figure 5-16). It allows the computer to monitor and control EGR valve movement and flow. Knowing the amount of EGR flow can have an effect on the air/fuel ratio and spark timing.

A Vane Air Flow Meter (VAF) or Linear Meter measures the amount of air entering the engine. The output signal from this sensor will vary depending on the volume of air flowing through it. This signal allows the computer to calculate the amount of fuel needed for the air entering the engine. This meter has been used for years on Bosch L-Jetronic EFI systems.

Resistive Network Sensors — Resistive Network Sensors can measure manifold pressure, vacuum, and barometric pressure. *A Manifold Absolute Pressure (MAP) sensor, Pressure Differential (vacuum)*

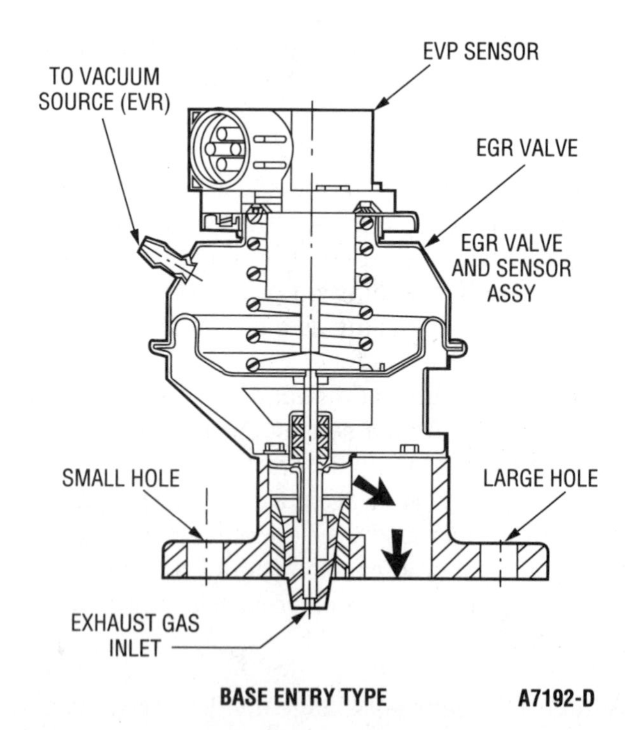

Figure 5-16 EGR position sensor mounted on top of the valve (courtesy OTC, a division of SPX Corporation).

sensor, or *Barometric (Baro) sensor* can be used to make these measurements. Figure 5-17 shows how the MAP and Baro change with pressure changes.

The Manifold Absolute Pressure (MAP) sensor provides information about intake manifold pressure while the engine is running. The MAP sensor needs an unrestricted pressure supply. It can be connected to the intake manifold by a hose or threaded directly into it. Monitoring intake manifold pressure allows the computer to know how much air is entering the engine, so it can calculate *engine load.*

Intake manifold pressure is a direct result of throttle position and engine speed. The computer uses this input to calculate:

1. Fuel delivery.
2. Ignition timing.
3. EGR flow.
4. Torque converter lockup and unlock.
5. Barometric pressure (when a separate barometric sensor is not used).

Each time the computer sends a spark output signal to the distributor, it looks for a voltage change from the MAP. If a voltage change occurs, it indicates to the computer that the MAP is receiving manifold pressure.

A *pressure differential* or *vacuum sensor* replaces the MAP sensor on General Motors carbureted engines. It is a pressure differential sensor that constantly compares engine vacuum to atmospheric pressure. It is connected to the intake manifold on one side and open to the atmosphere on the other side. The difference in pressure between the two sides is equal to engine vacuum. This allows the computer to monitor manifold vacuum and barometric pressure simultaneously. It does the same job as the MAP sensor, but output voltages will be opposite those of a MAP sensor.

The Barometric (Baro) Pressure (BP) sensor monitors barometric or atmospheric pressure any time the ignition key is on. At higher altitudes, less air is present because the pressure is lower, creating

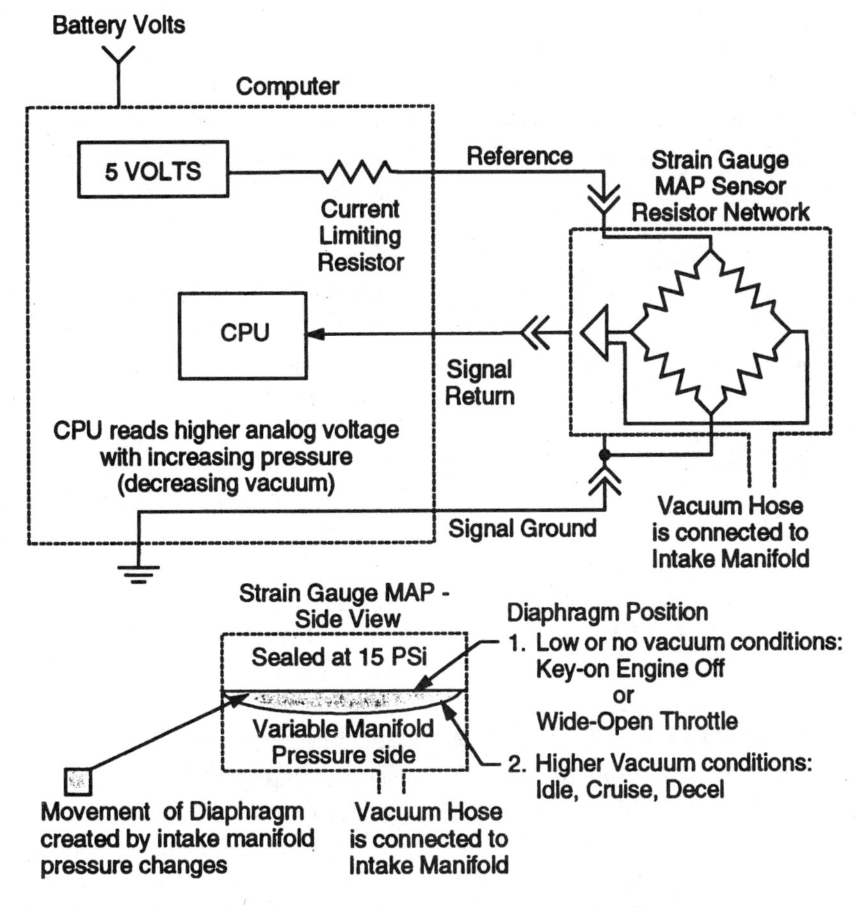

Figure 5-17 Typical resistor network (strain gauge) pressure sensor circuit.

a rich air/fuel ratio. Monitoring atmospheric pressure allows the computer to compensate by creating a leaner air/fuel ratio. In addition, a leaner air/fuel ratio requires more spark advance because it requires more time to burn.

Systems that do not use a separate Baro sensor still need to measure barometric pressure. This is accomplished using a *Barometric Manifold Absolute Pressure (BMAP) sensor.* A BMAP is actually a MAP sensor. The difference is how the computer uses the voltage from the MAP. When the key is turned to the start position, the computer reads the MAP and uses this voltage as a Baro reading. Once the engine is running, the MAP is again used to sense engine load. The computer updates the Baro reading through the MAP when the TPS is more than 60% open, which is equal to wide-open-throttle. At this time, manifold pressure is nearly the same as atmospheric pressure. Once the TPS drops below 60%, the computer uses the MAP to sense intake manifold pressure.

Mass Air Flow (MAF) sensors use a hot-wire or resistor type of system to measure the air mass of the incoming air. They provide the computer with information about the molecular structure of the incoming air. These sensors contain a sensing wire or resistor maintained at a calibrated temperature by controlling the amperage flow. As incoming air passes over the wire or resistor, it cools the wire/resistor. An increase in amperage to the wire/resistor raises the temperature to a calibrated temperature. The voltage changes with air flow, and the voltage changes are processed by the computer to calculate air mass for injector pulse-width and spark timing.

Variable DC Frequency

Sensors that fall into this category have the same purpose as previously mentioned sensors. The difference is the type of signal they generate. Variable DC frequency sensors generate a square wave (digital) depending on manifold, atmospheric pressure, or air flow.

Variable frequency sensors can be used as a MAP, Baro, or Mass Air Flow sensors. Frequency Baro and MAP sensors have the same three wires as other MAP, Baro, and MAF sensors. One wire is battery voltage, the signal return toggles 5-volts to ground, and the third wire is ground.

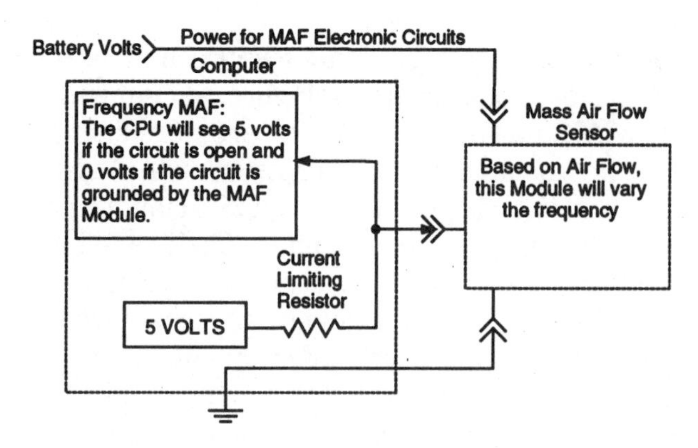

Figure 5-18 Typical DC frequency (Hertz) Mass Air Flow (MAF) sensor circuit.

The signal return line produces a frequency measured in *Hertz* (cycles per second) (Figure 5-18). As air flow increases, frequency increases because more air is entering the engine. As air flow increases, the sensor varies its on/off cycles per second, but voltage and amperage remain constant.

Variable DC Voltage Generators

Variable DC voltage generators make their own voltage. They do not need a power or reference voltage to generate a signal. The most common type is a zirconium oxygen sensor.

The *oxygen (O_2) sensor*, located in the exhaust stream, generates a variable voltage that the computer uses to detect a rich or lean exhaust. This sensor allows the computer to maintain an air/fuel ratio of 14.7:1 during closed-loop operation.

A zirconium oxygen sensor is a small galvanic battery that produces a low voltage signal. The heart of the oxygen sensor is a test tube-shaped piece of ceramic zirconium (Figure 5-19). The inner and outer surfaces of the ceramic tube are coated with platinum. Platinum provides the electrodes of the battery. Ambient (outside) air is always present inside the O_2 sensor. The O_2 sensor's outer plate is constantly exposed to the exhaust gases leaving the cylinders. A voltage is produced when there is a difference in oxygen content between the two plates. The voltage will vary between 100 millivolts (0.10v) and 900 millivolts (0.90v) on most applications.

When the engine is running rich, there is a shortage of oxygen in the exhaust gas, but there is plenty of oxygen in the ambient air side. Oxygen

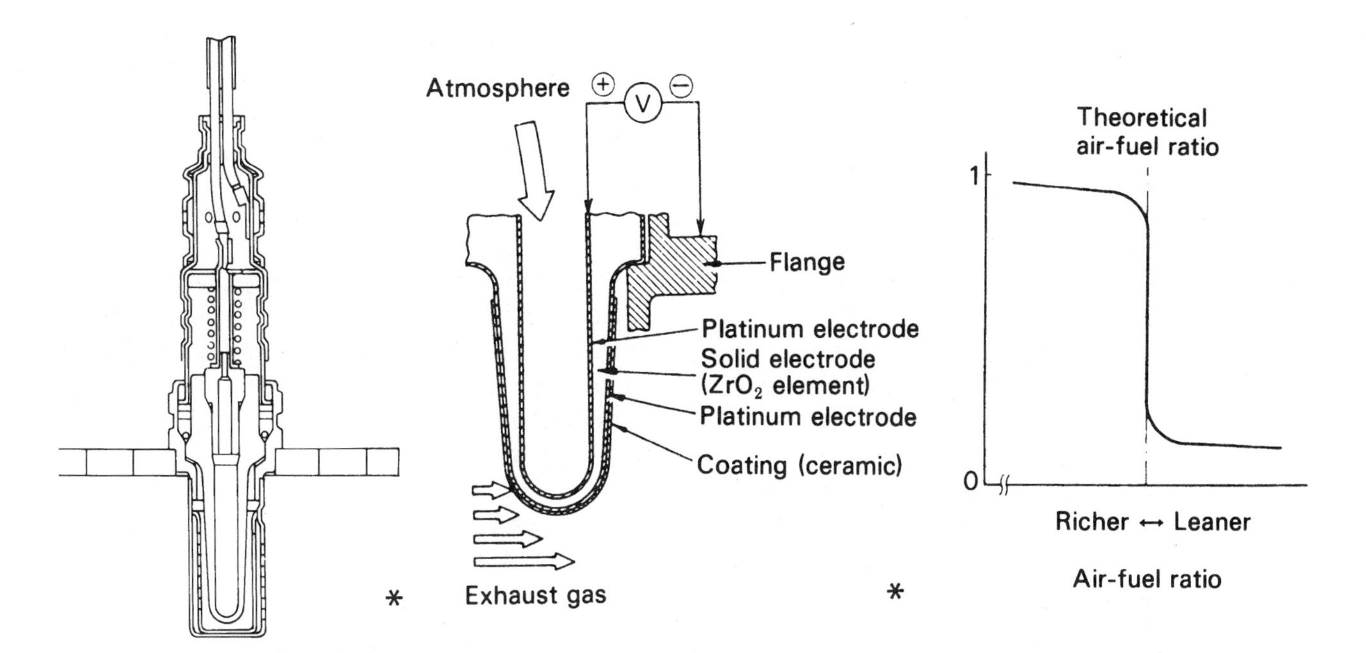

Figure 5-19 Zirconium oxygen sensor construction and voltage.

ions, a negatively charged atom, transfer from the atmospheric side of the sensor to the exhaust side, thus transferring a negative charge to the exhaust side of the sensor and generating a signal of 0.5 volts or more. The opposite occurs when the exhaust is lean. Since there is a high oxygen content in the exhaust side of the sensor, the number of oxygen ions transferred is less, causing less than 0.4 volts to be generated. The transition between rich and lean is almost instantaneous, and under ideal conditions this cycle should take approximately 1/10 of a second.

The ***higher*** the *oxygen* content in the exhaust, the ***leaner*** the *exhaust*. The *lower* the *oxygen* content, the *richer* the *exhaust*. It is important to remember that an oxygen sensor senses oxygen **only**. A non-firing cylinder, such as a bad spark plug or wire, can give a false lean reading, and drive the fuel system rich. Oxygen in the cylinder escapes into the exhaust manifold unburned, creating a lean exhaust condition. The computer receives a voltage signal from the O_2 sensor and interprets a lean condition. The computer compensates by enriching the air/fuel mixture.

The most serious cause of oxygen sensor failure is sensor poisoning. Clogging of the sensor's pores slows down response time and the sensor's output voltage will drop. Zirconium sensors can be poisoned from either side of the sensor since they are open on one side to ambient air. Some causes of sensor poisoning are listed below:

1. Tetraethyl lead found in regular gasoline is harmful if repeatedly used. An occasional misfueling with leaded gasoline probably won't hurt an oxygen sensor, because it will clean itself over a period of time. Continued usage of leaded fuel will clog an oxygen sensor and ruin a catalytic converter.
2. Methanol added to fuel is corrosive. It can dissolve some of the metallic lead put on the inside of the fuel tank to prevent rusting. The lead from this cannot only foul an oxygen sensor, it can ruin a catalytic converter.
3. Silicon is released from RTV gasket material and sealant as they cure in open air. The silicon deposits accumulate on the tip of the sensor and act as a filter, preventing the full flow of oxygen into the sensor. This will result in a delayed response by the sensor to changes in the air mixture. This condition can be identified by smooth, chalky white deposits on the tip of the oxygen sensor. Another sign of silicon poisoning is that the sensor output voltage actually goes negative.
4. Carbon from an overly rich mixture, engine oil leaking past defective rings, a blown turbocharger

turbine seal, and coolant leaking through cracked castings or leaking head gaskets can result in a failed sensor. Each of these contaminants can be identified by the residue it leaves on the sensor's tip.

Engine oil leaves a brown residue. These deposits may not kill the sensor right away, but they will reduce sensor efficiency and ruin it in the end. Rich exhaust conditions leave a black sooty coating that usually burns off within a few minutes of operation after the fault has been corrected. It can sometimes be cleaned by running the engine lean at 2000 RPM for a few minutes. Coolant leaves flaky white deposits that may have the "sweet" smell of ethylene glycol.

It is a good idea to make a habit of inspecting every sensor for contamination and making sure to correct the cause before installing a new sensor. Only RTV sealers that are specially formulated for use with oxygen sensor-equipped engines should be used.

5. Rust proofing, undercoating, solvents, brake fluid, and power steering fluid can alter the calibration of an oxygen sensor by blocking the reference (ambient air) entering it and slowing down response time.

Due to the location and construction of most oxygen sensors, they can be subjected to physical abuse. The wire(s) to the oxygen sensor have been found laying on hot exhaust manifolds, pinched between the transmission bell housing or starter housing and the engine, and with bad connections. They can be easily cracked if not handled properly or even shattered if blasted with cold or hot water. Although ceramic material is extremely hard, it's also brittle and a sharp blow will crack it.

Variable AC Voltage

Magnetic pickups consist of a wire coil surrounding a permanent magnet. As the magnetic field crosses through a coil, an electrical voltage is generated in the wire coils. They do not need power or a reference voltage to operate.

A magnetic pickup generates an AC analog positive and negative sine wave. The analog signal is converted to a digital signal by the computer. It then counts the number of cycles per second. Magnetic pickups can be used to calculate:

1. Engine RPM.
2. Crankshaft position.
3. Vehicle speed.

A distributor pickup coil generates an AC sine wave as it rotates. The computer uses this input to calculate a spark output to the ignition module. As the ignition module is cycled on and off, the primary ignition system collapses, producing a secondary spark.

A magnetic pickup used on a distributor-less ignition system is called a crankshaft sensor or reluctor. They can usually be found in one of three places:

1. In the bell housing next to the flywheel.
2. On the front of the engine next to the harmonic balancer.
3. In the side of the engine block.

Vehicle speed sensors produce pulses the computer counts to calculate:

1. Torque converter lockup and unlock.
2. Idle bypass air on deceleration to prevent stalling.
3. EGR flow on some applications.

The detonation (knock) sensor is a piezoelectric crystal that senses spark detonation. When detonation occurs, it generates a DC voltage the computer uses to retard spark timing under load. Cylinder detonation must reach a minimum level before the computer modifies the spark advance timing. Figure 5-20 shows a Toyota knock sensor generating a signal.

Switches

Switches are simple in theory; they are either on or off. There are three types of switches:

1. Hall-Effect.
2. Switch to power.
3. Switch to ground.

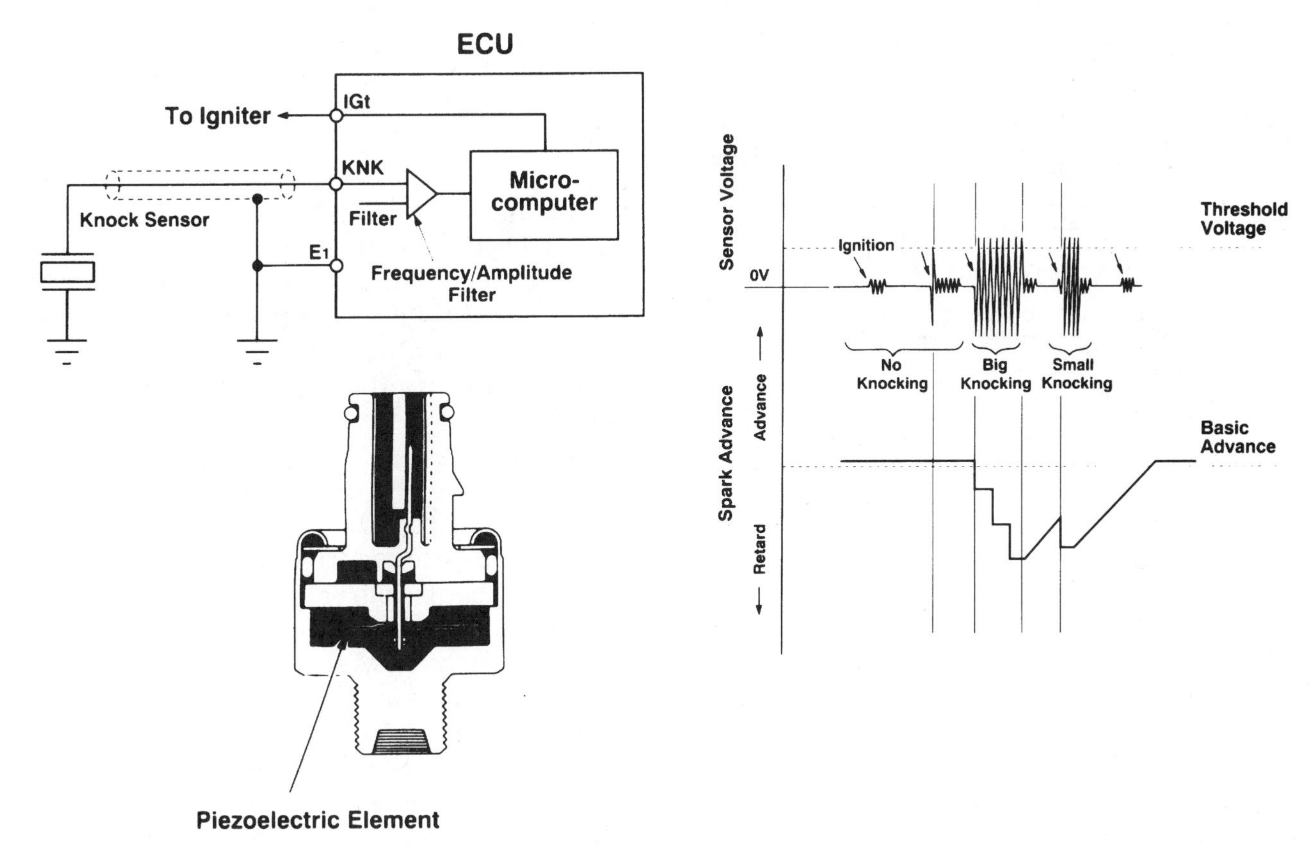

Figure 5-20 Typical detonation (knock) sensor circuit that shows the relationship between sensor voltage and spark advance.

Hall-Effect Switches — A Hall-Effect switch is a magnetic switch with electronic circuits that create a signal the computer uses to:

1. Control ignition primary.
2. Calculate engine RPM.
3. Identify crankshaft position.
4. Identify cylinder number one through a camshaft position sensor.

All Hall-Effect switches consist of three wires (Figure 5-21). One wire has battery voltage that powers the sensor's electronics. The second wire is the signal line that the Hall-Effect switch grounds and ungrounds. This creates the digital switching that is necessary to run the engine. The third wire is ground.

Switch to Power Switches — A switch to power (Figure 5-22) circuit allows the computer to see zero volts until the switch is closed. This is because the power feed is through the ignition switch to a control switch. The air conditioning request switch is an example of this type of switch. With the ignition on, the driver pushes the air conditioning on/off switch.

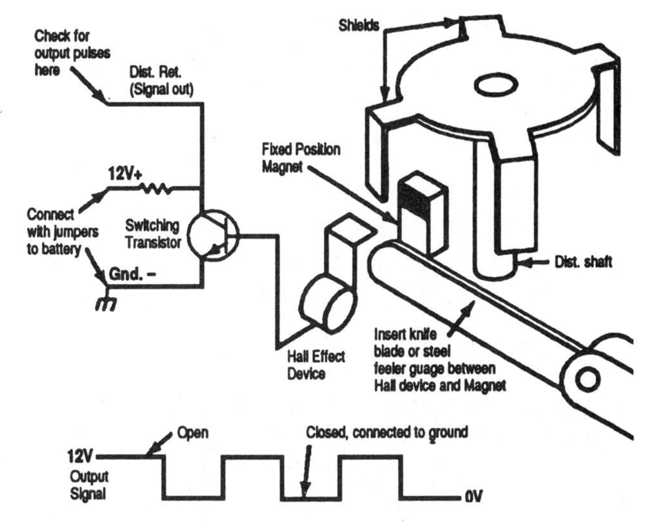

Figure 5-21 Checking Hall-Effect switches (courtesy Fluke Corporation).

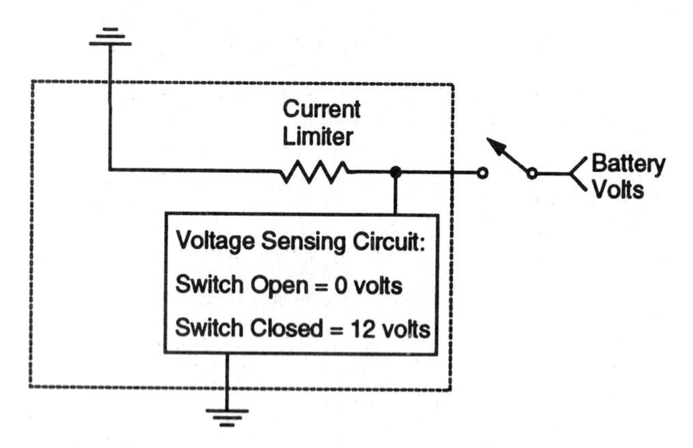

Figure 5-22 Typical switch to power circuit and voltages.

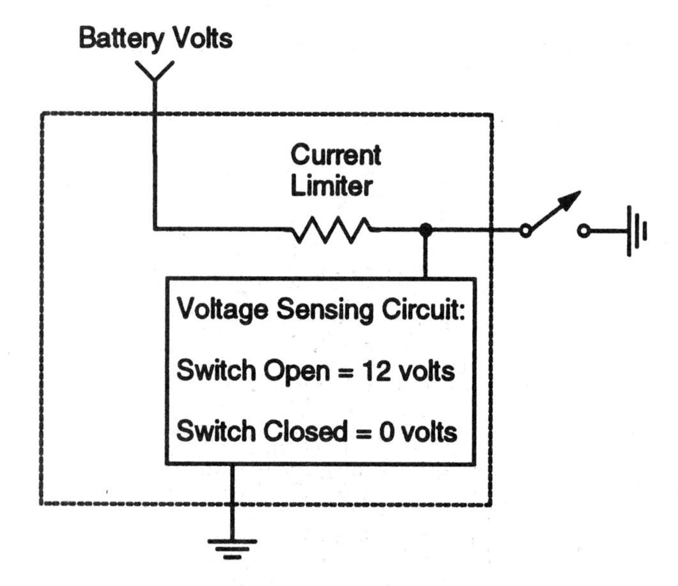

Figure 5-23 Typical switch to ground circuit and voltages.

Once the computer sees a voltage from this switch, it can make necessary adjustments for the load placed on the engine by the air conditioning compressor.

Switch to Ground Switches — A switch to ground (Figure 5-23) circuit allows the computer to see battery voltage until the switch is grounded. The Park/Neutral, brake, power steering, idle contact, and transmission gear switches are examples of this type of switch. For example, shifting the transmission gear selector into reverse creates a voltage drop across the Park/Neutral switch and the computer sees zero volts.

Importance of Sensors and Switches — Now that we have covered the theory, operation, and testing groups of inputs, we need to learn what importance they play in providing a smooth running engine. The farther down the list sensors and switches are, the less effect they have on driveability:

1. *RPM* — This signal must be present for a fuel injected engine to run. This input can be a variable AC magnetic pickup or a digital DC voltage from a Hall-Effect switch or optical sensor. The computer uses the RPM signal to time each ignition pulse and injector pulse.
2. *Load Sensor* — This sensor can be a variable resistor or a variable DC frequency signal. It can be a MAP, Vacuum, Vane Air Flow Meter, or Mass Air Flow sensor. The computer uses this input and the RPM signal to establish the necessary air/fuel ratio.
3. *Throttle Position* — It can be a variable resistor potentiometer or an on-off-on contact switch. The computer uses this input to establish acceleration enrichment or deceleration lean-out.
4. *Coolant Temperature* — This sensor is a thermistor, and it affects open-loop air/fuel, spark advance, idle speed, and the on-off of various emission control systems.
5. *Barometric Pressure* — It informs the computer of the altitude and compensates for less air by leaning the fuel and advancing the spark.
6. *Inlet Air Temperature* — This sensor is used to fine tune fuel delivery and idle speed based on air density.
7. *Oxygen* — It is used by the computer to correct the air/fuel ratio to 14.7:1 in a closed-loop system.
8. *Switches* — The final inputs have an effect on fuel delivery and idle speed. They are the Park/Neutral and throttle contact switches. Both switches inform the computer of engine load and will affect fuel delivery and idle speed depending on the gear selected and if the throttle is closed or open.

Remember that the farther down the list the sensor or switch is, the less effect it will have on driveability. Often the sensors and switches farther down the list are less obvious and often cause intermittent driveability complaints that can be hard to find.

OUTPUT/ACTUATOR CONTROLS

The computer sends commands to actuators to control various mechanical, pneumatic, and hydraulic functions. Computer output voltages to most actuators are on/off or high/low.

Actuators are electromechanical devices that use an electrical current to produce a mechanical action. By controlling current flow through these devices, the computer can control the engine operation within its designed parameters.

Most actuators have battery voltage applied to them. The computer controls the ground side of the circuit, which is called *ground side switching* (Figure 5-24). This allows the computer to control high current devices by grounding the low current side of the circuit. Ground side switching is controlled by the computer through a transistor, called a *driver.*

Transistor Switching

The common device used in the control of output signal processing is a *PNP transistor.* PNP transistors act like a switch between the collector and emitter when a voltage is applied at the base. A small amount of current will flow from the control voltage source through the base and out the emitter to ground when voltage is applied to the base of the transistor. This will forward bias (energize) the transistor, allowing current to flow from the collector to the emitter. When the voltage is removed from the transistor base, the circuit is reverse biased (de-energized).

Quad Driver Modules

Transistors can be in groups of four in a single module called a *Quad Driver Module* (QDM). This increases space available inside the computer and provides a *fault line* for improved diagnostics (Figure 5-25).

Quad Driver Modules are sensitive to current flow. Excessive current flow can destroy a driver and prevent the computer from controlling the

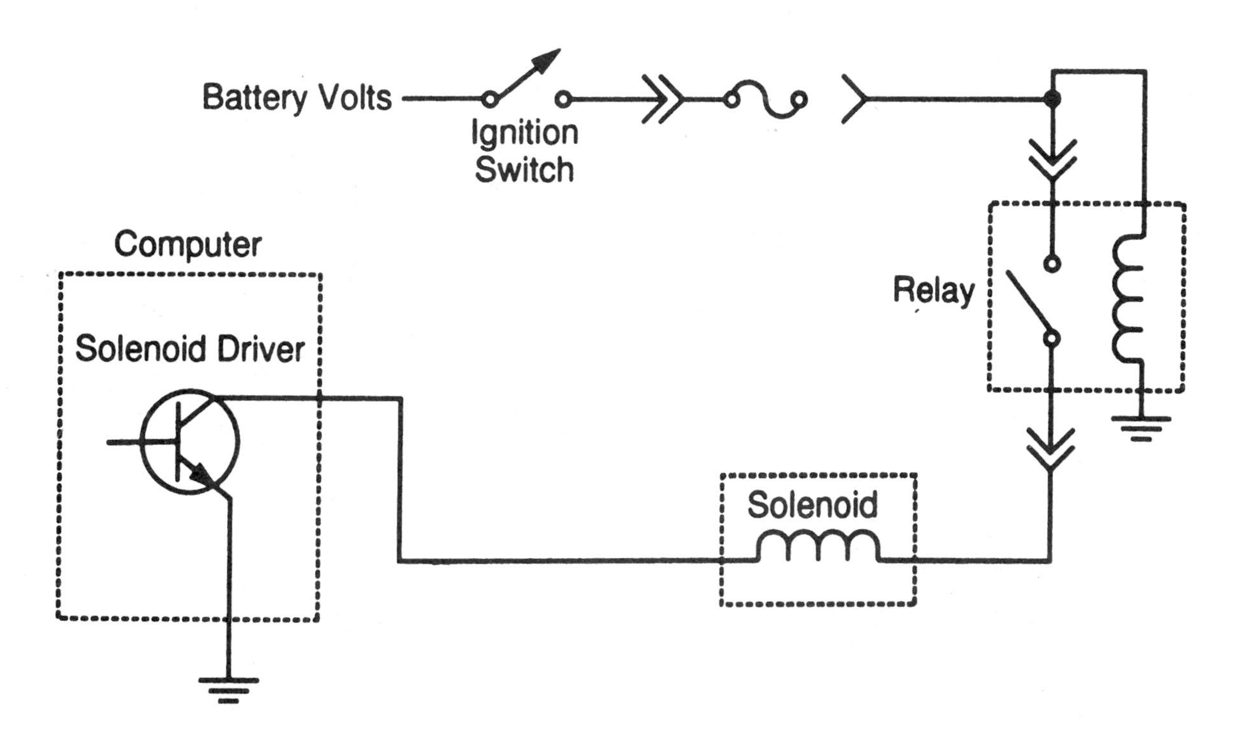

Figure 5-24 A *driver* (transistor) is used to control the ground side of the circuit.

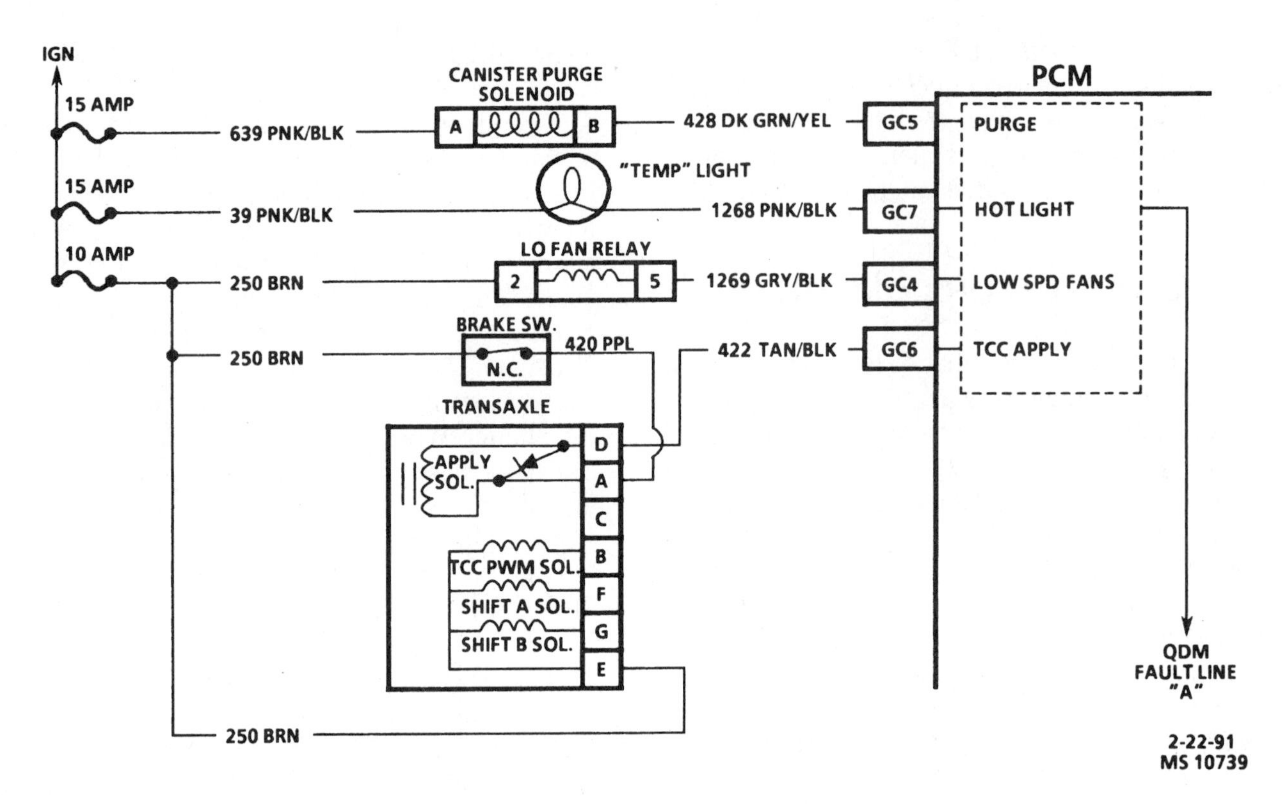

Figure 5-25 A Quad Driver Module (QDM) can control up to four actuators and includes a *fault line* to detect a circuit failure (courtesy GM).

actuator. Excessive current flow through a Quad Driver Module can cause more than one actuator to malfunction.

With the circuit energized, the computer monitors the fault line for current flow. If no current flow is detected, the computer will set a *hard fault* code.

Types of Output/Actuators

Outputs can be divided into five categories:

1. Solenoids.
2. Relays.
3. DC motors.
4. Control modules.
5. Diagnostic readouts.

Solenoids — A solenoid is an electromagnetic device that can control the flow of vacuum, fuel vapor, and liquid fuel. Solenoids are used to control:

1. A feedback carburetor mixture control solenoid.
2. Fuel injectors.
3. Torque converter lockup.
4. Idle speed.
5. Emissions control devices such as: EGR, Evaporative Canister Purge, Early Fuel Evaporation, and Air Injection.

A solenoid consists of two wires (Figure 5-26). One terminal receives fused battery power and the other terminal is grounded by the computer. Grounding the circuit energizes the actuator. Ungrounding the circuit de-energizes the actuator.

Solenoids are normally closed or normally open. Normally closed solenoids have to be grounded (energized) to open. Once open, vapor, liquid, or vacuum can be controlled by the solenoid. Fuel injectors are a normally closed solenoid and must be grounded for fuel to flow. Normally open (de-energized) solenoids have to be grounded to stop the flow of vapor, liquid, or vacuum. For example, a mixture control solenoid in a feedback carburetor is a normally open solenoid. A de-energized solenoid forces the carburetor to go full rich. An energized solenoid forces the carburetor to go full lean.

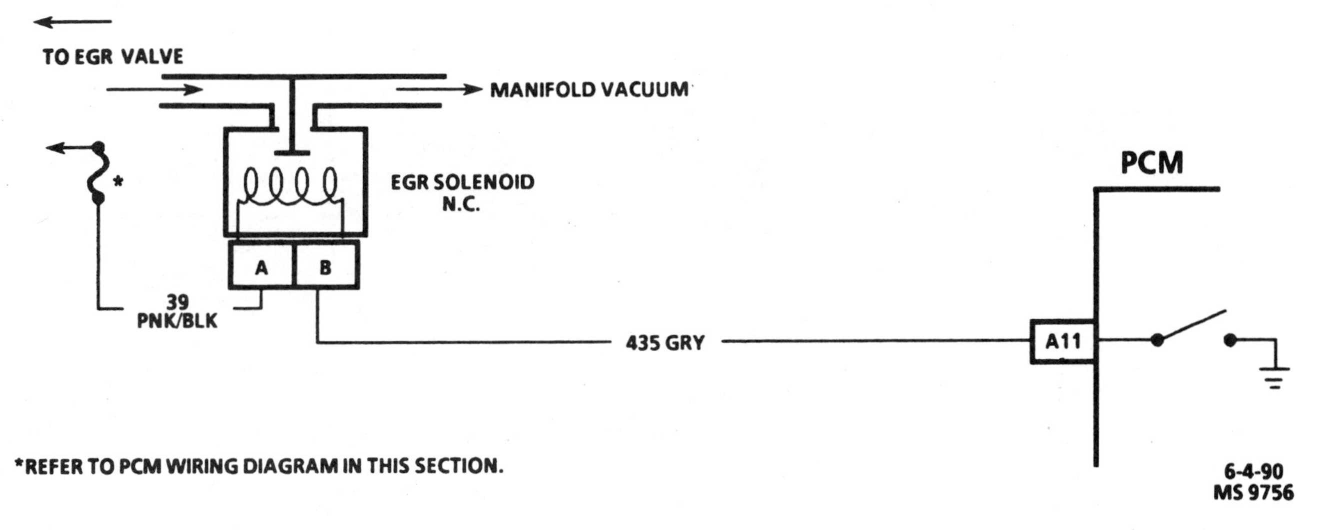

Figure 5-26 Manifold vacuum is controlled by a solenoid that requires battery power and a computer to ground circuit (courtesy GM).

Voltage Signals

Solenoid *on-time* (energized) can be controlled three ways:

1. A fixed duty-cycle.
2. A variable pulse-width modulation.
3. A variable pulse-width.

An example of a fixed *duty-cycle* solenoid is a mixture control solenoid in a feedback carburetor. It is cycled at a rate of ten times per second. The duty-cycle will vary depending on whether the system is in *closed loop* or *open loop*. When in closed loop, the duty-cycle will vary depending on the signal from the oxygen sensor. If the system is in open loop, the duty-cycle varies based on a preset program in the computer.

Solenoids controlled by a square wave that are not cycled at a fixed rate and vary the on/off time are called *pulse-width modulated.* A pulse-width modulated solenoid is often used to control the EGR valve. When the computer wants exhaust gases recycled into the intake manifold, it energizes a solenoid at a variable rate. A 0% duty-cycle indicates the solenoid is closed and no EGR flow is occurring. A 100% duty-cycle indicates that the EGR valve is fully open, allowing maximum flow.

Figure 5-27 shows a Hitachi Idle Air Bypass solenoid used to regulate idle speed on many Ford fuel-injected engines. It consists of two wires and is duty cycled to regulate the amount of air bypassing the closed throttle plates. The computer regulates the duty from 0% to 100%, depending on the RPM needed for a stable idle speed.

A fuel injector is an example of a pulse-width solenoid measured in milliseconds. As the fuel demand increases, the milliseconds of pulse-width should increase.

Relays — Relays control high-current devices when rapid switching is not critical. They can control the:

1. Electric fuel pump.
2. Air conditioning compressor clutch.
3. Engine cooling fan(s).
4. Power to the computer, oxygen sensor heater, ignition coil positive terminal, fuel injectors, and alternator field.

The computer controls the low current coil side of the relay, which when grounded, allows higher current to pass through the switch to operate the device. Most relays are normally open, which means the computer must ground the coil side to allow the switch side to close.

De-Spiking Diodes (Clamping Diodes) or Resistors — Many relays and solenoids have a de-spiking diode or resistor across their load coil (Figure 5-28).

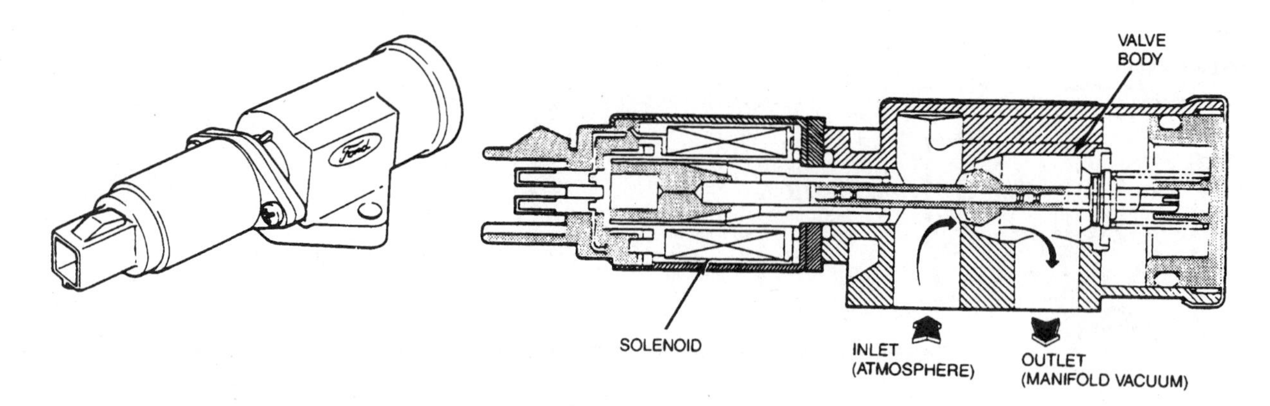

Figure 5-27 A solenoid can be used to regulate idle speed by allowing air to bypass the closed throttle plates (courtesy Ford).

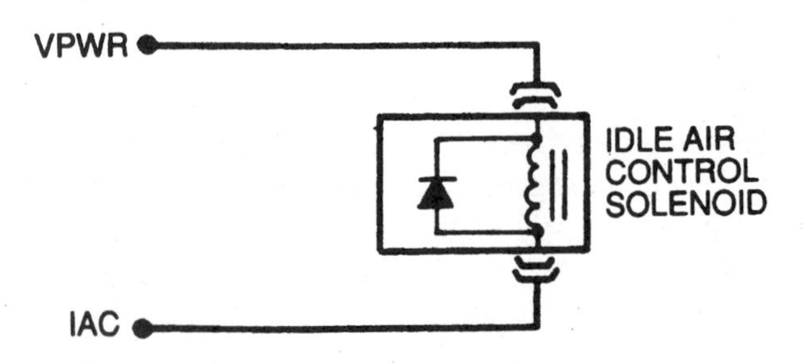

Figure 5-28 Diodes are connected across a coil to prevent voltage spikes that could damage the computer (courtesy Ford).

They are installed to:

1. Reduce radio interference.
2. Prevent voltage surges from affecting computer operation.
3. Prevent damage to the computer from induced voltage surges when solenoids and relays are switched off.

De-spiking diodes limit voltage surges when a coil is turned off. The computer de-energizes the circuit to stop current flow. The magnetic field of the coil collapses and produces a high voltage, trying to keep the current flowing. A reverse polarity will forward bias the diode, allowing current to flow. A de-spiking diode acts as a short circuit across the coil. This limits the spike to a voltage level that cannot damage the computer.

A de-spiking diode is connected in parallel across the 20Ω solenoid inside the transaxle as shown in Figure 5-29. If the computer energizes the solenoid or relay, the diode is reverse biased (turned off).

When the solenoid or relay circuit is de-energized, current flow stops, causing the magnetic field to collapse. The magnetic lines of force cut through the coil and induce a voltage. The voltage builds until it reaches approximately 0.7 volts, enough to forward bias (energize) the diode, completing the circuit to the other end of the coil. Current flows through the diode and coil circuit until the voltage is dissipated.

There are three disadvantages to the diode:

1. If subjected to high temperatures, it can short out and damage the transistor inside the computer.
2. Diodes can be damaged by a reverse polarity because the diode would be connected in the conducting direction and because it can handle very little current, it would immediately burn out, usually in an open condition.
3. If a voltage spike occurs from something else in the system, the diode acts as a short across the winding and reduces the voltage drop across the winding with a resultant higher voltage across the transistor in the computer. This higher voltage could damage the computer.

Solenoids and relays sometimes use a *resistor* across the coil windings rather than a diode. The advantages of the resistor are:

1. It is less likely to short with high temperatures.
2. It is not affected by reverse polarity.

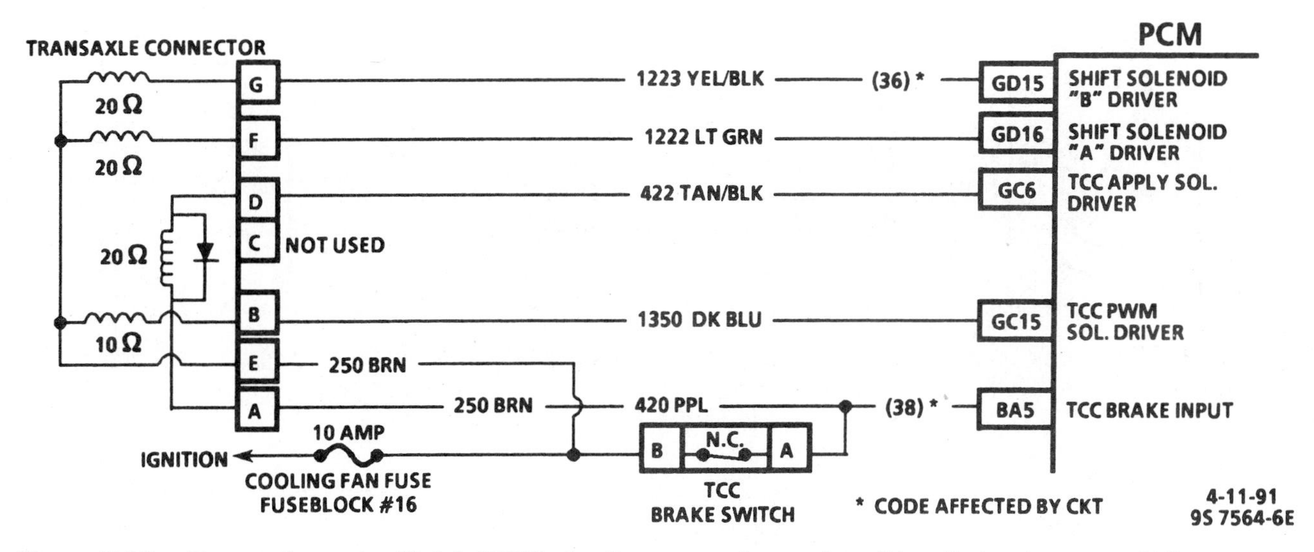

Figure 5-29 Torque Converter Clutch (TCC) circuits commonly use de-spiking diodes (courtesy GM).

3. It reduces the possibility of damage to the computer by an induced voltage other than from the winding itself.

The disadvantage of a resistor is when the coil is connected across the source of the induced voltage. It does not do as good a job of keeping the voltage down because it has more resistance (approximately 400-600 ohms) than the diode does in the forward direction. Because of the resistor's high resistance, they are not quite as efficient at suppressing a voltage spike.

DC Motors — Direct current (DC) reversible motors usually control engine idle speed. The computer internally reverses the polarity of the motor to achieve the desired idle speed. There can be two-wire or four-wire motors. Two types of motors are in use:

1. Idle Speed Control (ISC).
2. Idle Air Control (IAC).

An ISC motor consists of four wires and is externally mounted to the carburetor or throttle body. The computer controls the polarity to move the throttle plates for idle speed control. The other two wires connect the computer to an on/off *grounding switch* inside the motor (Figure 5-30). When the throttle is closed, the switch is closed and the computer sees a low voltage signal. A closed throttle switch signals the computer that it is time to control the idle speed.

The IAC controls the idle speed on fuel-injected engines only. It is typically a stepper motor consisting of four wires that control two sets of electromagnetic

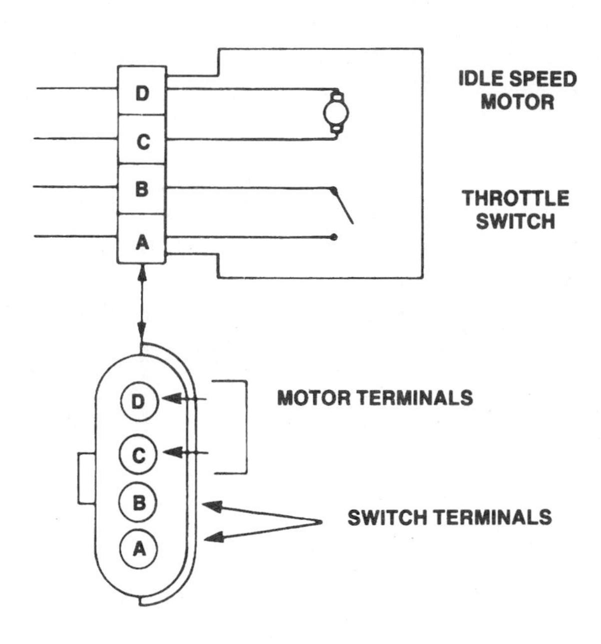

Figure 5-30 Terminals A and B allow the computer to identify a closed throttle. Terminals C and D are for the DC motor controlled by the computer (courtesy GM).

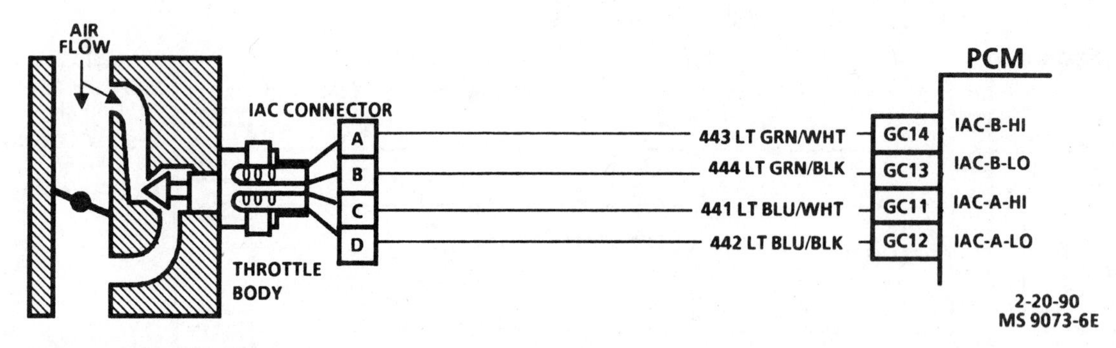

Figure 5-31 The Idle Air Control (IAC) moves in steps to regulate idle speed and to compensate for loads placed on the engine (courtesy GM).

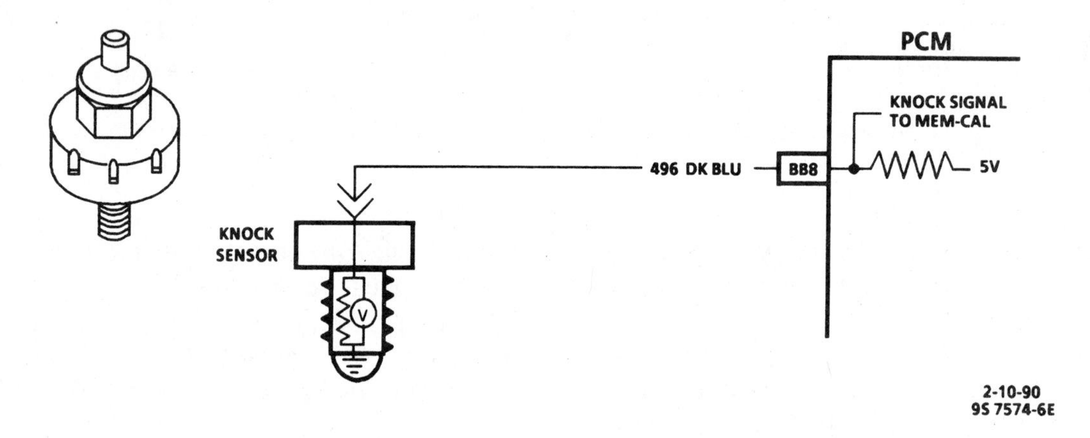

Figure 5-32 The detonation (knock) sensor control module is located inside the computer in the MEM-CAL chip (courtesy GM).

circuits inside the IAC. With this type of arrangement, the computer moves the motor in steps. Movement of the motor inward reduces the amount of air that bypasses the throttle plates (Figure 5-31), thus reducing the idle speed. Increasing the steps increases air bypassing the throttle plates, increasing the idle speed.

An ISC or IAC is necessary to maintain curb idle speed when a load is placed on the engine, and to provide cold fast idle for smooth engine operation and rapid warm-up. The idle speed control system regulates idle speed under at least one or more of the following conditions, depending on application:

1. Cold fast idle.
2. Warm curb idle.
3. Air conditioner load.
4. Electrical load.
5. Transmission load.
6. Power steering load.
7. Deceleration dashpot control.

Control Modules Used as Actuators — Separate computer-controlled modules are used to control:

1. Detonation control and spark output (which controls ignition timing for best engine operation under differing conditions).
2. Air conditioning compressor clutch operation under various loads and temperatures.
3. Cruise control engagement and disengagement.

In some applications, these devices may be built into the computer as shown in Figure 5-32.

Diagnostic Readouts — Diagnostic readouts can be a *check engine* light on the dash, flashing LED (Light Emitting Diode) on the computer, or data sent to a scan tool. The important thing to remember is the information is a "clue" about what the problem might be. Use it wisely, and remember to think of what inputs are responsible for controlling each output.

CHAPTER 6

Computer-Controlled Emission Control Systems

PRE-COMPUTER EMISSIONS CONTROLS

To meet air quality standards, a variety of emissions systems have been in use for many years. The main systems in use are:

1. Air Injection Reaction (AIR).
2. Evaporative emissions controls.
3. Early fuel evaporation.
4. Exhaust gas recirculation.
5. Catalytic converters.

Before the use of computer controls, thermo vacuum switches, vacuum hoses, and ported vacuum signals were used for control of the various emissions systems. These systems relied on coolant temperature to regulate thermo vacuum switches. Thermo vacuum switches were calibrated to open or close vacuum passages at a specific temperature. Ported vacuum signals were the main source of vacuum because they could be regulated by the position of the throttle plates.

Each additional thermo vacuum switch required more hoses and ported vacuum passages in the carburetor or throttle body to assure a strong vacuum signal. All of this made for a very cluttered engine compartment. The under-hood vacuum label looked like a map of the Los Angeles freeway system. Any technician who has worked on a carbureted Honda engine can understand this problem.

While these systems were effective in controlling emissions, they tended to have a negative effect on driveability. They could not calculate the amount of EGR flow or compensate for fuel vapors purged from the evaporative charcoal canister. For example, to reduce the effect of the EGR system, automotive engineers have tried:

1. Positive and negative backpressure EGR valves.
2. Backpressure transducers.
3. Primary and secondary EGR valves.
4. Vacuum-modulated EGR valves.
5. Vacuum amplifiers.

These attempts to improve driveability sometimes resulted in reduced fuel economy, added cost to the consumer, and wasted hours by the technician diagnosing the system.

COMPUTER-CONTROLLED SYSTEMS

The introduction of computerized engine controls has not changed the purpose of the various emission control systems. What has changed with these systems are: the type of vacuum used, the way the vacuum is controlled, the elimination of vacuum as the source of control, and system or component failure diagnosis.

The computer uses input sensors, switches, and actuators to regulate system operation. The sensors, switches, and actuators most commonly used are the:

1. Engine RPM signal.
2. Engine load sensor.
3. Throttle position sensor.
4. Coolant temperature sensor.
5. EGR valve position or pressure sensor.
6. Oxygen sensor.
7. Vehicle speed sensor.
8. Park/Neutral switch.
9. Solenoid.

The coolant temperature sensor and solenoid replace the thermo vacuum switch shown in Figure 6-1. Controlling vacuum through a solenoid can eliminate the need for ported vacuum passages, because the manifold is a better source of vacuum. Some late model EGR systems eliminate the need for a vacuum source and vacuum hoses by building everything into one valve.

The remaining sensors and switches allow the computer to calculate EGR flow and compensate for fuel vapors purged from the charcoal canister.

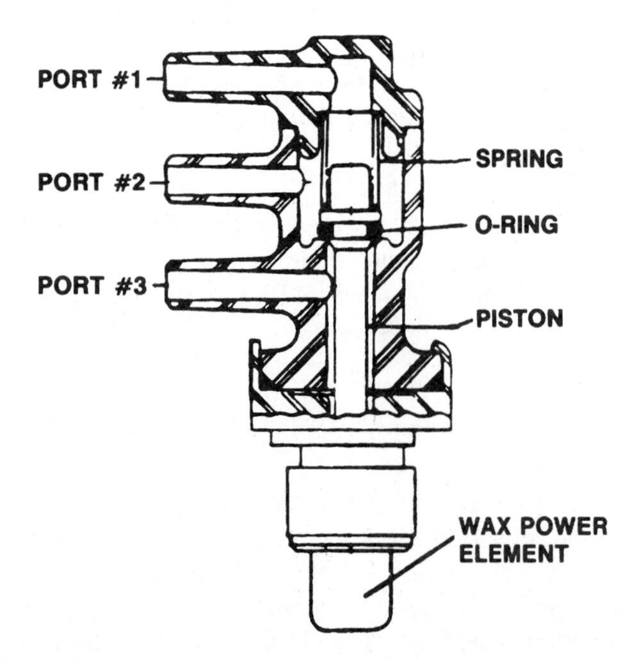

Figure 6-1 Thermo vacuum switches are used to regulate vacuum flow with engine temperature (courtesy GM).

They allow automotive engineers to restore driveability, improve engine performance, and increase fuel economy, while lowering exhaust and evaporative emissions.

On some engines, the computer has allowed automotive engineers to discontinue the use of Early Fuel Evaporation devices and Air Management systems. Over the last few years, some engines have been designed without EGR valves or systems. This trend has been reversed because of stringent NO_x regulations and the need for detonation (knock) control.

AIR INJECTION REACTION (AIR) SYSTEMS

Air injection systems allow the computer to control the flow of air from the air pump to either the exhaust manifold, catalytic converter, or atmosphere. Fresh air injected into the exhaust manifold allows for complete burning of the air/fuel mixture as it leaves the cylinder. The additional air also helps to heat up the exhaust system, oxygen sensor, and catalytic converter. Faster warm-up allows for closed loop to happen faster on a cold engine.

This air pump switching action is controlled by the computer through one of the following: switching valve, bypass valve, diverter valve, or a combination of valves. Vehicles not equipped with an air line to the catalytic converter bypass AIR to the air cleaner or atmosphere during all engine modes except open loop (Figure 6-2).

Regulated by the computer, control valves stop exhaust manifold air injection during closed loop by sending the air to the converter or atmosphere, depending on the engine calibration (Figure 6-3).

Air injection switching downstream to the catalytic converter is not a requirement for entering closed loop, but a result of the closed-loop requirements being met. The computer does not know that the air pump actually switched the air downstream. An air injection component failure can be the source of lean exhaust conditions because it induces false air into the exhaust stream ahead of the oxygen sensor. During closed-loop operation, the introduction of air into the exhaust manifold would create a false lean signal from the oxygen sensor, causing the computer to issue a rich command to the carburetor or fuel injectors.

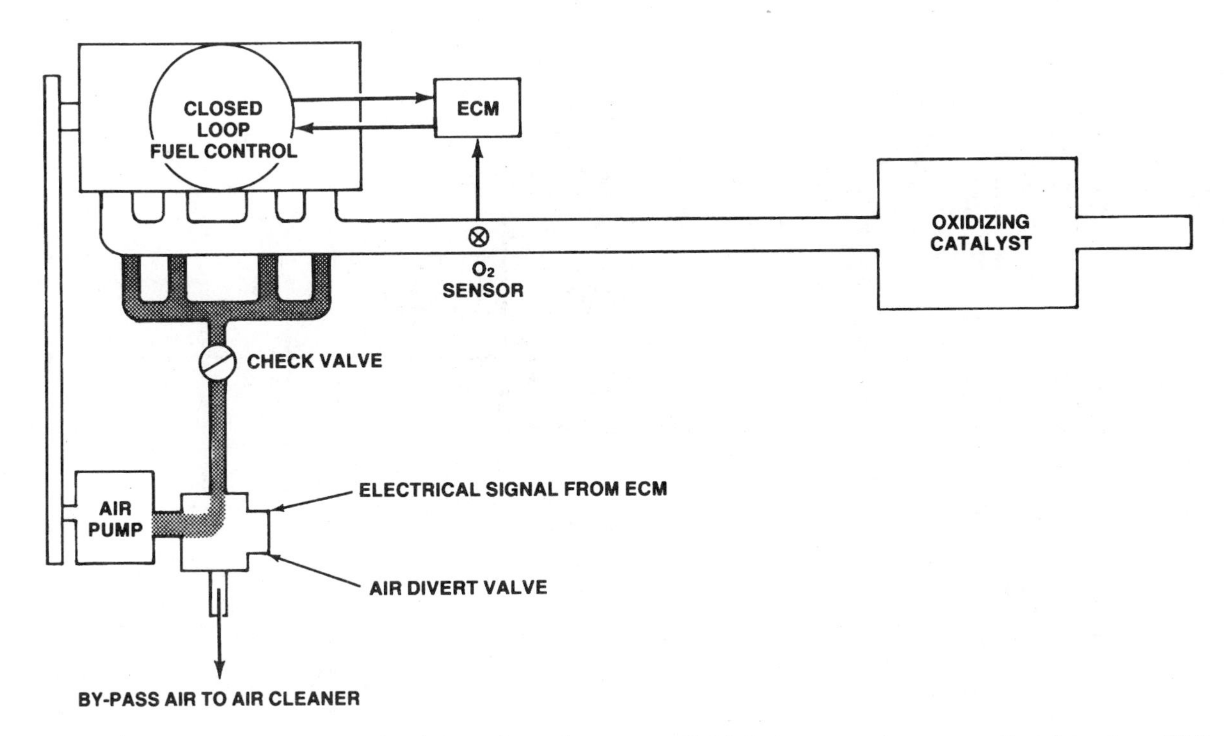

Figure 6-2 A single solenoid sends air into the exhaust manifold during open-loop operation (courtesy GM).

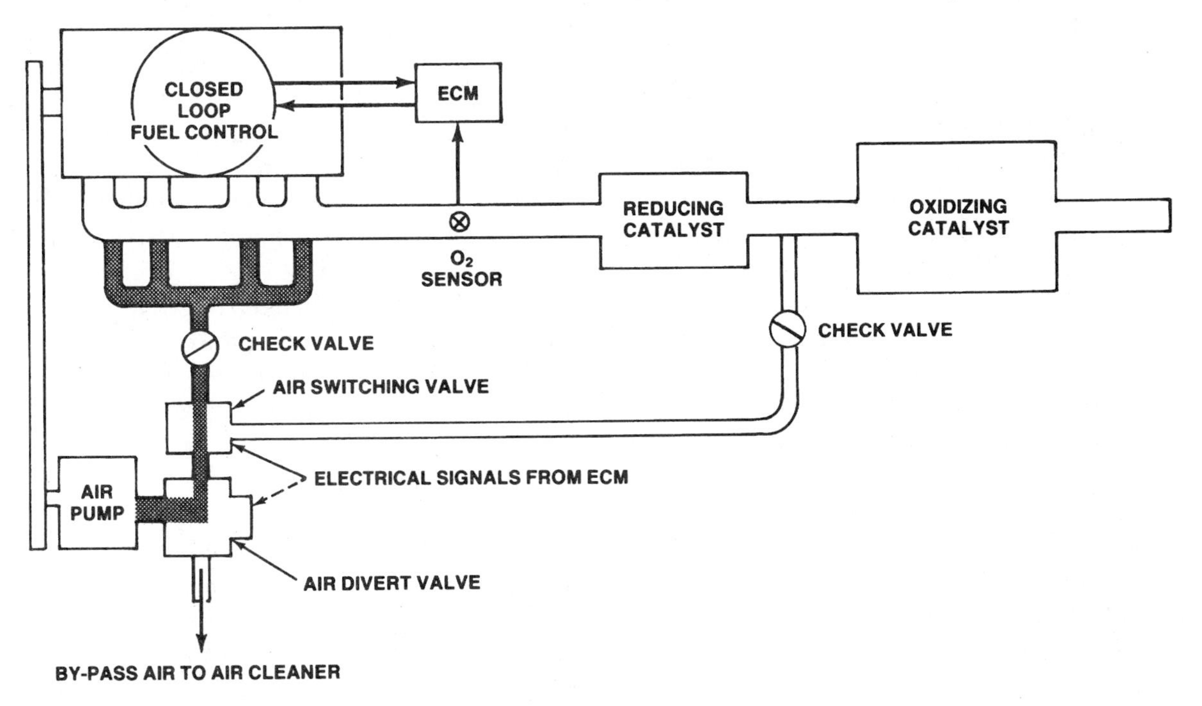

Figure 6-3 A dual solenoid can switch the air upstream (exhaust manifold) or downstream (catalytic converter) depending on the engine mode (courtesy GM).

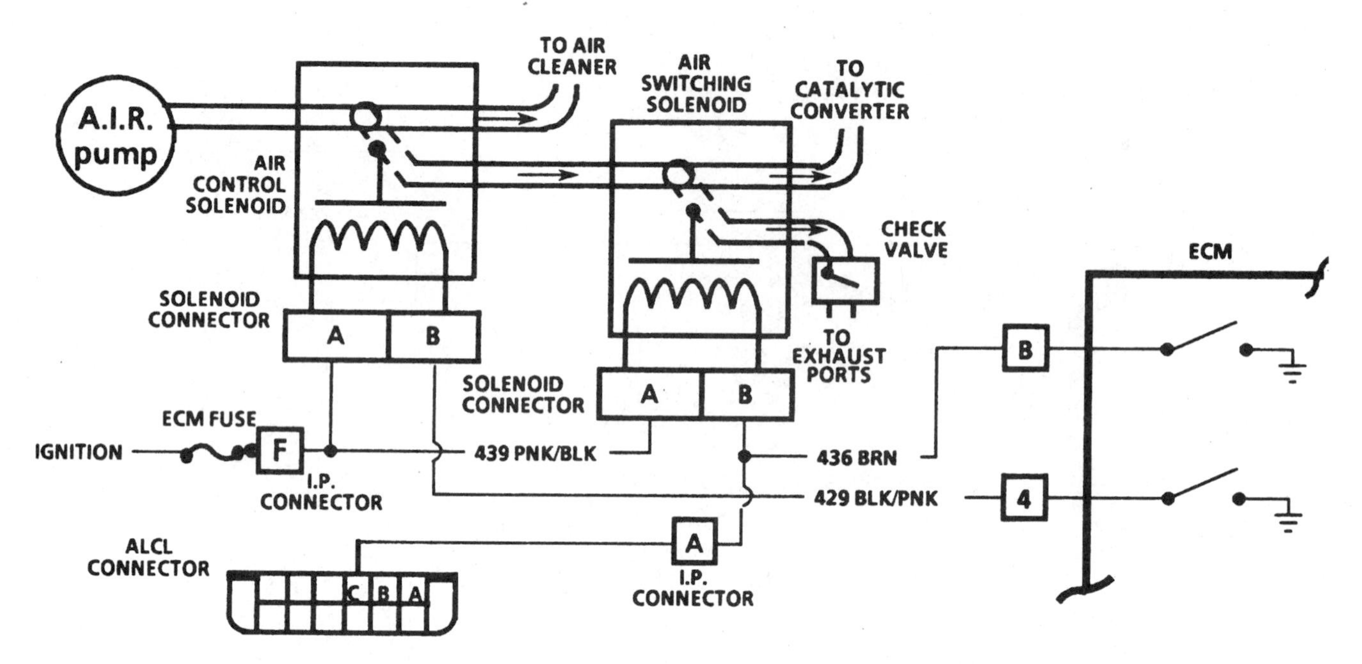

Figure 6-4 Dual (normally closed) solenoids are controlled by the computer switching the ground side of the circuit (courtesy GM).

Troubleshooting AIR Systems

Computer-controlled AIR systems experience the same problems as those that are not computer-controlled. System failures might include: seized AIR pumps, broken belts, burned-out check valves and divert/bypass valves, and rusted AIR tubes.

Testing the electrical operation of a solenoid can be performed with a voltmeter. Figure 6-4 shows a typical dual solenoid AIR system being controlled on the ground side by the computer.

For the AIR to flow into the exhaust manifold (open loop), both solenoids must be energized. When the voltmeter is placed in Terminal B to ground of each solenoid, voltage should be less than 0.3 volts.

During closed loop, the AIR must reach the catalytic converter. For the AIR to flow to the catalytic converter, the computer keeps the air control solenoid energized. Battery voltage can be read on the voltmeter when the air switching valve is de-energized.

If the AIR is being diverted to the air cleaner, both solenoids will be de-energized. The voltmeter will read battery voltage on both terminals.

Other checks of the system should include:

1. Check for the presence of manifold vacuum.
2. Check for battery voltage at Terminal A.
3. Check solenoid resistance (every solenoid has a specification for minimum resistance).
4. Place an ammeter in series and measure current flow with the solenoid energized (0.75 amp maximum).

If the system is equipped with only the air control solenoid (Figure 6-2), it should be energized in open loop. This indicates that the AIR flow is into the exhaust manifold. Under all other engine conditions, the air control solenoid should be de-energized with the AIR flow going into the air cleaner or atmosphere.

AIR flow can be checked by squeezing an air rail hose or just disconnecting it to feel for air flow. To understand which hose to test for flow, the technician must understand when and to which location(s) AIR may be diverted. (Refer to Figure 6-5.)

EVAPORATIVE EMISSION CONTROL (EEC) SYSTEM

The basic *evaporative emission control system* is the charcoal canister storage method. It transfers fuel vapor from the fuel tank and/or carburetor float bowl to a canister of activated carbon (charcoal). The canister stores vapors while the engine is not

Air is Diverted to:	Computer Mode:
• Exhaust Manifold • Catalytic Converter or Atmosphere • Atmosphere	• Open Loop • Closed Loop • Limp Mode

Figure 6-5 AIR flow locations and conditions.

operating. When the engine is running, fuel vapor is purged from the canister by intake air flow, and is consumed in normal combustion.

The computer controls an electric solenoid valve that controls a ported or manifold vacuum, depending on the system. The solenoid valve is a normally closed or open solenoid. A *normally closed* (N.C.) solenoid valve must be energized by the computer to allow the system to purge. In the event the computer fails to energize the solenoid, the canister will not purge, resulting in a loaded canister.

A *normally open* (N.O.) solenoid valve must be energized by the computer to prevent canister purging. In the event the computer fails to energize the solenoid, the canister will purge all the time, resulting in a rich condition at idle or during warm-up.

The computer allows canister purging under different engine conditions that vary between manufacturers. Some engine conditions include:

1. Engine coolant temperature above a minimum.
2. Engine run time more than a specified minimum.
3. Vehicle speed is above a minimum speed.
4. TPS above a minimum voltage.
5. Engine has entered closed loop.

Other than the IM240 purge flow test, there is no effective way to measure the actual flow of fuel vapors. Most manufacturers provide diagnostic charts to pinpoint problems with this circuit. If one is not available, the following tests can be used (Figure 6-6):

1. Check for the presence of manifold vacuum.
2. Check for battery voltage at Terminal A.
3. Backprobe Terminal B and watch for the voltage to drop below 0.3 volts with the solenoid energized.
4. Check solenoid resistance (every solenoid has a specification for minimum resistance).
5. Place an ammeter in series and measure current flow with the solenoid energized (0.75 amp maximum).
6. Check for fault codes if the computer is equipped with a fault protected quad-driver.

While conditions and specifications will vary, the above tests should allow the technician to quickly pinpoint a canister purge system failure.

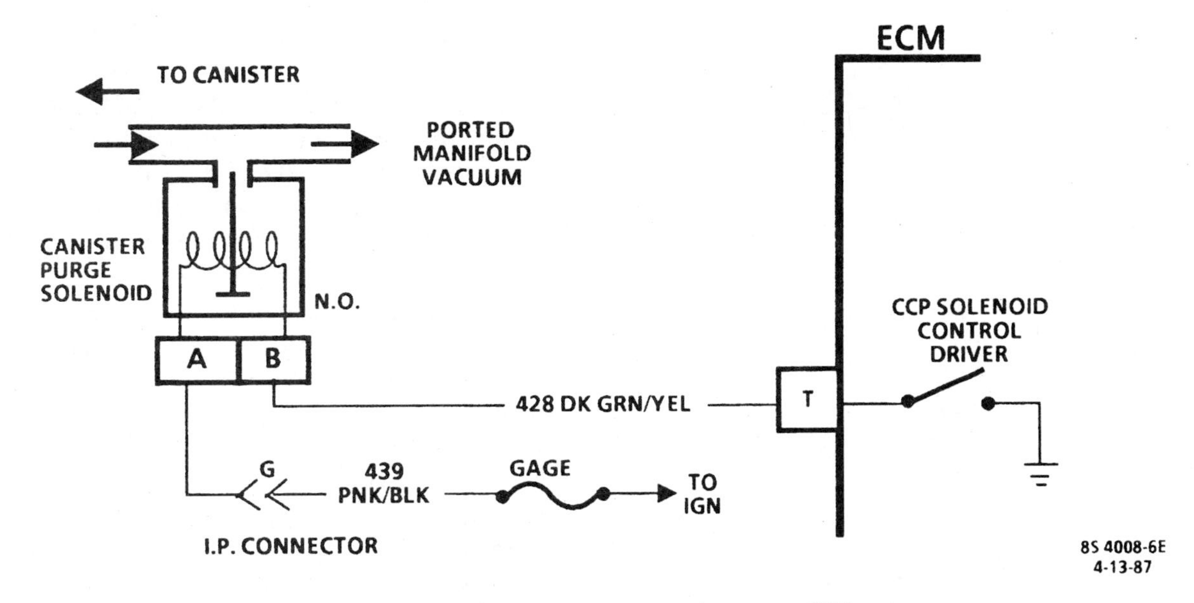

Figure 6-6 Typical computer-controlled canister purge system (courtesy GM).

EARLY FUEL EVAPORATION (EFE)

The purpose of Early Fuel Evaporation is to provide a source for rapid heat to the intake manifold during the warm-up cycle. Rapid heating of the intake system is needed for quick fuel vaporization for uniform fuel distribution and to improve cold driveability. To further reduce HC and CO emissions, EFE reduces the length of time the carburetor's automatic choke remains on. This system can be found on some carbureted and throttle body-injected engines.

There are two types of computer-controlled EFE systems:

1. One system uses a solenoid and vacuum servo mounted in the exhaust stream (Figure 6-7). The computer energizes a normally closed solenoid to operate the valve. In the event the computer fails to energize the solenoid, manifold vacuum is blocked off and the valve will not operate. This can result in engine stalling or stumble during warm-up. If manifold vacuum continues to reach the valve after the engine reaches operating temperature, it can cause high intake manifold temperature, fuel percolation in the carburetor, difficult hot engine restarts, and pre-ignition.

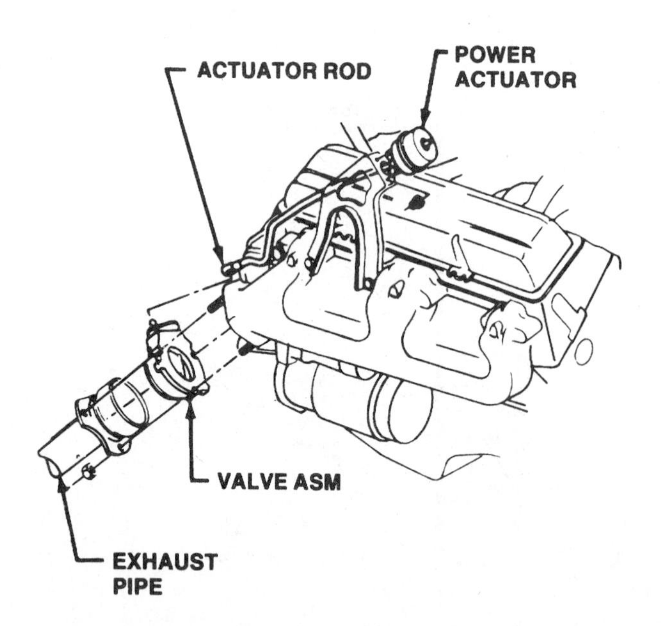

Figure 6-7 Vacuum-operated Early Fuel Evaporation (EFE) assembly (courtesy GM).

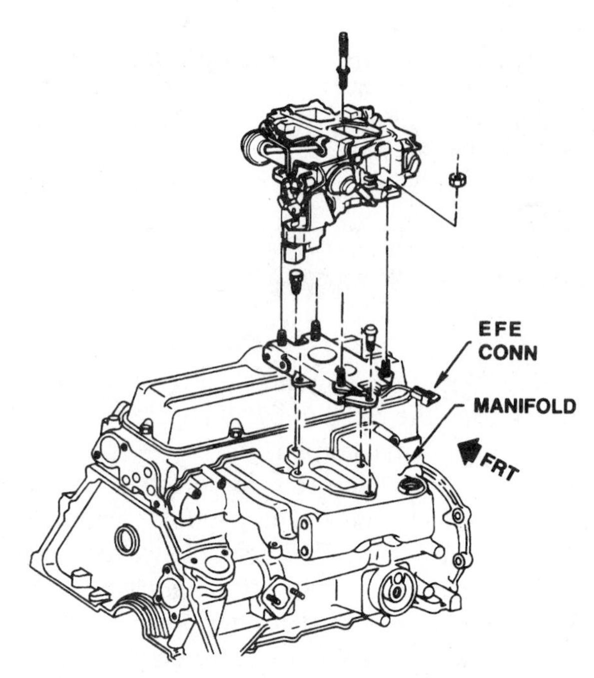

Figure 6-8 Electrically-controlled Early Fuel Evaporation (EFE) ceramic heater (courtesy GM).

2. The second type is a ceramic heater grid, mounted underneath the carburetor or throttle body (Figure 6-8). Operation of the relay is controlled by the computer. Two computer inputs detect if the relay contacts are closed. A signal from the coolant temperature sensor must indicate that the temperature is below 140°F. Another signal from the ignition coil must indicate that the engine has started and is operating above 500 RPM. (On some systems an RPM signal is not required; the ignition switch just needs to be in the on position.) When both signals indicate the need for the heater, the computer grounds the relay coil circuit.

   The heater contains a PTC thermistor that is self-regulating. That is, it does not need any special device to regulate the temperature. As the heater reaches its maximum temperature, its resistance rises to the point where it becomes so high that electrical flow is reduced, lowering the heat output. The results of a failed system are similar to the vacuum servo type.

   Figure 6-9 shows a typical electrically-heated circuit. Notice that the circuit has two power feeds and that the computer controls the low current side of the relay. Specific heater resistances will vary between manufacturers.

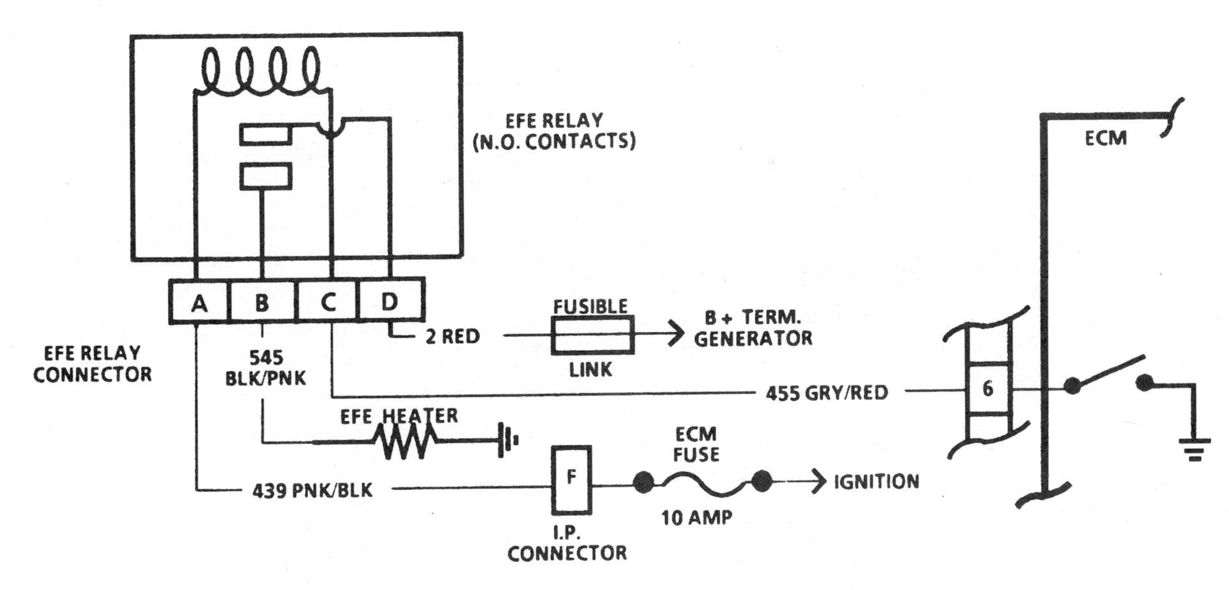

Figure 6-9 Typical Early Fuel Evaporation (EFE) heater circuit (courtesy GM).

EXHAUST GAS RECIRCULATION (EGR) SYSTEM

The purpose of an *Exhaust Gas Recirculation* (EGR) system is to control oxides of nitrogen (NO_x) emissions caused by high combustion chamber temperatures. The EGR valve recirculates small amounts of inert (inactive) exhaust gases back into the cylinder. EGR will have several effects on the combustion process:

1. Combustion temperatures are lowered, so EGR is not added until the engine has had a chance to warm up.
2. The combustion process is slowed, which requires additional spark advance to allow enough time for the air/fuel to burn completely.
3. Exhaust gases introduced into the intake manifold displace a certain amount of air by replacing it with inert gases. This reduces the amount of oxygen available for combustion, making the air/fuel ratio rich. The computer will then adjust the fuel injector pulse-width to balance the air/fuel ratio.

The EGR valve should operate only after the engine warms up and the throttle is opened beyond idle. Certain inputs will be monitored by the computer to calculate EGR operation. These inputs will vary between manufacturers and might include:

1. Coolant temperature above approximately 130°F.
2. Throttle position sensor above idle voltage.
3. Engine load: MAP, VAC, MAF, etc.
4. Park/Neutral switch (must be in gear).
5. EGR valve position sensor.
6. EGR backpressure sensor.
7. TCC and 4th (and 5th) gear.
8. Vehicle speed.

Monitoring and Calculating EGR Flow

Three types of computer-controlled EGR input sensors are commonly used today. One sensor measures EGR valve movement and the other two measure EGR exhaust pressure:

1. An EGR Valve Position (EGR-EVP) sensor (Figure 6-10) is located on top of or inside of the EGR valve. It is a three-wire potentiometer consisting of a five-volt reference, signal return, and ground wire. This input is used by the computer to calculate the amount of EGR gases flowing and to identify if the valve is fully closing at idle.
2. A Pressure Feedback EGR (PFE) sensor (Figure 6-11) is a ceramic capacitive-type pressure transducer that operates electrically the same as a potentiometer. The PFE system serves only as a pressure regulator rather than a flow metering device. It converts exhaust system pressure or vacuum into an analog electrical

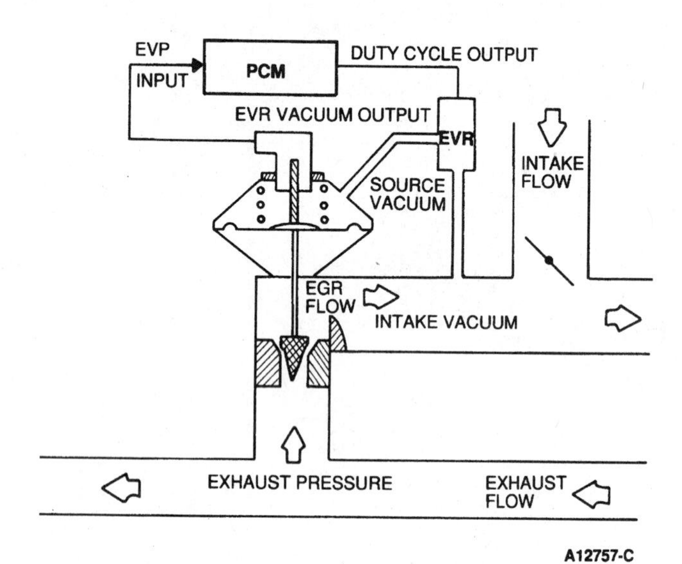

Figure 6-10 Typical EGR Valve Position (EGR-EVP) system. The computer regulates the duty-cycle based on input from the EVP (courtesy Ford).

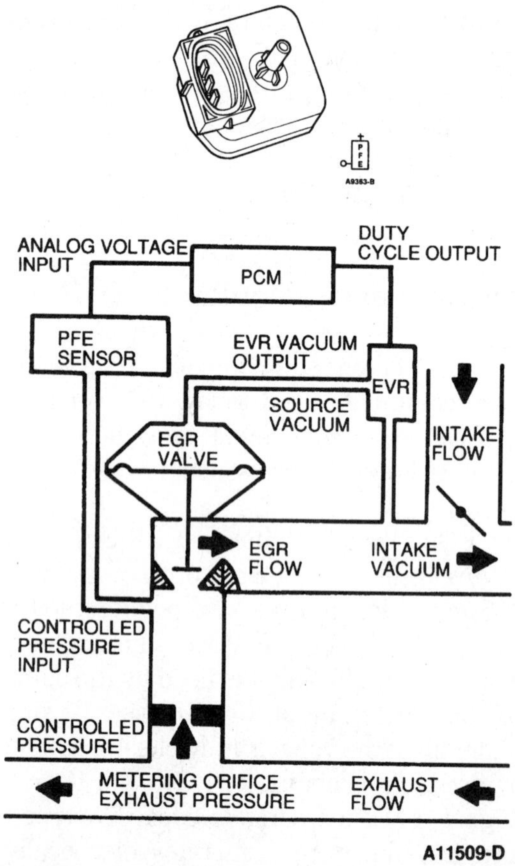

input signal of 0.2 to 4.75 volts DC. A high-voltage signal output indicates a minimum or no EGR flow. With the engine running at idle, an output signal of approximately 3.0 to 3.5 volts should be seen.

3. A Differential Pressure Feedback EGR (DPFE) sensor (Figure 6-12) operates in the same manner as a PFE, except it monitors the pressure drop (differential) across the metering orifice. This allows for a more accurate assessment of EGR flow requirements. Voltage ranges are similar to that of the sensor.

With the addition of these sensors, the computer can closely monitor the movement, flow, and overall operation of the EGR valve. These systems allow for improved diagnostics because the computer can set a code if the system should fail. When diagnosing these sensors, always inspect the computer for codes to pinpoint the problem.

Figure 6-11 The controlled pressure input signal is used by the computer to calculate EGR operation (courtesy Ford).

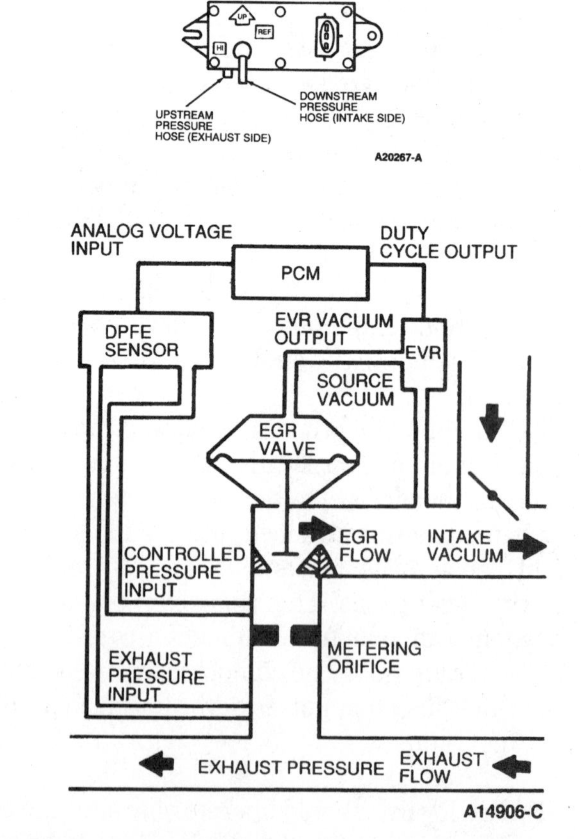

Figure 6-12 Monitoring the pressure differential gives a more accurate measurement of EGR flow requirements (courtesy Ford).

EGR Output Controls

To control the EGR valve, the computer regulates ported or manifold vacuum through an electric solenoid(s). Depending on the manufacturer, this solenoid can be normally closed or normally open. A normally closed solenoid must be energized by the computer to allow the valve to operate. If a circuit failure should occur, the EGR valve will not open. This condition will result in high combustion chamber temperatures and possible pinging. A normally open solenoid must be energized to prevent EGR valve operation. If a circuit failure should occur, the EGR valve will operate continuously. If the system uses manifold vacuum, the results are stalling or stumble during cold and hot idle and possible stalling on deceleration. Systems using ported vacuum can result in stalling during warm-up, or fast idle or stumble during cold acceleration.

Solenoids are designed with a vacuum vent (bleed-off) for when the vacuum source is closed off to the EGR valve. If a vent is not included, vacuum is trapped inside the vacuum line between the solenoid and the EGR valve. The trapped vacuum will result in stalling on deceleration or rough idle after deceleration.

A variety of solenoid configurations are used to regulate the operation of the EGR valve:

1. Ported vacuum source with a single solenoid.
2. Manifold vacuum source with single or dual solenoids.
3. Ported vacuum source, internal EGR valve position sensor, and an electrically-controlled bleed-off solenoid.
4. Single, dual, and triple digital EGR solenoids.

Figure 6-13 shows an EGR system using a computer-controlled solenoid that pulses ported vacuum to the EGR valve. The computer controls the EGR vacuum by energizing a normally closed and vented solenoid. The EGR is pulse-width modulated, which means the computer varies the amount of on or off time. Part of the control includes a vacuum diagnostic control switch that is normally open and is closed by vacuum. The purpose of the switch is to signal the computer when vacuum is being applied to the EGR valve. A code will be set if the EGR diagnostic switch senses vacuum when the computer signals no EGR or does not sense vacuum when the computer signals for EGR.

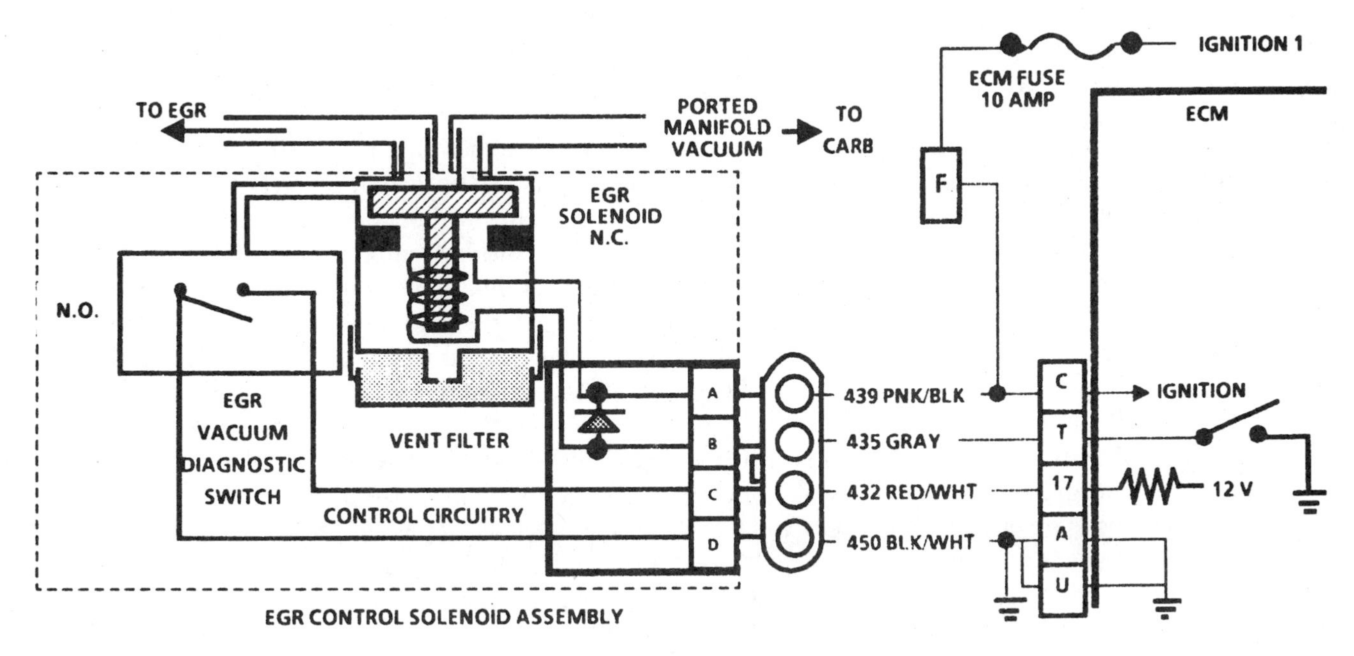

Figure 6-13 EGR Vacuum Regulator (EVR) solenoid with built-in diagnostic vacuum switch (courtesy GM).

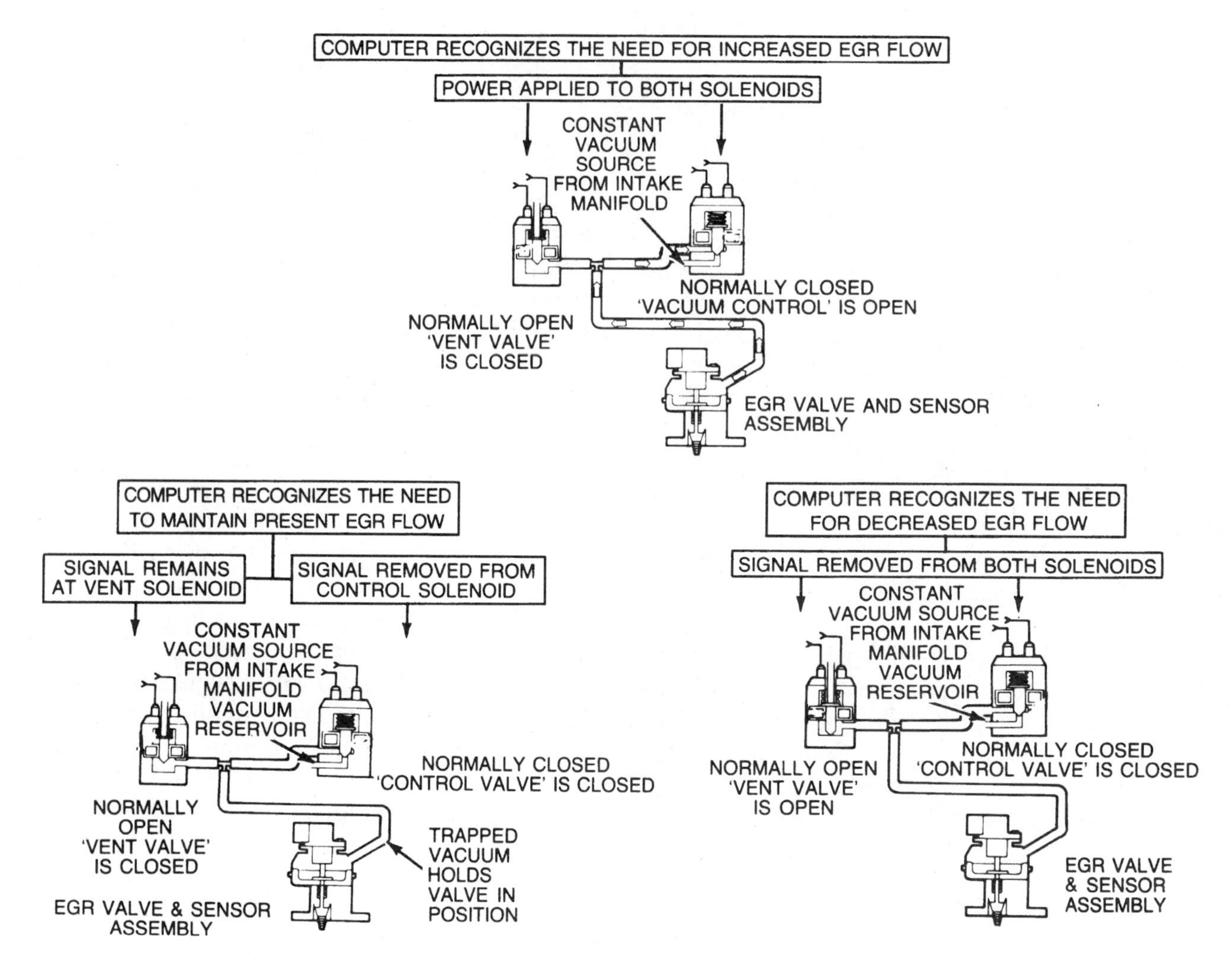

Figure 6-14 Dual solenoid EGR systems include a vacuum control solenoid and vent solenoid (courtesy Motorcraft – Ford).

On some applications (Figure 6-14), the computer controls two solenoids (one is normally closed and one is normally open) to control the flow EGR. This system separates the single solenoid (control and vent) into individual control and vent solenoids. To prevent the EGR from operating, both solenoids are de-energized, which means battery voltage will be seen at both terminals. Both solenoids must be energized for EGR flow to occur; this is indicated by a ground side voltage of less than 0.3 volts. If the vent solenoid is energized (less than 0.3 volts) and the control solenoid is de-energized (battery volts), the flow of EGR gases is maintained.

An Integrated EGR valve combines the EGR valve, solenoid vent, and EGR Valve Position sensor into a single valve (Figure 6-15). The internal solenoid is normally open, which causes the ported vacuum signal to be vented off to the atmosphere when the EGR is de-energized. This valve includes a voltage regulator, which converts the computer signal to provide different amounts of EGR flow by regulating current to the solenoid. EGR flow is controlled by the computer using a pulse-width modulated signal based on intake airflow, throttle position, and engine RPM.

The system also contains a pintle position (EGR-EVP) sensor, which works like a TPS sensor. As EGR flow is increased, sensor output voltage increases. The EGR position sensor voltage is used to determine that the pintle is moving. When no EGR is commanded (0% duty-cycle), the position sensor should read between 0.05 and 1.9 volts, and increase with the increase in duty-cycle as the valve opens. Failure conditions can result in a code.

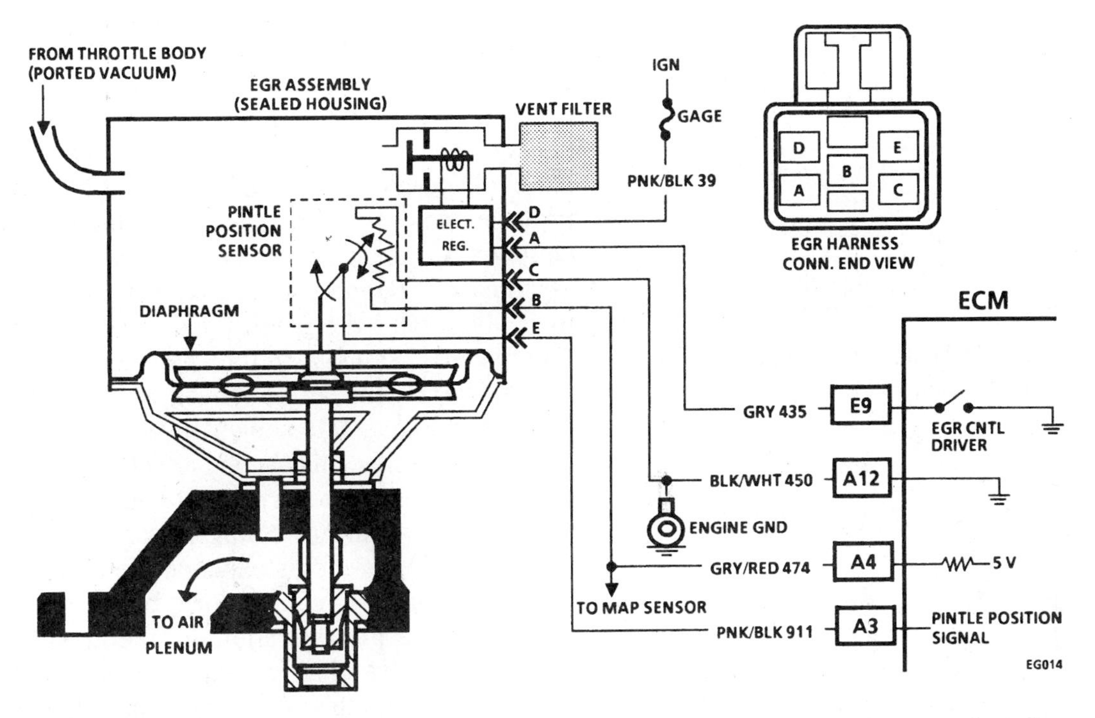

Figure 6-15 An integrated EGR valve includes a vent filter, voltage regulator, and pintle position sensor (courtesy GM).

The digital EGR valve is designed to accurately supply EGR to an engine, independent of the intake manifold vacuum. The response speed of this valve is about ten times faster than similar vacuum-operated EGR valves. It controls EGR flow to the intake through:

1. Single solenoid linear EGR valve (Figure 6-16).
2. Dual solenoid digital EGR.
3. Triple solenoid digital EGR.

EGR flow to the intake manifold is regulated by one to three solenoids. Depending on the number of solenoids, EGR flow rates can be varied (Figure 6-17). Using fixed orifices, as many as seven different solenoid combinations are used to regulate EGR flow.

When a solenoid is energized, its armature is lifted, opening the orifice. Exhaust gas flow through the digital EGR valves differ from other valve designs because the pintle(s) are opened to allow exhaust gas to exit the valve rather than enter it (Figure 6-18). With most EGR valve designs, the pintle valve is opened to allow exhaust gas to pass through the valve. The digital EGR valve allows exhaust gas to enter an open passage in the base and remain inside the valve until the pintle valve is opened. The swivel pintle insures good sealing of exhaust gases because the shaft and seals are exposed to exhaust pressure instead of intake manifold vacuum.

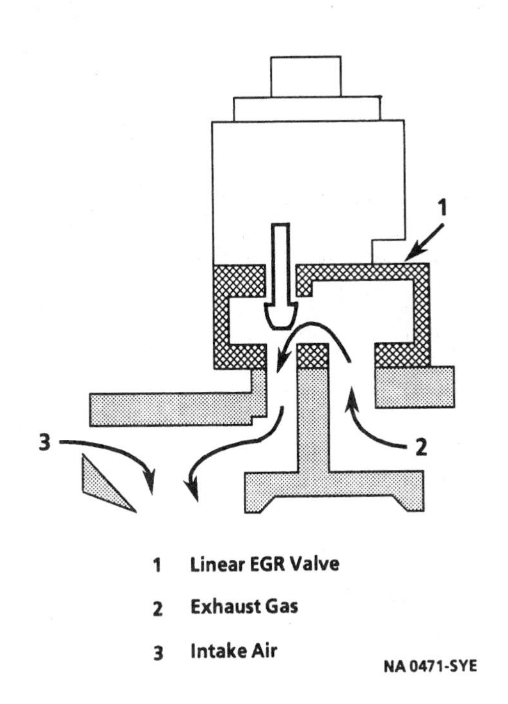

Figure 6-16 Electronically-controlled linear EGR valve (courtesy GM).

Solenoid(s) Actuated	Flow (g./sec.)
# 1	1.00 ±0.20
# 2	2.00 ±0.20
# 1 and # 2	3.00 ±0.30

Figure 6-17 Digital EGR solenoids can be energized individually or in pairs to achieve the desired EGR flow rate (courtesy GM).

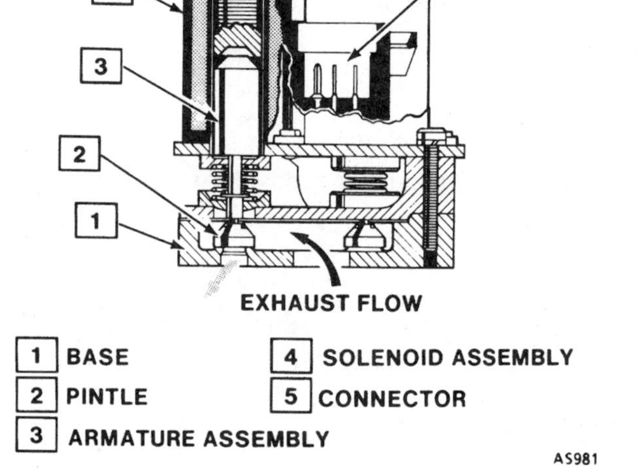

Figure 6-18 Digital EGR construction and flow (courtesy GM).

Built-in diagnostics are used by the computer on a closed throttle coast-down. It cycles the solenoid(s) individually on and off, and looks for a change in engine RPM and O_2 sensor activity. If the computer does not see the correct change in RPM and O_2 sensor activity, it will set a trouble code. The code indicates a lack of EGR gas flow from that particular solenoid/orifice combination.

While each of these digital EGR valves has built-in diagnostics, the solenoids can be tested for available voltage and resistance. Figure 6-19 shows a triple solenoid digital EGR valve. Terminal D is the common terminal and it receives battery power. If Terminals A, B, and C are tested with a voltmeter, the voltage will switch high (de-energized) and low (energized). Removing the harness connector allows the technician to check each solenoid's resistance. Figure 6-19 shows the terminals and resistance values for this particular valve.

In summary, the purpose of each system has not changed, what has changed is how the job is completed. The systems explained in this chapter were not named by manufacturer because the theory and concept of operation are the most important things. The technician with an open mind and good understanding of these systems will be more knowledgeable, versatile, and in higher demand in the job market.

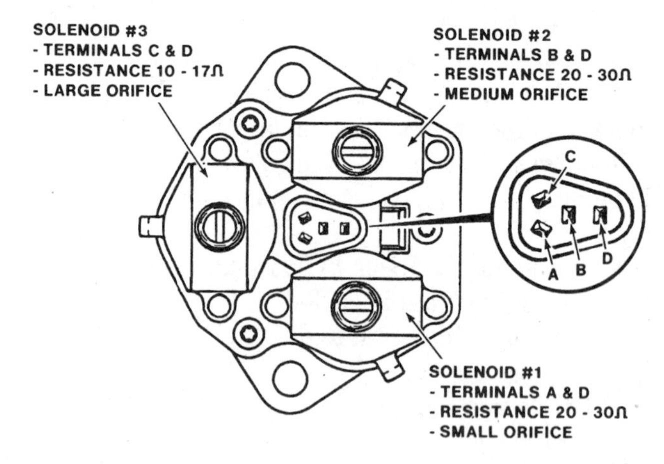

Figure 6-19 Testing a three-solenoid digital EGR valve (courtesy GM).

CHAPTER 7

Introduction to Scan Tools

An effective way to look at a lot of information entering and exiting the vehicle's on-board computer is to use a scan tool. A scan tool is a computer that reads data supplied by the vehicle's on-board computer(s). Figure 7-1 shows some typical input/output data that can be read from a scan tool. By making data available to the technician, the scan tool provides valuable diagnostic information, much of which would be very difficult or impossible to obtain by other means.

Keep in mind that a scan tool is only as good as the user's understanding of the data. When used with the proper reference materials, a scan tool can be a great help to a knowledgeable, experienced technician performing computer-related diagnostics.

```
1993 FORD TRUCK                    A/C

3.0L V6 - EECIV EFI                 A/T

  ** SCROLL FOR DATA. OK TO DRIVE. **

    RPM_2544 02S1(mV)_558 02S2(mV)___133
TP=TPS(V)________1.77  TP MODE____________P/T
TPCT______________0.9  ECT(V)____________0.48
ECT(dF)___________214  IAT=ACT(V)________2.19
IAT=ACT(dF)_______104  IDLE AIR(%)_______98.8
MASS AIR(V)______2.93  DPFE(V)___________2.81
EVR(%)___________63.2  INJ PW1(mS)______11.60
INJ PW2(mS)_____11.68  SFTRIM 1(%)________2.5
SFTRIM 2(%)_______0.5  LFTRIM 1(%)_______-6.2
LFTRIM 2(%)______-6.2  VPWR=BATT(V)______14.0
VREF(V)__________4.97  SPARK ADV(d)________29
WAC=WOT A/C________ON  FP=FUEL PUMP________ON
CANP=PURGE_________ON  VEH SPEED(MPH)______30
PARK/NEU POS___-R-DL   BOO=BRAKE SW_______OFF
OPEN/CLSD LOOP_CLSD    ACCS=A/C___________OFF
```

Figure 7-1 Input/Output data from a 1993 Ford truck using a Snap-On Scan Tool and DCL cartridge.

Often, the tool is more than a help; it's a necessity. However, it can't solve problems by itself. Understanding the data and accurately interpreting the information displayed is the key. The technician, not the tool, is the diagnostician.

ADVANTAGES OF A SCAN TOOL

A computer processes a tremendous amount of information in a very short time. For the technician to collect this data with a multimeter or lab scope would require a great deal of time. A scan tool, when connected to a vehicle, can provide the technician with several advantages:

First, the scan tool allows quick access to the computer's data through a diagnostic connector (Figure 7-2) located in the passenger or engine compartment of the vehicle. This quick access to computer data can significantly reduce diagnostic time if the technician understands the displayed data.

Second, aftermarket scan tools made by Snap-On, OTC, MPSI, etc.(Figure 7-3) give a technician access to many domestic (and some imported) vehicle data streams. By changing cartridges, the technician can use the same scan tool to repair a Ford, Toyota, Chrysler, etc. Since these systems are cartridge-based they can be updated annually by purchasing new cartridges or PROM chips.

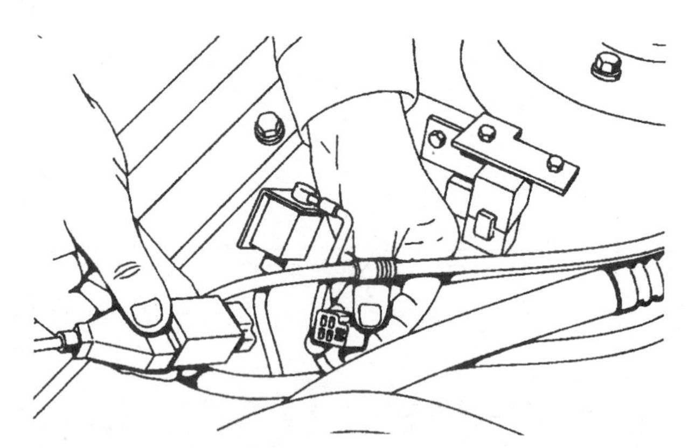

Figure 7-2 Chrysler diagnostic connector.

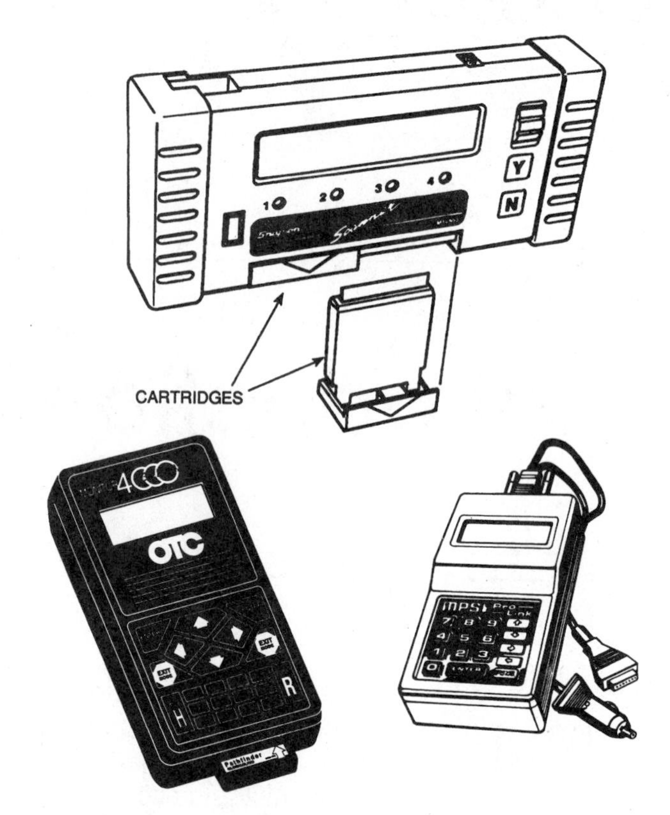

Figure 7-3 Aftermarket scan tools.

Third, a scan tool can be taken on a road test. On the road, data can be displayed live or recorded and viewed back at the shop. This feature is explained later in this chapter.

Fourth, a scan tool can be connected to a printer which allows the technician to print out a "hard copy." This documentation can be used to build a database of vehicle parameters, to explain the problem to the customer, and to provide the technician with documentation that the repair has been corrected and verified.

Fifth, software is available that allows scan tool data to be transferred to a Personal Computer (PC). With this software, data can be displayed "live" on a PC or saved and recalled at a later time. These features are explained later in this chapter.

LIMITATIONS OF A SCAN TOOL

As with any tool or piece of equipment, it is important to know its advantages and at the same time understand its limitations. Though the scan tool can be a very effective tool, it has several important limitations:

First, the scanner can only be used to diagnose computer-related systems, since it relies on the on-board computer to supply data. It is not very useful in diagnosing some types of fuel, ignition, charging, or emissions system problems, because many of the components in these systems are not controlled or monitored by the computer.

Second, since the scan tool reads data supplied to the on-board computer by the various circuits, it can only indicate in which circuit(s) to look for possible problems. A trouble code or abnormal reading does not necessarily mean that a particular sensor, switch, or actuator has failed. Something else in the circuit may be faulty (or the code may have been set accidentally during the previous repair or inspection). When a scan tool indicates a possible problem in a circuit, it's important to make sure the wiring, connectors, splices, and grounds are good before replacing a sensor, switch, actuator, or computer.

Third, what is displayed for an input component or actuator controlled by the on-board computer is not always accurate. The scan tool reads processed data from the computer and displays what the computer "thinks" is occurring. For example, in the case of actuators, it indicates that the on-board computer has sent a command to operate the device — not whether, in fact, the device is operating properly. For instance, an EGR signal reading that goes from "off" to "on" indicates only that the on-board computer is sending an electrical command to operate the EGR solenoid. The solenoid may actually be stuck open or closed. If the symptoms indicate a possible problem with the solenoid, further testing is necessary to determine if the solenoid is functioning correctly.

Fourth, OBD regulations do not require vehicle manufacturers to provide a diagnostic connector with a serial data stream. If a diagnostic connector is provided, in many cases the scan tool can only serve as a code readout box. There are usually cheaper alternative methods to pulling codes; e.g., a test light, logic probe, voltmeter, etc. These items should already be in a technician's toolbox.

Fifth, a limitation occurs when the manufacturer does not provide a diagnostic connector with serial data stream. Under OBD, vehicle manufacturers are not required to provide the aftermarket with access to their data stream. Japanese cars are a good example: many mid 1980's to early 1990's Mitsubishi and Hyundai products are equipped with a serial data stream, but the manufacturers chose not to allow the aftermarket access to this information. However, if the vehicle is returned to a Mitsubishi or Hyundai dealer, the dealership technician can access the data stream with the vehicle manufacturer's scan tool. Import manufacturers are addressing this problem by providing access protocols for vehicles with serial data streams. Scan tool developers are adding these systems to their new cartridges (See Figure 7-4.)

```
1992 HYUNDAI SONATA          A/C

3.0L V6 MFI                  A/T

**  CODES & DATA.  OK TO DRIVE.  **

(NO CODES PRESENT)

     RPM_____0 02(mV)_____20 INJ(mS)_____2.0
TPS(V)__________0.488  MOTOR POS STEPS__100
AIRFLOW(Hz)_________0  IDLE SW__________CLSD
COOLANT(dF)_______178  AIR TEMP(dF)______108
BARO("Hg)________29.4  EGR TEMP(dF)______158
SPK ADV(dBTDC)______4  CRANKING___________NO
P/N SWITCH_____P-N--
```

```
1990 MITSUBISHI              A/C

3.0L V6 MFI                  A/T

**  CODES & DATA.  OK TO DRIVE.  **

(NO CODES PRESENT)

     RPM__5094 02(mV)____918 INJ(mS)_____3.8
TPS(V)__________1.523  MOTOR POS STEPS_____7
AIRFLOW(Hz)_______363  IDLE SW__________OPEN
COOLANT(dF)_______160  AIR TEMP(dF)_______68
BARO("Hg)________29.2  SPK ADV(dBTDC)_____50
CRANKING___________NO  BATTERY(V)______14.28
P/N SWITCH_____-R-DL   HI PS PRESSURE_____NO
A/C REQUEST________NO  A/C RELAY_________OFF
```

```
1992 HYUNDAI SONATA          A/C

3.0L V6 MFI                  A/T

**  CODES & DATA.  OK TO DRIVE.  **

(NO CODES PRESENT)

     RPM__1719 02(mV)____879 INJ(mS)_____3.1
TPS(V)__________0.488  MOTOR POS STEPS___57
AIRFLOW(Hz)________56  IDLE SW__________CLSD
COOLANT(dF)_______178  AIR TEMP(dF)______109
BARO("Hg)________29.4  EGR TEMP(dF)______156
SPK ADV(dBTDC)_____17  CRANKING___________NO
P/N SWITCH_____P-N--
```

```
1990 MITSUBISHI              A/C

3.0L V6 MFI                  A/T

**  CODES & DATA.  OK TO DRIVE.  **

(NO CODES PRESENT)

     RPM___750 02(mV)____195 INJ(mS)_____3.6
TPS(V)__________0.684  MOTOR POS STEPS_____2
AIRFLOW(Hz)________63  IDLE SW__________CLSD
COOLANT(dF)_______184  AIR TEMP(dF)_______68
BARO("Hg)________29.4  SPK ADV(dBTDC)_____26
CRANKING___________NO  BATTERY(V)______14.28
P/N SWITCH_____P-N--   HI PS PRESSURE_____NO
A/C REQUEST________NO  A/C RELAY_________OFF
```

Figure 7-4 Hyundai and Mitsubishi data stream readings using a Snap-On Scanner with an import cartridge.

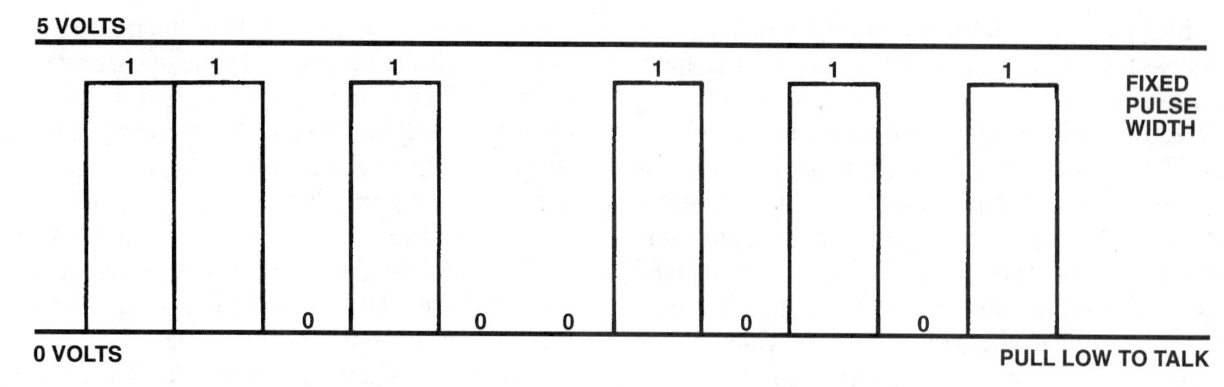

Figure 7-5 Two computers can talk to each other by rapidly pulsing or switching the voltage from 5 volts to 0 volts.

SERIAL DATA

A computer is made up of several components. It contains chips to store information, a clock generator to measure the time of the pulsed voltage signals, and a central processing unit. The computer performs three functions:

1. Receives information from sensors/switches.
2. Processes that information.
3. Transmits commands to actuators.

The computer makes this processed input and output information available to the technician through a scan tool. The scan tool picks up the variable readings from the computer. The inputs and outputs are continuous, creating what is called a "serial data stream."

Serial data is how computers communicate with each other. It is actually a series of rapidly-changing voltage signals pulsed from high voltage to low voltage. These voltage signals are shown in Figure 7-5 as a binary code with 1 representing a high voltage, typically 5 volts, and 0 representing a low voltage, which is always 0 volts. These voltages are transmitted from the computer through a wire to the diagnostic connector. When data is transmitted in this way it is referred to as a serial data line.

Digital computers can only understand signals that are either high or low, on or off. Because of this, all input and output signals must be converted to a binary code represented by the number 1 or 0. The computer strings together eight of the *bits* to make one *byte* or *word.* A word can represent any of the numerous inputs and outputs, such as an oxygen sensor voltage, a throttle position angle, a switch on/off status, or a diagnostic trouble code. Figure 7-6 shows the relationship between a digital signal and the converted binary code or word.

Serial data refers to information that is transferred in a linear fashion over a single line, one bit at a time. A typical data stream, as shown in Figure 7-7A, can have as few as 10 words or as many as 70 words. The numbers of words on a data stream varies from vehicle to vehicle.

A scan tool receives serial data through a Data Communications Link (DCL) or an Assembly Line Data Link (ALDL). The words to be looked at are selected by the technician using a scan tool. The communications link between the on-board computer and the scan tool can be uni-directional (one way) as seen in Figure 7-7A or bi-directional (two way) as in Figure 7-7B.

DIGITAL SIGNALS ARE USED TO TRANSMIT BINARY CODE

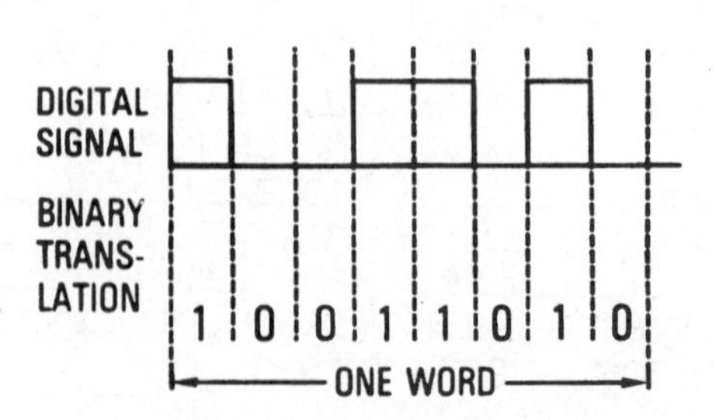

Figure 7-6 A digital signal converted to an eight-bit binary code.

Uni-directional means the scan tool can only receive data from the on-board computer. Bi-directional means the scan tool can receive data and can send commands to the computer, instructing it to perform system-specific tests on the actuators. These tests might include: raising and lowering the idle speed; performing an injector balance test; activating the fuel pump; turning on/off output solenoids for the EGR, purge, etc.

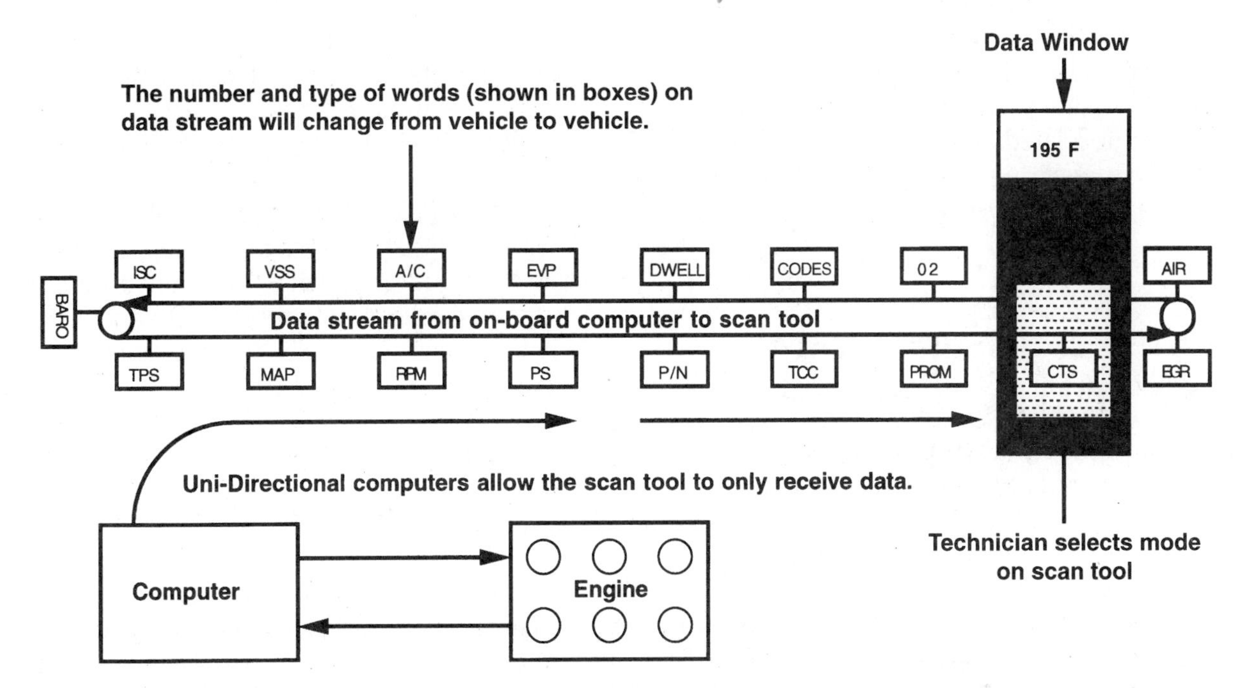

A. TYPICAL *WORDS* ON THE DATA STREAM THAT CAN BE READ USING A SCAN TOOL.

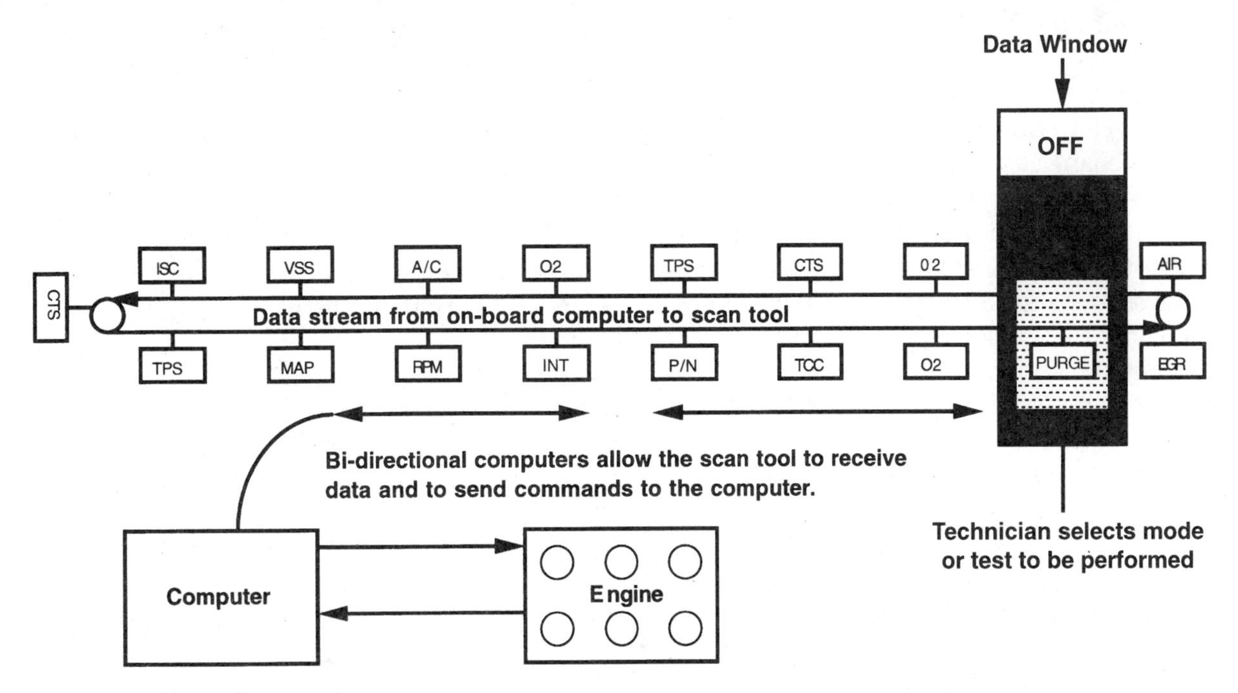

B. A BI-DIRECTIONAL COMPUTER ALLOWS THE TECHNICIAN TO *OVERRIDE* THE COMPUTER AND PERFORM SPECIALIZED DIAGNOSTIC TESTS.

Figure 7-7 Serial data transmission.

The advantage of the bi-directional computer is that it allows the technician to override (bypass) input to the computer through the scan tool. The computer and the actuators can then be controlled through a pre-set program and not by the input data from the sensors and switches. If the system now works correctly, the problem must be on the input side of the computer. If the system does not work correctly, then the problem must be computer- and/or actuator-related. The technician should always consider the wiring, connectors, and grounds if these tests to not perform correctly.

COMPUTER BAUD/SERIAL DATA REFRESH RATE

The speed at which bits are transmitted is called the *baud rate.* A computer with a baud rate of 160 (one of the slowest) transmits serial data at the rate of 160 bits per second. The higher the rate, the faster the computer. This also allows for more data to be placed on the data stream, which in turn allows more information to be displayed on the scan tool.

The *refresh rate*, sometimes referred to as the sampling rate, is the time it takes for the computer to update the words in the data stream. If the stream were a clothes line (see Figure 7-7), each updated data would be placed on the line each time the loop completed a cycle. The time depends on the baud rate of the on-board computer and the number of words in the serial data stream. A computer with a 160-baud rate has approximately 20 to 30 words, or parameters, on the serial data line. It takes approximately 1.25 seconds to refresh the serial data line information on the scan tool display. For example, a Chrysler computer with a baud rate of 62,500 (one of the fastest) has approximately 50 words, or parameters, in the serial data line. It takes approximately 11 milliseconds to refresh the serial data line information. Generally, the faster the refresh rate of the on-board computer, the better the diagnostics.

Data is transmitted through the data stream at a speed much too fast for the human brain to process. To allow the technician to see the necessary information, serial data is transmitted in three ways:

1. The computer sends serial data to the scan tool at a reduced baud rate from that used to process input/output data. (Because of the slower baud rate, intermittent problems can be missed when watching the data live.)
2. Most scan tools default to a refresh rate that does not display every change of serial data. Most scan tools default to updating data once every fifth loop of the data stream. The slower update rate makes it easier for the technician to read. Because of the slower refresh rate, intermittent problems can be missed when watching the data live.
3. The MPSI Pro-Link scan tool defaults to updating the data stream every fifth loop. However, it also allows the technician the option of changing the refresh rate to a slower or faster update ratio. A one-to-one ratio can be set, but rapidly changing data may be hard to read.

RECORD/PLAYBACK

Wouldn't it be helpful to have a dynamometer, or an extra pair of eyes or hands on a road test? The fastest, easiest substitute is to use the record feature found in all scan tools. This is commonly called a *snapshot*. Snap-On calls it a *movie*. It helps isolate intermittent problems by storing data stream samples in memory immediately before, during, and after the problem occurs.

A snapshot (Figure 7-8) is called an *event* or *movie*, much like a roll of film, with each picture as a frame. Each frame contains data for each input and output during one loop of the data stream.

The number of frames that can be recorded depends on the refresh rate of the computer and the individual scan tool. Each scan tool varies in the number of pictures that can be recorded. The total number of frames captured depends on the baud rate of the on-board computer, the number of words in the data stream, and the available memory of the scan tool cartridge.

When testing a GM fuel-injected engine, the technician should make sure the scan tool is in the *ROAD* mode as seen in Figure 7-9 or indicates it is *OK TO DRIVE* as in Figure 7-4.

If the scan tool indicates any other test mode, the computer is in a special diagnostic state. Driving the vehicle in this mode could create a false driveability complaint or cause damage to the engine.

91VR3301 E-Z EVENT PLUS
CHRYSLER 1991 3.3/3.8 MFI VIN
EVENT 1 TAG-- FRAME -30
FAULT CODES: NO

	-15	-14	-13	-12	-11	-10	-9	-8	-7	-6	-5	-4	-3	-2	-1	0
COOLANT TEMP	120	120	120	120	120	120	120	120	120	120	120	120	120	120	120	120
RPM	0000	0000	0000	0000	0000	0000	0000	0000	0000	0000	0000	0000	0000	0000	0000	0000
INJ PULSE WIDTH	14.7	14.7	14.7	14.7	14.7	14.7	14.7	14.7	14.7	14.7	14.7	14.7	14.7	14.7	14.7	14.7
MAP	29.4	29.4	29.4	29.4	29.4	29.4	29.4	29.4	29.4	29.4	29.4	29.4	29.4	29.4	29.4	29.4
O2 VOLTS	0.50	0.50	0.50	0.50	0.50	0.50	0.50	0.50	0.50	0.50	0.50	0.50	0.50	0.50	0.50	0.50
THROT POS SENSOR	0.66	0.66	0.66	0.66	0.66	0.66	0.66	0.66	0.66	0.66	0.66	0.66	0.66	0.66	0.66	0.66
ADAPTIVE FUEL FAC	-03	-03	-03	-03	-03	-03	-03	-03	-03	-03	-03	-03	-03	-03	-03	-03
BARO PRESSURE	29.4	29.4	29.4	29.4	29.4	29.4	29.4	29.4	29.4	29.4	29.4	29.4	29.4	29.4	29.4	29.4
BARO PRESSURE VOL	4.76	4.76	4.76	4.76	4.76	4.76	4.76	4.76	4.76	4.76	4.76	4.76	4.76	4.76	4.76	4.76
BATTERY TEMP	065	065	065	065	065	065	065	065	065	065	065	065	065	065	065	065
BATTERY VOLTS	12.2	12.2	12.2	12.2	12.2	12.2	12.2	12.2	12.2	12.2	12.2	12.2	12.2	12.2	12.2	12.2
CHARGE TEMP	105	105	105	105	105	105	105	105	105	105	105	105	105	105	105	105
CHARG SYS TARG VOL	13.9	13.9	13.9	13.9	13.9	13.9	13.9	13.9	13.9	13.9	13.9	13.9	13.9	13.9	13.9	13.9
CRUISE CNTROL CUT	KEY	KEY	KEY	KEY	KEY	KEY	KEY	KEY	KEY	KEY	KEY	KEY	KEY	KEY	KEY	KEY
CRUISE CONTROL	OFF	OFF	OFF	OFF	OFF	OFF	OFF	OFF	OFF	OFF	OFF	OFF	OFF	OFF	OFF	OFF
CRUISE CONTROL STA	CRSW	CRSW	CRSW	CRSW	CRSW	CRSW	CRSW	CRSW	CRSW	CRSW	CRSW	CRSW	CRSW	CRSW	CRSW	CRSW
CRUISE SET SPEED	000	000	000	000	000	000	000	000	000	000	000	000	000	000	000	000
DIS CAM STATUS	NO	NO	NO	NO	NO	NO	NO	NO	NO	NO	NO	NO	NO	NO	NO	NO
DIS CRANK STATUS	NO	NO	NO	NO	NO	NO	NO	NO	NO	NO	NO	NO	NO	NO	NO	NO
ELECTRONIC ADV/RE	016	016	016	016	016	016	016	016	016	016	016	016	016	016	016	016
IDLE AIR CONTROL	041	041	041	041	041	041	041	041	041	041	041	041	041	041	041	041
IDLE RPM X 10	078	078	078	078	078	078	078	078	078	078	078	078	078	078	078	078
KNOCK ACTIVE	NO	NO	NO	NO	NO	NO	NO	NO	NO	NO	NO	NO	NO	NO	NO	NO
KNOCK VOLTS	0.00	0.00	0.00	0.00	0.00	0.00	0.00	0.00	0.00	0.00	0.00	0.00	0.00	0.00	0.00	0.00
MAP VOLTS	4.56	4.56	4.56	4.56	4.56	4.56	4.56	4.56	4.56	4.56	4.56	4.56	4.56	4.56	4.56	4.56
TPS-MINIMUM	0.84	0.84	0.84	0.84	0.84	0.84	0.84	0.84	0.84	0.84	0.84	0.84	0.84	0.84	0.84	0.84
VACUUM	00 I	00 I	00 I	00 I	00 I	00 I	00 I	00 I	00 I	00 I	00 I	00 I	00 I	00 I	00 I	00 I
VEHICLE SPEED SEN	000	000	000	000	000	000	000	000	000	000	000	000	000	000	000	000

Figure 7-8 Snapshot captured using an OTC Monitor 2000/4000.

01 BASIC INFO. 92 CHEVY TRUCK TBI

RPM	0650	O2	0.85 V
COOL	185° F	FUEL	RICH
TPS	0.52 V	LOOP	CLOSED
DIAG STATE	9 ROAD	BATTERY	14.0 V
TROUBLE CODES	NO	EGR	NU
MAP	09 IHG	MAP V	1.21 V
KNOCK SIGNAL	NO	BARO VOLTS	NU
M/C DWELL	NU	PULSE WID	NU

Select Diag State:

6-FIELD SERVICE

7-BACKUP FUEL

8-ALCL

9-ROAD TEST

Figure 7-9 Diagnostic States – ROAD Mode *(OKAY TO DRIVE).*

See the manual for the specific scan tool for instructions about changing diagnostic modes. Snap-On displays *OK TO DRIVE* for most vehicles, but GM is the only manufacturer that allows the user to select other modes.

When the vehicle is ready to test drive, prepare the scan tool to record before leaving. Some scan tools allow for the selection of different methods for activating the snapshot trigger once the problem is experienced:

1. *Quick Trigger* is the standard method used by most scan tool manufacturers. It activates manually by either pushing any numeric key or by pushing the record button.
2. Any *Fault Code Trigger* allows the scan tool to take a snapshot when any fault code sets during the test drive.
3. *Specific Fault Code Trigger* allows the technician to specify which code activates the trigger.

The number of snapshots and frames that can be captured on one cartridge varies between the different scan tools. Some allow multiple snapshots to be recorded, while others allow only one.

Once the problem is experienced and the snapshot data is captured, the technician should return to the shop and review the data. The playback of the snapshot displays the data in the data stream at the time the problem was experienced and the trigger activated. No frames are missed and diagnostics are greatly enhanced.

REVIEWING SNAPSHOT DATA

When reviewing a scan tool snapshot, the minus (-) frames represent data stored in memory *before* the trigger or record button was pushed. Frame "0" is the point in time the technician experienced the problem and pushed the trigger or record button. The positive frames represent data captured *after* pushing the trigger or record button.

When studying data from a snapshot, it is best to concentrate first on frames -10 to 0 (Figure 7-10), since by the time the trigger or record button was pushed, the problem had begun. If the problem is not seen in these frames, then the technician should look at the 0 to +10 frames. Most driveability complaints should be found in the -10 to +10 frames. Systems equipped with a higher baud rate computer might require the technician to look further into the minus or plus frames to locate the problem.

PRIORITIZE INPUT/OUTPUT DATA

When diagnosing with a scan tool, an order of importance should be established for sensors and switches. Outputs should also be included because they are an indication of what the computer is attempting to do, based on the input data. (Refer to Chapter 5.) If the output does not correspond to the input data, there may be a problem with the onboard computer.

The first step to effective scan tool diagnosis is: don't be too hasty to start the vehicle. The technician should:

1. Always take a look at key-on, engine off (KOEO) data.
2. Always remember that not all data is available from all vehicles.

E-Z EVENT PLUS
G.M. 1991 TRUCK TBI VIN K
EVENT 2 TAG -- FRAME -30
DIAGNOSTIC STATE 9 ROAD
TROUBLE CODES: NO

	-15	-14	-13	-12	-11	-10	-9	-8	-7	-6	-5	-4	-3	-2	-1	0
COOLANT TEMP	174	174	174	174	174	174	174	174	174	174	174	174	174	174	174	174
MAP	10 I	09 I	08 I	10 I	11 I	09 I	09 I	09 I	09 I	11 I	08 I	09 I	08 I	08 I	10 I	09 I
THROT POS SENSOR	0.62	0.62	0.62	0.62	0.62	0.62	0.62	0.62	0.62	0.62	0.62	0.62	0.62	0.62	0.62	0.62
RPM	0725	0775	0825	0700	0675	0775	0775	0800	0750	0675	0825	0750	0825	0825	0675	0850
O2 VOLTS	0.23	0.24	0.25	0.23	0.23	0.23	0.24	0.24	0.23	0.23	0.25	0.23	0.25	0.26	0.23	0.24
O2 CROSS COUNTS	000	000	000	000	000	000	000	000	000	000	000	000	000	000	000	000
AIR DIVERT	OFF	OFF	OFF	OFF	OFF	OFF	OFF	OFF	OFF	OFF	OFF	OFF	OFF	OFF	OFF	OFF
AIR SWITCH	OFF	OFF	OFF	OFF	OFF	OFF	OFF	OFF	OFF	OFF	OFF	OFF	OFF	OFF	OFF	OFF
BACKUP FUEL	OFF	OFF	OFF	OFF	OFF	OFF	OFF	OFF	OFF	OFF	OFF	OFF	OFF	OFF	OFF	OFF
BATTERY VOLTS	14.2	14.2	14.2	14.2	14.2	14.2	14.3	14.3	14.2	14.2	14.2	14.2	14.2	14.2	14.2	14.3
BLOCK LEARN MULT	142	143	135	144	144	144	145	135	147	147	147	148	150	135	150	135
IDLE AIR CONTROL	039	039	039	039	039	039	039	039	039	039	039	039	039	039	039	039
INTEGRATOR	132	134	137	129	129	131	133	135	129	129	131	133	135	128	128	128
KNOCK SIGNAL	NO	NO	NO	NO	NO	NO	NO	NO	NO	NO	NO	NO	NO	NO	NO	NO
LOOP STATUS	CLOS	CLOS	CLOS	CLOS	CLOS	CLOS	CLOS	CLOS	CLOS	CLOS	CLOS	CLOS	CLOS	CLOS	CLOS	CLOS
MAP VOLTS	1.27	1.13	1.05	1.29	1.48	1.13	1.13	1.11	1.11	1.52	1.07	1.19	1.01	1.03	1.37	1.11
O2 FUEL STATE	LEAN	LEAN	LEAN	LEAN	LEAN	LEAN	LEAN	LEAN	LEAN	LEAN	LEAN	LEAN	LEAN	LEAN	LEAN	LEAN
PARK/NEUTRAL	ON	ON	ON	ON	ON	ON	ON	ON	ON	ON	ON	ON	ON	ON	ON	ON
TORQUE CONV CLUT	OFF	OFF	OFF	OFF	OFF	OFF	OFF	OFF	OFF	OFF	OFF	OFF	OFF	OFF	OFF	OFF
VEHICLE SPEED SEN	000	000	000	000	000	000	000	000	000	000	000	000	000	000	000	000
A/C REQUEST	NO	NO	NO	NO	NO	NO	NO	NO	NO	NO	NO	NO	NO	NO	NO	NO
A/C CLUTCH	OFF	OFF	OFF	OFF	OFF	OFF	OFF	OFF	OFF	OFF	OFF	OFF	OFF	OFF	OFF	OFF
PROM ID	6301	6301	6301	6301	6301	6301	6301	6301	6301	6301	6301	6301	6301	6301	6301	6301

Figure 7-10 Analyzing snapshot data.

3. Know that various inputs/outputs may have different names.

See Figures 7-11 and 7-12. As a starting point, some information to look at might include:

1. *Coolant Temperature* and *Intake Air Temperature* — If the vehicle has sat long enough for the engine to cool down, the sensors should display similar readings. Do they correspond to the surrounding ambient air temperature?
2. *Baro Sensor* — If the system is equipped with a separate barometric sensor, this information is then used to establish barometric air pressure to calculate spark timing and air/fuel.
3. *Map Sensor* — If the vehicle is equipped with a separate Baro sensor, are the MAP and Baro readings similar? If the vehicle is equipped with

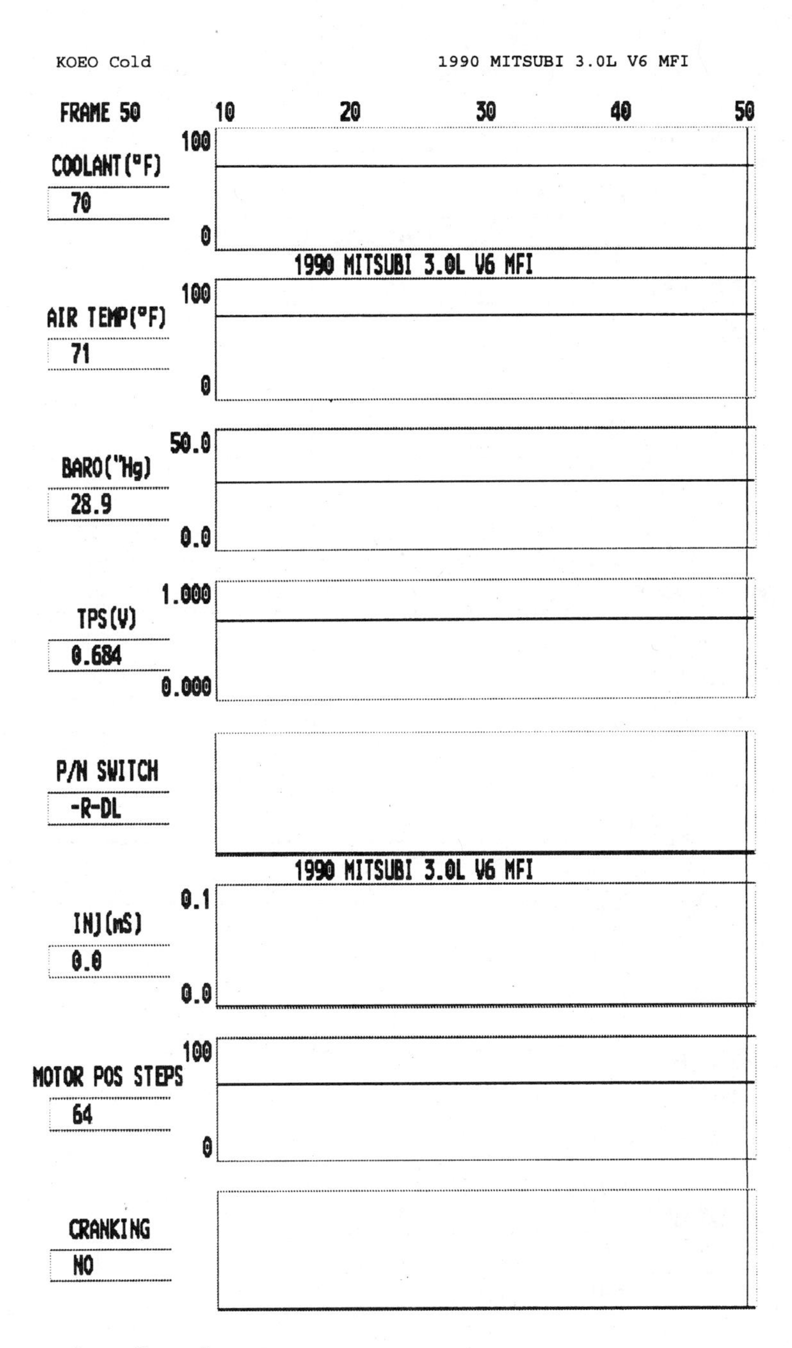

Figure 7-11 Key-on, engine-off readings.

only a MAP sensor (no Baro sensor), this is a Barometric/Manifold Absolute Pressure (BMAP) sensor. In the KOEO position, the computer uses this sensor to obtain the barometric pressure reading needed to calculate spark timing and air/fuel.

4. *Throttle Position Sensor* — If the vehicle is fuel-injected, the TPS voltage should be approximately 1 volt or under. If TPS voltage exceeds approximately 3 volts in this mode, a clear flood mode may be set, resulting in a no start.
5. *Park/Neutral Switch* — It should indicate that the transmission selector is in Park/Neutral. If out of adjustment, this could affect idle quality.
6. *Fuel Injector Pulse-Width* — A number may or may not be present. If a number is present, it

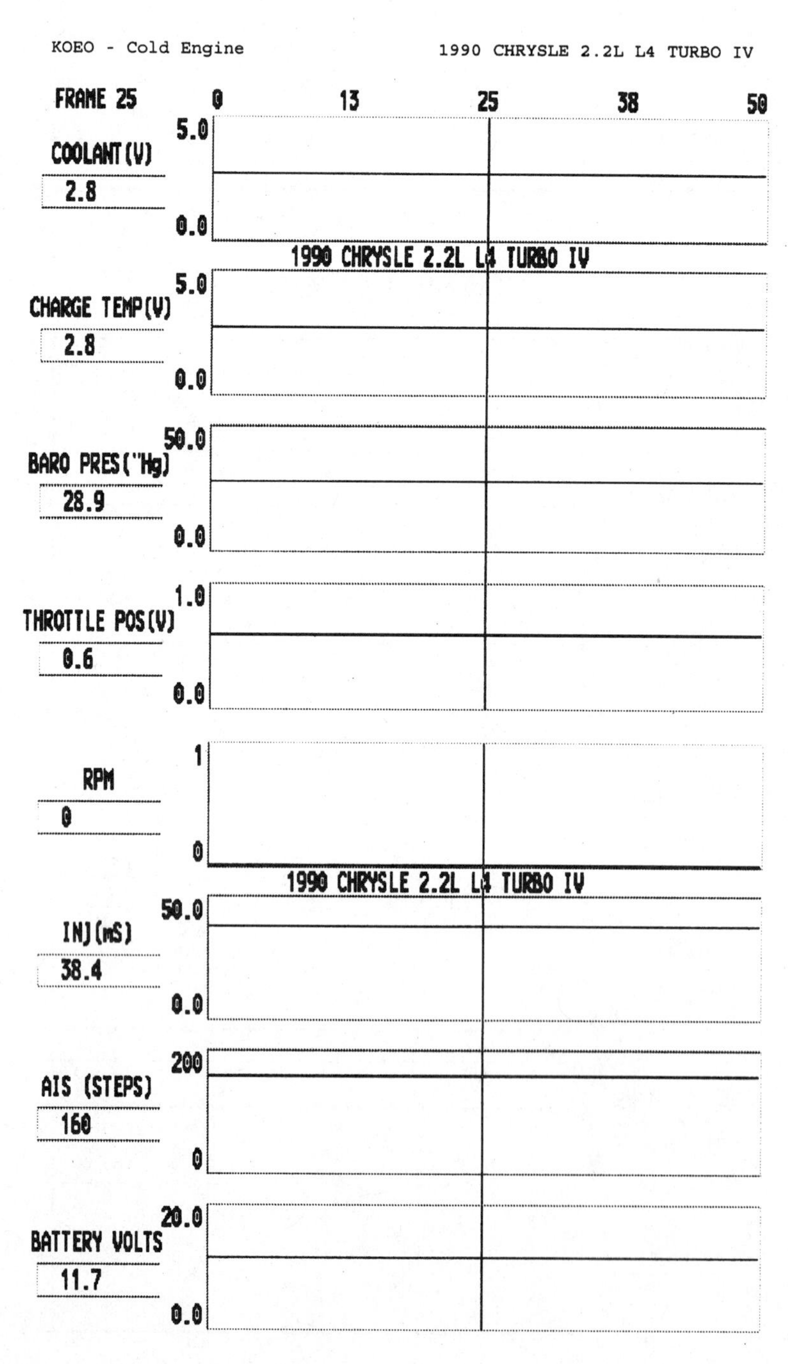

Figure 7-12 Key-on, engine-off readings.

indicates that the computer is anticipating the next start-up. If no pulse-width is indicated, the vehicle may be in a clear flood mode, or a pulse-width may not be available until the key is turned to the crank position.

7. *Idle Air Control (IAC)* — Verify that a number is present because this indicates that the computer has positioned the IAC for start-up. (Each manufacturer uses a different step count to prepare the engine for the next restart.)

Once the engine is running, inputs should be prioritized based on the effect they have on driveability. These inputs that have the most pronounced effect on engine operation (in descending order):

1. *RPM* — If the vehicle is fuel-injected and this signal is not being generated, the engine cannot start. If the signal disappears while the engine is running, the engine will stall. If the vehicle is carbureted, it will continue to run with the *check light* on.
2. *Load Sensor* — Map, Vacuum, or Mass Air Flow sensors tell the computer how much air is in the engine. This sensor has a major affect on air/fuel ratios and ignition timing.
3. *Throttle Position Sensor* — Can be a sensor or switch that informs the computer if the vehicle is at a closed throttle, steady throttle, accelerating, or decelerating. Some systems display throttle position as an angle or a mode. If the throttle is closed, does the throttle angle indicate 0°? If throttle position is displayed as a mode, it shows C/T for a closed throttle. This indicates that the computer has recognized that the throttle is closed.
4. *Coolant Temperature Sensor* — Affects a variety of engine operations; for that reason it should always be monitored.
5. *Barometric Sensor* — Compensating for increases in altitude (less pressure = less air) is important because less air requires less fuel, and leaner air/fuel ratios require additional spark advance.
6. *Manifold Air Temperature Sensor* — It allows the computer to fine-tune the air/fuel ratio based on the density of the air in the intake air stream.
7. *Oxygen Sensor* — Indicates the condition of exhaust oxygen content. It can be read in volts or cross counts. Once the vehicle is operating in closed loop, does the sensor rapidly cycle between 0.2 and 0.8 volts? If the voltage is cycling between 0 and 0.6 volts, the computer is correcting for a lean exhaust. If the voltage is cycling between 0.4 and 1.0 volts, the computer is correcting for a rich exhaust.
8. *Battery Sense Input* — Most computers operate effectively when battery/charging voltage is from 10 to 16 volts. If voltage is not in this range, the computer might revert to a *limp-in* mode.
9. *Switch Circuits* — They can be in one of two states, on/off or high/low. Most switches indicate that a load is being placed on the engine.

Listed below are the various outputs that are good indications of how the computer is "thinking." Remember that the computer sends commands to the various outputs based on the data it receives from the inputs. Placing a priority on the outputs is not important.

1. *Idle Air Control (IAC)* — This should be checked with the key-on engine off, at idle, during acceleration and deceleration. The number should increase when the throttle is opened and decrease as the throttle is closed. Additional engine loads should increase the IAC number: i.e., when the air conditioning is turned on, the vehicle is shifted into gear, or when a heavy electrical demand is placed on the charging system.
2. *Fuel Injector Pulse-Width* — The pulse-width should increase with fuel demand or decrease under deceleration or lean air/fuel ratio.
3. *Fuel Control* (Refer to Chapter 8) — Different terms are used for this, such as: Mixture Control Dwell, Block Learn/Integrator, Short-Term/Long-Term Fuel Trim, and Adaptive Fuel Factor. Whatever the name, they indicate the amount of correction the computer is attempting to make to the air/fuel ratio to achieve a stoichiometric ratio.
4. *Spark Advance* — It should change with coolant temperature, throttle opening, engine load, and spark knock.
5. *Emissions Controls* (Refer to Chapter 6) — These systems are controlled by the computer, based on input from the various sensors. They indicate what the computer is attempting to do, not necessarily what is actually occurring.

This covers the most important inputs and outputs that help the technician to diagnose driveability complaints or emissions failures using a scan tool.

DEFAULT VALUES

A default value may be substituted for a failed sensor that has set a hard fault. For example, when the Mass Air Flow sensor on a GM vehicle (Figure 7-13) sets a hard fault code, the computer uses the RPM, CTS, IAC, and TPS signals to calculate a value that allows the customer to *limp-in* to the shop. The scan tool displays what appears to be a normal value for the circuit involved. Actually, the display is a number the computer has calculated to be correct based on other inputs.

Another example is a Chrysler coolant sensor (Figure 7-14, Sheets 1 and 2). In this case, the computer identifies a problem in the coolant sensor circuit, indicated by *Fault Code 22* at the top of the figure. The vehicle had just returned from a 15-minute test drive. The coolant temperature

90VC3803 E-Z EVENT PLUS
G.M. 1990 3800 VIN C
EVENT 2 TAG-- FRAME -25
DIAGNOSTIC STATE 9 ROAD
TROUBLE CODES: 34
CURRENT CODES: 34

	-25	-24	-23	-22	-21	-20	-19	-18	-17	-16	-15	-14	-13	-12	-11	-10
RPM	1025	1025	0975	0950	0925	0900	0900	0875	0875	0875	0875	0900	0900	0875	0875	0900
MASS AIR FLOW	007	007	007	007	007	007	007	007	007	007	007	007	007	007	007	007
THROT POS SENSOR	0.42	0.42	0.42	0.42	0.42	0.42	0.42	0.42	0.42	0.42	0.42	0.42	0.42	0.42	0.42	0.42
COOLANT TEMP	058	058	058	058	058	058	058	058	058	058	058	059	059	059	059	059
O2 VOLTS	0.49	0.49	0.49	0.49	0.50	0.50	0.50	0.50	0.50	0.50	0.50	0.50	0.50	0.50	0.50	0.50
BATTERY VOLTS	14.2	14.2	14.2	14.2	14.2	14.2	14.2	14.2	14.2	14.2	14.2	14.2	14.2	14.2	14.2	14.2
INJ PULSE WIDTH	04.8	04.8	04.9	05.0	05.1	05.2	05.2	05.3	05.3	05.3	05.3	05.1	05.2	05.2	05.2	05.1
BLOCK LEARN MULT	128	128	128	128	128	128	128	128	128	128	128	128	128	128	128	128
INTEGRATOR	128	128	128	128	128	128	128	128	128	128	128	128	128	128	128	128

	-10	-9	-8	-7	-6	-5	-4	-3	-2	-1	0	1	2	3	4	5
RPM	0900	0875	0850	0850	0850	0850	0875	0850	0875	0850	0850	0850	0875	1250	1675	1950
MASS AIR FLOW	007	007	007	007	007	007	007	007	007	007	007	007	013	014	017	018
THROT POS SENSOR	0.42	0.42	0.42	0.42	0.42	0.42	0.42	0.42	0.42	0.42	0.42	0.42	0.58	0.58	0.62	0.64
COOLANT TEMP	059	059	059	059	059	059	059	059	059	059	059	059	059	059	059	059
O2 VOLTS	0.50	0.50	0.51	0.51	0.51	0.51	0.50	0.51	0.51	0.51	0.51	0.51	0.51	0.51	0.50	0.50
BATTERY VOLTS	14.2	14.1	14.2	14.1	14.1	14.0	14.1	14.2	14.2	14.1	14.2	14.2	14.2	14.2	14.1	14.2
INJ PULSE WIDTH	05.1	05.3	05.4	05.3	05.3	05.3	05.2	05.3	05.2	05.3	05.3	05.3	09.1	07.0	06.4	05.8
BLOCK LEARN MULT	128	128	128	128	128	128	128	128	128	128	128	128	128	128	128	128
INTEGRATOR	128	128	128	128	128	128	128	128	128	128	128	128	128	128	128	128

	5	6	7	8	9	10	11	12	13	14	15	16	17	18	19	20
RPM	1950	2100	2200	2275	2325	2400	2475	2525	2575	2600	2600	2625	2625	2625	2650	2650
MASS AIR FLOW	018	018	019	020	021	022	022	022	022	022	022	022	022	022	022	022
THROT POS SENSOR	0.66	0.66	0.66	0.66	0.66	0.66	0.66	0.66	0.66	0.66	0.66	0.66	0.66	0.66	0.66	0.66
COOLANT TEMP	059	059	059	059	059	059	059	059	059	059	059	059	059	059	059	059
O2 VOLTS	0.50	0.50	0.51	0.51	0.51	0.51	0.51	0.51	0.51	0.51	0.51	0.51	0.51	0.52	0.51	0.52
BATTERY VOLTS	14.2	14.2	14.1	14.1	14.1	14.2	14.1	14.1	14.1	14.1	14.1	14.1	14.1	14.1	14.1	14.1
INJ PULSE WIDTH	05.8	05.5	05.4	05.7	05.6	05.7	05.5	05.5	05.4	05.4	05.3	05.3	05.3	05.3	05.3	05.3
BLOCK LEARN MULT	128	128	128	128	128	128	128	128	128	128	128	128	128	128	128	128
INTEGRATOR	128	128	128	128	128	128	128	128	128	128	128	128	128	128	128	128

This vehicle has a "Check Engine" light on and a Hard Fault Code 34. Code 34 represents a Mass Air Flow Voltage Low problem.

The Mass Air Flow sensor is measured in Grams per Second and should increase with RPM and throttle position. Look at the readings and you will see that the Mass Air Flow does increase with RPM and throttle position. Is it real or is it Memorex?

What you are seeing is a calculated Mass Air Flow value with a computer substituted default value. The only way to be sure of this is to go directly to the sensor and test the output signal from the sensor. Remember, you can't always believe what you see.

Figure 7-13 General Motors MAF sensor in default.

DEFAULT E-Z EVENT PLUS
CHRYSLER 1987 3.0 MFI VIN 3 1 of 2 pages
EVENT 1 TAG -- FRAME -30
FAULT CODES: 22

	-30	-29	-28	-27	-26	-25	-24	-23	-22	-21	-20	-19	-18	-17	-16	-15
COOLANT TEMP	102	102	102	102	102	102	102	102	102	102	102	102	102	102	102	102
RPM	2158	2151	2151	2238	2238	2316	2338	2338	2351	2351	2376	2337	2337	2356	2356	2351
INJ PULSE WIDTH	02.8	02.9	02.9	03.2	03.2	03.0	02.9	02.9	02.9	02.9	03.1	02.9	02.9	03.0	03.0	03.0
MAP	08.9	09.1	09.1	09.3	09.3	09.3	09.3	09.3	09.1	09.1	09.1	09.2	09.2	09.1	09.1	09.1
O2 VOLTS	0.82	0.60	0.60	0.34	0.34	0.76	0.76	0.76	0.60	0.60	0.52	0.72	0.72	0.72	0.72	0.10
O2 FUEL STATE	RICH	RICH	RICH	LEAN	LEAN	RICH	RICH	RICH	RICH	RICH	RICH	RICH	RICH	RICH	RICH	LEAN
THROT POS SENSOR	1.24	1.26	1.26	1.26	1.26	1.26	1.26	1.26	1.28	1.28	1.26	1.26	1.26	1.26	1.26	1.26
BARO PRESSURE	29.8	29.8	29.8	29.8	29.8	29.8	29.8	29.8	29.8	29.8	29.8	29.8	29.8	29.8	29.8	29.8
BARO PRESSURE VOL	4.84	4.84	4.84	4.84	4.84	4.84	4.84	4.84	4.84	4.84	4.84	4.84	4.84	4.84	4.84	4.84
BATTERY TEMP	068	068	068	068	068	068	068	068	068	068	068	068	068	068	068	068
BATTERY VOLTS	14.5	14.1	14.1	14.3	14.3	14.4	14.2	14.2	14.2	14.2	14.6	14.5	14.5	14.7	14.7	14.2
CHARGE TEMP	099	099	099	099	099	099	099	099	099	099	099	099	099	099	099	099
ELECTRONIC ADV/RE	002	004	004	003	003	003	003	003	003	003	002	003	003	003	003	002
IDLE RPM X 10	080	080	080	080	080	080	080	080	080	080	080	080	080	080	080	080
MAP VOLTS	1.06	1.14	1.14	1.22	1.22	1.18	1.16	1.16	1.16	1.16	1.14	1.16	1.16	1.16	1.16	1.16
TPS-MINIMUM	1.12	1.12	1.12	1.12	1.12	1.12	1.12	1.12	1.12	1.12	1.12	1.12	1.12	1.12	1.12	1.12
VACUUM	21 I	21 I	21 I	21 I	21 I	21 I	21 I	21 I	21 I	21 I	21 I	21 I	21 I	21 I	21 I	21 I
VEHICLE SPEED SEN	000	000	000	000	000	000	000	000	000	000	000	000	000	000	000	000

	-15	-14	-13	-12	-11	-10	-9	-8	-7	-6	-5	-4	-3	-2	-1	0
COOLANT TEMP	102	102	102	102	102	102	102	102	102	102	102	102	102	102	102	102
RPM	2351	2351	2328	2328	2367	2367	2337	2337	2358	2358	2385	2391	2391	2406	2374	2374
INJ PULSE WIDTH	03.0	03.0	02.8	02.8	03.1	03.1	02.5	02.5	02.9	02.9	03.0	02.8	02.8	03.1	02.9	02.9
MAP	09.1	09.1	09.1	09.1	09.1	09.1	09.1	09.1	09.3	09.3	09.3	09.1	09.1	09.1	09.1	09.1
O2 VOLTS	0.10	0.10	0.64	0.64	0.18	0.18	0.20	0.20	0.74	0.74	0.62	0.70	0.70	0.56	0.40	0.40
O2 FUEL STATE	LEAN	LEAN	RICH	RICH	LEAN	LEAN	LEAN	LEAN	RICH	RICH	RICH	RICH	RICH	RICH	LEAN	LEAN
THROT POS SENSOR	1.26	1.26	1.26	1.26	1.26	1.26	1.28	1.28	1.28	1.28	1.28	1.28	1.28	1.28	1.28	1.28
BARO PRESSURE	29.8	29.8	29.8	29.8	29.8	29.8	29.8	29.8	29.8	29.8	29.8	29.8	29.8	29.8	29.8	29.8
BARO PRESSURE VOL	4.84	4.84	4.84	4.84	4.84	4.84	4.84	4.84	4.84	4.84	4.84	4.84	4.84	4.84	4.84	4.84
BATTERY TEMP	068	068	068	068	068	068	068	068	068	068	069	068	068	068	068	068
BATTERY VOLTS	14.2	14.2	14.8	14.8	14.3	14.3	14.3	14.3	14.3	14.3	14.4	14.6	14.6	14.6	14.1	14.1
CHARGE TEMP	099	099	099	099	099	099	099	099	099	099	099	099	099	099	099	099
ELECTRONIC ADV/RE	002	002	002	002	002	002	002	002	003	003	004	003	003	001	003	003
IDLE RPM X 10	080	080	080	080	080	080	080	080	080	080	080	080	080	080	080	080
MAP VOLTS	1.16	1.16	1.14	1.14	1.16	1.16	1.10	1.10	1.14	1.14	1.16	1.14	1.14	1.12	1.12	1.12
TPS-MINIMUM	1.12	1.12	1.12	1.12	1.12	1.12	1.12	1.12	1.12	1.12	1.12	1.12	1.12	1.12	1.12	1.12
VACUUM	21 I	21 I	21 I	21 I	21 I	21 I	21 I	21 I	21 I	21 I	21 I	21 I	21 I	21 I	21 I	21 I
VEHICLE SPEED SEN	000	000	000	000	000	000	000	000	000	000	000	000	000	000	000	000

Figure 7-14 Chrysler coolant sensor in default (Sheet 1 of 2).

sensor is only 102° F, while the charge temperature (intake air) is 99° F. After testing the coolant sensor circuit, it was found to have an open in the sensor. The coolant sensor value displayed in the figure is a default value calculated from data stored in memory and input from the intake air temperature sensor.

Default values change from vehicle to vehicle and manufacturer to manufacturer. Not every system displays a default value on the data stream. Some manufacturers show the actual value. For example, on a GM, an open coolant sensor circuit displays -40° F and not the default value. Internally, the computer has calculated a default value of 119° F which allows the vehicle to continue to run in *limp-in* mode.

If the technician is not aware of this feature, a default value can be misleading. The key to identifying a default value is to see if a ***hard fault*** code exists for the particular sensor in question. If so, the value observed may be a default value. The

DEFAULT E-Z EVENT PLUS
CHRYSLER 1987 3.0 MFI VIN 3 2 of 2 pages
EVENT 1 TAG -- FRAME -30
FAULT CODES: 22

	0	1	2	3	4	5	6	7	8	9	10	11	12	13	14	15
COOLANT TEMP	102	102	102	102	102	102	102	102	102	102	102	102	102	102	102	102
RPM	2374	2415	2415	2392	2392	2392	2380	2380	2380	2408	2408	2408	2437	2396	2396	2419
INJ PULSE WIDTH	02.9	03.0	03.0	02.9	02.9	02.9	03.0	03.0	03.0	02.8	02.8	02.8	03.0	02.6	02.6	02.9
MAP	09.1	09.1	09.1	09.1	09.1	09.1	08.9	08.9	08.9	08.8	08.8	08.8	09.1	09.1	09.1	08.8
O2 VOLTS	0.40	0.60	0.60	0.30	0.30	0.30	0.26	0.26	0.26	0.44	0.44	0.44	0.54	0.14	0.14	0.44
O2 FUEL STATE	LEAN	RICH	RICH	LEAN	LEAN	LEAN	LEAN	LEAN	LEAN	LEAN	LEAN	LEAN	RICH	LEAN	LEAN	LEAN
THROT POS SENSOR	1.28	1.28	1.28	1.28	1.28	1.28	1.28	1.28	1.28	1.28	1.28	1.28	1.28	1.28	1.28	1.28
BARO PRESSURE	29.8	29.8	29.8	29.8	29.8	29.8	29.8	29.8	29.8	29.8	29.8	29.8	29.8	29.8	29.8	29.8
BARO PRESSURE VOL	4.84	4.84	4.84	4.84	4.84	4.84	4.84	4.84	4.84	4.84	4.84	4.84	4.84	4.84	4.84	4.84
BATTERY TEMP	068	068	068	068	068	068	068	068	068	068	068	068	068	068	068	068
BATTERY VOLTS	14.1	14.2	14.2	14.3	14.3	14.3	14.3	14.3	14.3	14.3	14.3	14.3	14.6	14.3	14.3	14.8
CHARGE TEMP	099	099	099	099	099	099	099	099	099	099	099	099	099	099	099	099
ELECTRONIC ADV/RE	003	003	003	002	002	002	002	002	002	002	002	002	003	003	003	002
IDLE RPM X 10	080	080	080	080	080	080	080	080	080	080	080	080	080	080	080	080
MAP VOLTS	1.12	1.16	1.16	1.14	1.14	1.14	1.10	1.10	1.10	1.10	1.10	1.10	1.12	1.16	1.16	1.10
TPS-MINIMUM	1.12	1.12	1.12	1.12	1.12	1.12	1.12	1.12	1.12	1.12	1.12	1.12	1.12	1.12	1.12	1.12
VACUUM	21 I	21 I	21 I	21 I	21 I	21 I	21 I	21 I	21 I	21 I	21 I	21 I	21 I	21 I	21 I	21 I
VEHICLE SPEED SEN	000	000	000	000	000	000	000	000	000	000	000	000	000	000	000	000

	15	16	17	18	19	20	21	22	23	24	25	26	27	28	29	30
COOLANT TEMP	102	102	102	102	102	102	102	102	102	102	102	102	102	102	102	102
RPM	2419	2419	2408	2408	2424	2424	2428	2428	2417	2417	2463	2463	2407	2407	2423	2423
INJ PULSE WIDTH	02.9	02.9	02.9	02.9	03.0	03.0	03.5	03.5	02.9	02.9	02.9	02.9	02.5	02.5	03.0	03.0
MAP	08.8	08.8	09.1	09.1	08.8	08.8	09.2	09.2	08.9	08.9	09.1	09.1	08.8	08.8	09.3	09.3
O2 VOLTS	0.44	0.44	0.76	0.76	0.14	0.14	0.22	0.22	0.38	0.38	0.72	0.72	0.46	0.46	0.08	0.08
O2 FUEL STATE	LEAN	LEAN	RICH	RICH	LEAN	LEAN	LEAN	LEAN	LEAN	LEAN	RICH	RICH	RICH	RICH	LEAN	LEAN
THROT POS SENSOR	1.28	1.28	1.28	1.28	1.28	1.28	1.28	1.28	1.28	1.28	1.28	1.28	1.28	1.28	1.26	1.26
BARO PRESSURE	29.8	29.8	29.8	29.8	29.8	29.8	29.8	29.8	29.8	29.8	29.8	29.8	29.8	29.8	29.8	29.8
BARO PRESSURE VOL	4.84	4.84	4.84	4.84	4.84	4.84	4.84	4.84	4.84	4.84	4.84	4.84	4.84	4.84	4.84	4.84
BATTERY TEMP	068	068	068	068	068	068	068	068	068	068	068	068	068	068	068	068
BATTERY VOLTS	14.8	14.8	14.4	14.4	14.1	14.1	14.1	14.1	14.5	14.5	14.4	14.4	14.2	14.2	14.2	14.2
CHARGE TEMP	099	099	099	099	099	099	099	099	099	099	099	099	099	099	099	099
ELECTRONIC ADV/RE	002	002	003	003	002	002	003	003	002	002	002	002	002	002	004	004
IDLE RPM X 10	080	080	080	080	080	080	080	080	080	080	080	080	080	080	080	080
MAP VOLTS	1.10	1.10	1.12	1.12	1.14	1.14	1.14	1.14	1.10	1.10	1.14	1.14	1.08	1.08	1.14	1.14
TPS-MINIMUM	1.12	1.12	1.12	1.12	1.12	1.12	1.12	1.12	1.12	1.12	1.12	1.12	1.12	1.12	1.12	1.12
VACUUM	21 I	21 I	21 I	21 I	21 I	21 I	21 I	21 I	21 I	21 I	21 I	21 I	21 I	21 I	21 I	21 I
VEHICLE SPEED SEN	000	000	000	000	000	000	000	000	000	000	000	000	000	000	000	000

Figure 7-14 Chrysler coolant sensor in default (Sheet 2 of 2).

technician should proceed to the circuit and the component in question to perform additional tests.

TROUBLESHOOTER CARTRIDGES

Troubleshooter cartridges are available to assist the technician with computer-related driveability problems. Snap-On has a separate cartridge called *Fast Track*; OTC's is included in their master cartridge called *Pathfinder*. Troubleshooter cartridges provide diagnostic help information (Figure 7-15) on the scan tool screen. They can greatly minimize the time-consuming task of digging through repair manuals, technical service bulletins, and GM PROM identification books to find the information needed to solve the problem.

2. Press 2-QUICK PATH. The next menu and the sub-menus will present a number of choices. Each vehicle manufacturer may not support all of the choices shown.

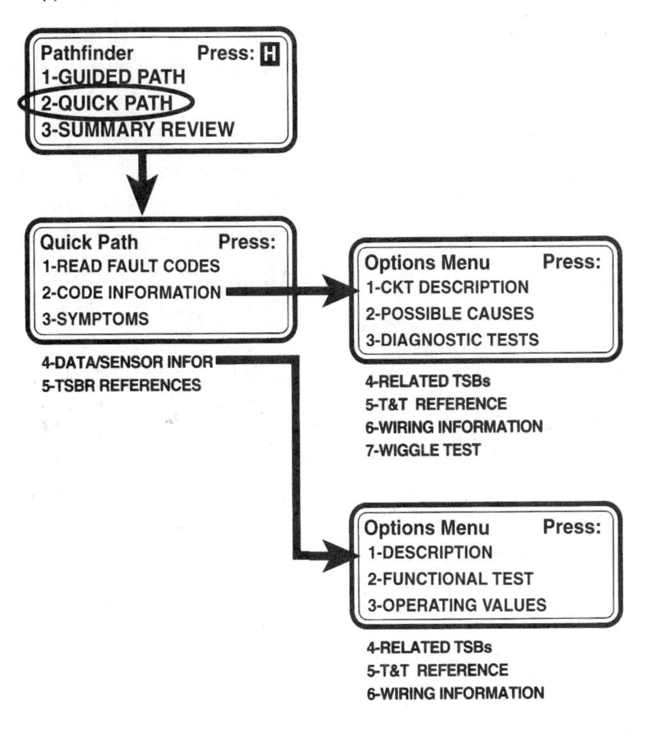

Figure 7-15 OTC Pathfinder II troubleshooter menu.

Troubleshooter cartridges contain information on many common input parameters and adjustments, trouble code problems, and driveability complaints for vehicles covered by the cartridge (Figure 7-16). They do not, however, contain information for every possible code and problem that could occur on all vehicles.

The checks in each Troubleshooter cartridge begin with the most likely cause of a problem or with the tests that should be performed first. The checks then progress through other possible causes and tests. All checks are based on common causes of a problem or important basic tests, generally in order of importance. For the most effective use of this cartridge, follow the checks in the order in which they are given.

When using a Troubleshooter cartridge, references are made to the manuals that are shipped with it. The technician should always consult the references as directed by the Troubleshooter. Trying to use the references by themselves may cause the technician to miss information or to perform some test or adjustment out of sequence.

4-Data/Sensor Information

This function displays sensor information. The tester can be either connected or disconnected to the vehicle. It serves as an "on-line" or "off-line" reference source for the following options:

1-Descriptions-Definition or description of the sensor.

This display helps diagnose trouble code 41, including tacho-meter

2-Functional Tests-loads the diagnostic test specified for the vehicle.

A properly functioning TPS should read between 0.2 & 1.2 volts with ign on or at idle. The voltage ↑ ↓

3-Operating Values-Shows the specified value and the typical range of the sensor.

TPS
Spec: 1.7 volts
Typical: 1.2 – 2.5v
0 – 5 volts

4-Related TSBs-The Technical Service Bulletins for the vehicle symptoms and fault code(s) are identified.

TSBR GM168
See TSBR manual for more information
TSBR 03 of 16

5-Tool & Tech References-Reference pages, charts, and steps for the fault code are displayed.

See Ford Tool & Techniques, 1986. #103120
Refer to Page 36, chart 2, step 2.

6-Wiring Information-shows pin number and wire color.

	PIN	Color
VREF	B26	Orange
SIG	B28	Drk Grn
GND	B23	Black

Press the ↑ or ↓ arrow keys to scroll through the list.

Figure 7-16 OTC Pathfinder II troubleshooter menu.

VT-100 EMULATION

The easiest way to view scan tool data on a larger screen is to use a Video Display Terminal (VT-100). A patch cable is needed, but no software is required. Once the terminal's protocol is set up, the VT-100 can display Snap-On, MPSI, or OTC scan tools.

When the Snap-On and MPSI scan tools are displayed on the terminal, the same information that appears on the scan tool appears on the terminal. The Snap-On scan tool can be displayed on the VT-100 as a full page of information.

When ZSTEM is used with the OTC Monitor 2000 or 4000, additional data can be displayed on the VT-100. (Figure 7-17.)

The larger screen allows the technician to create custom pages of data and to organize and

VT-100
01 - BASIC INFO.
1992 CHEVY TRUCK TBI

RPM	0650	COOL	185 F	O2	0.85 V
TPS	0.52 V	FUEL	RICH	LOOP	CLOS
DIAG STATE	9 ROAD	BATTERY			14.1 V
TROUBLE CODES	NO	EGR			NU
MAP	09 IHG	MAP V			1.21 V
KNOCK SIGNAL	NO	BARO VOLTS			NU
M/C DWELL	NU	PULSE WID			NU

Figure 7-17 ZSTEM basic information screen using a Monitor 2000/4000.

prioritize the data. Reducing the amount of data on the screen and concentrating on specific inputs and outputs makes it easier for the technician to diagnose problems. Figure 7-18 shows the inputs and outputs available on the data stream that can be displayed on a custom screen.

Another option is to use the ZSTEM terminal emulation software. This software is needed to connect an IBM PC or compatible to a Snap-On, MPSI, or OTC scan tool. This feature allows the scan tool information to be placed on a larger screen for easier viewing. The Hamilton (ESP) or Bear BAR '90 can be used as a VT-100 because they have an OBD connector on the back of the machine. In some cases, the ZSTEM software is already installed in these BAR '90 machines under the *Utilities* menu and identified as VT-100 (Figure 7-19). It might have to be activated by the equipment manufacturer.

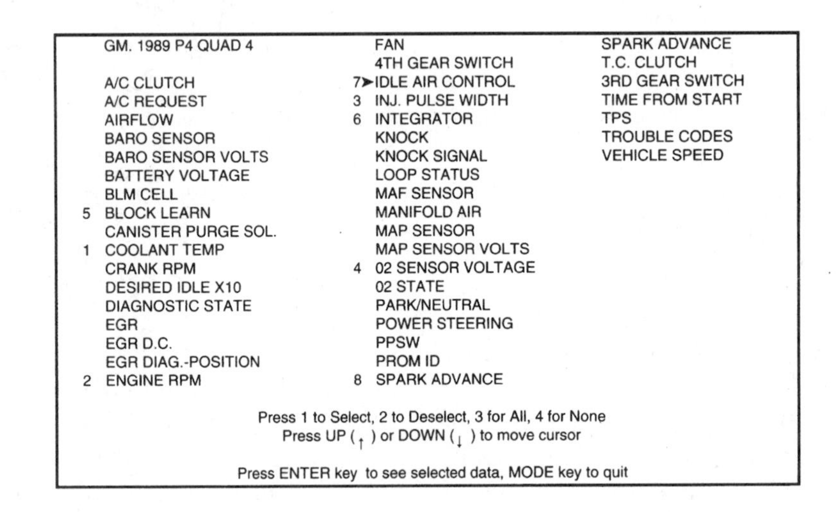

CUSTOM		GM. 1989 P4 QUAD 4	
COOLANT TEMP	092C	SPARK ADVANCE	23 DEG
ENGINE RPM	1100	A/C CLUTCH	ON
INJECTOR PW #1	1.6ms	EGR	OFF
02 SENSOR VOLT	0.140	LOOP STATUS	CLOSED
BLOCK LEARN	124	PROM ID	2309
INTEGRATOR	125	BATTERY VOLT	11.9
IDLE AIR CONTROL	032	VEHICLE SPEED	42

Figure 7-18 Building a custom data screen using a Monitor 2000/4000 and a VT-100 or ZSTEM with a PC.

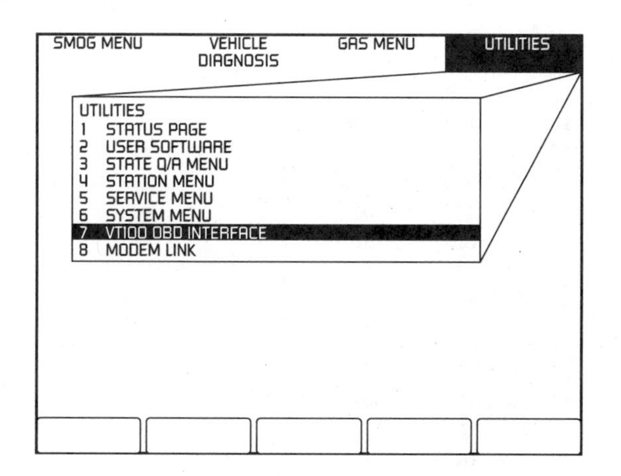

Figure 7-19 Selecting the VT-100 from the *Utilities* menu.

SCAN TOOL TO PC SOFTWARE

Software that allows data transfer to a PC is available from a variety of sources. The two most common software packages are EZ-Events Plus (OTC) and Scan-Gra-Fix (Snap-On). They run on most IBM PC or compatible computers and are available for both 3-1/2" and 5-1/4" disk drives.

EZ-Events Plus (OTC) (Figure 7-20) software was developed for use with the Monitor 2000 or 4000. This software makes the Monitor 2000 a more versatile tool. Once the data is displayed on the larger screen it's just like having a Monitor 4000. Features of this software include: graphing data live, downloading recorded events to an IBM PC or compatible computer for future reference, graphing recorded data, and printing some or all of the data recorded.

Scan-Gra-Fix (Snap-On) allows data to be to displayed and analyzed using an IBM PC or PC compatible and the Snap-On MT 2500 Scanner. It can display or print scanner data in text or graphic formats (Figure 7-21). Scan-Gra-Fix can be used to display live engine data, review a movie, or recall a previously saved file.

Scan-Gra-Fix can display engine system serial data for all vehicles supported by the scanner. It must be used with one of the following cartridges: 1994 and later Domestic Combination Primary Cartridge; 1993 GM Primary Cartridge; or 1992 and later Asian Import Cartridge.

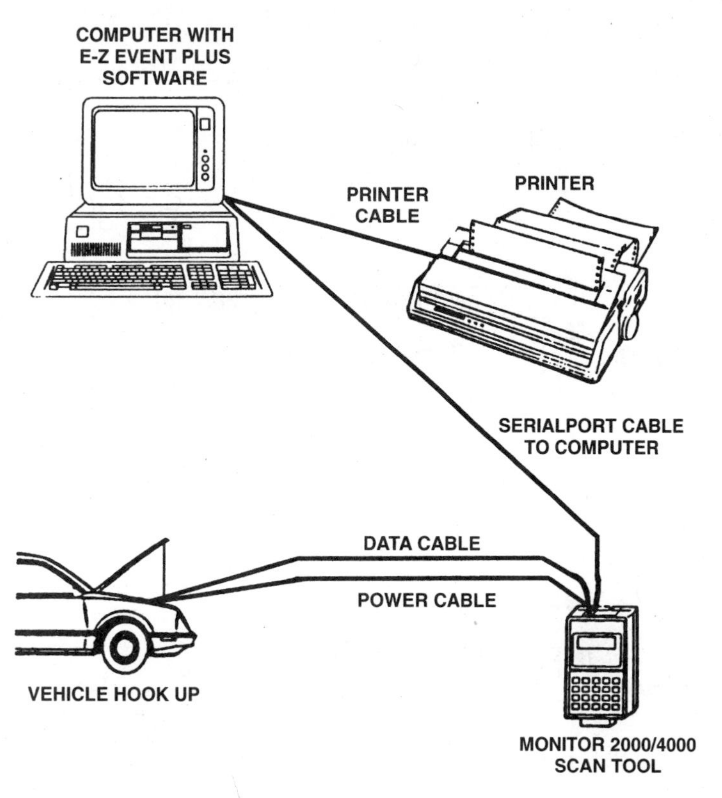

Figure 7-20 Connecting a scan tool to a PC.

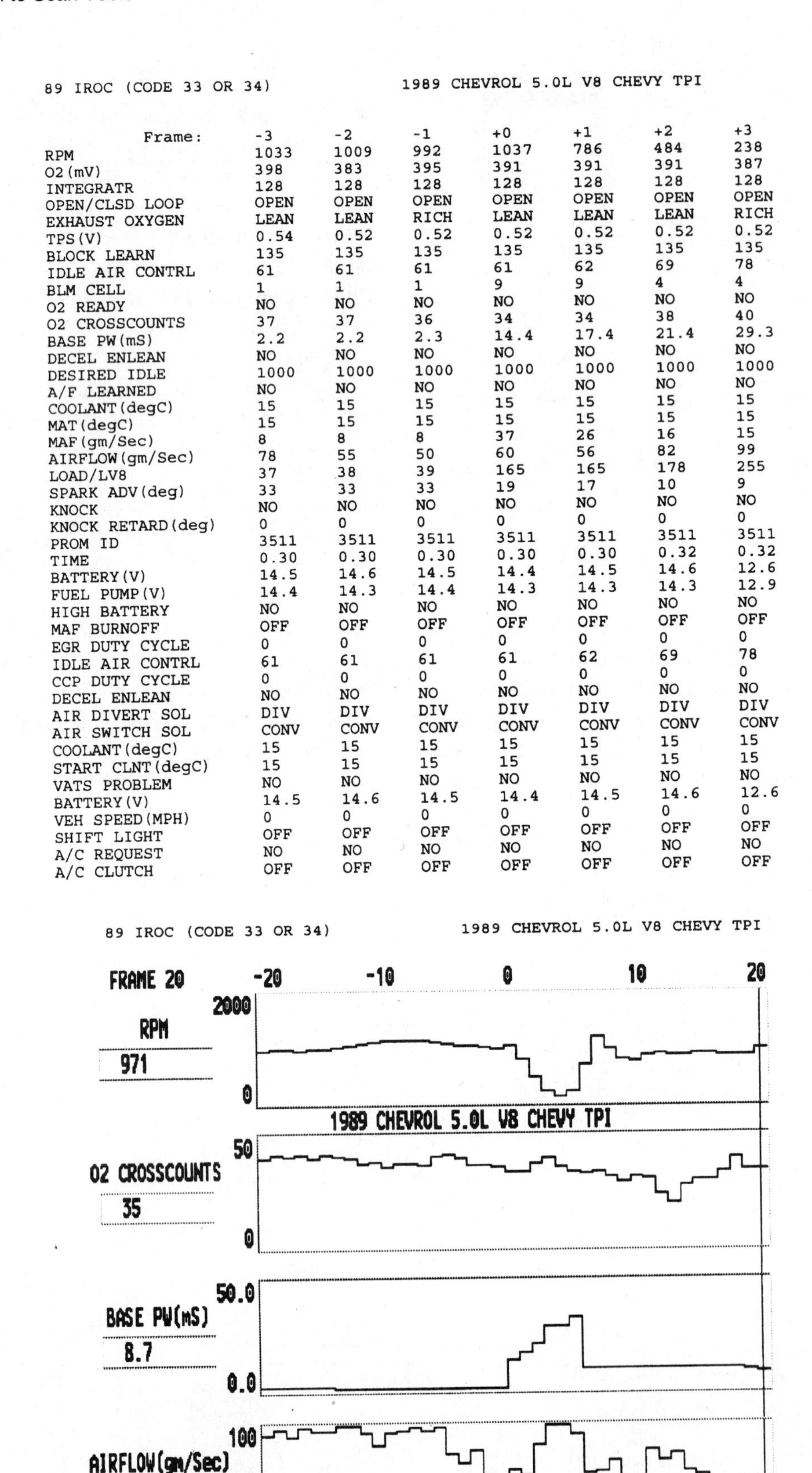

89 IROC (CODE 33 OR 34) 1989 CHEVROL 5.0L V8 CHEVY TPI

Frame:	-3	-2	-1	+0	+1	+2	+3
RPM	1033	1009	992	1037	786	484	238
O2 (mV)	398	383	395	391	391	391	387
INTEGRATR	128	128	128	128	128	128	128
OPEN/CLSD LOOP	OPEN	OPEN	OPEN	OPEN	OPEN	OPEN	OPEN
EXHAUST OXYGEN	LEAN	LEAN	RICH	LEAN	LEAN	LEAN	RICH
TPS (V)	0.54	0.52	0.52	0.52	0.52	0.52	0.52
BLOCK LEARN	135	135	135	135	135	135	135
IDLE AIR CONTRL	61	61	61	61	62	69	78
BLM CELL	1	1	1	9	9	4	4
O2 READY	NO	NO	NO	NO	NO	NO	NO
O2 CROSSCOUNTS	37	37	36	34	34	38	40
BASE PW (mS)	2.2	2.2	2.3	14.4	17.4	21.4	29.3
DECEL ENLEAN	NO	NO	NO	NO	NO	NO	NO
DESIRED IDLE	1000	1000	1000	1000	1000	1000	1000
A/F LEARNED	NO	NO	NO	NO	NO	NO	NO
COOLANT (degC)	15	15	15	15	15	15	15
MAT (degC)	15	15	15	15	15	15	15
MAF (gm/Sec)	8	8	8	37	26	16	15
AIRFLOW (gm/Sec)	78	55	50	60	56	82	99
LOAD/LV8	37	38	39	165	165	178	255
SPARK ADV (deg)	33	33	33	19	17	10	9
KNOCK	NO	NO	NO	NO	NO	NO	NO
KNOCK RETARD (deg)	0	0	0	0	0	0	0
PROM ID	3511	3511	3511	3511	3511	3511	3511
TIME	0.30	0.30	0.30	0.30	0.30	0.32	0.32
BATTERY (V)	14.5	14.6	14.5	14.4	14.5	14.6	12.6
FUEL PUMP (V)	14.4	14.3	14.4	14.3	14.3	14.3	12.9
HIGH BATTERY	NO	NO	NO	NO	NO	NO	NO
MAF BURNOFF	OFF	OFF	OFF	OFF	OFF	OFF	OFF
EGR DUTY CYCLE	0	0	0	0	0	0	0
IDLE AIR CONTRL	61	61	61	61	62	69	78
CCP DUTY CYCLE	0	0	0	0	0	0	0
DECEL ENLEAN	NO	NO	NO	NO	NO	NO	NO
AIR DIVERT SOL	DIV	DIV	DIV	DIV	DIV	DIV	DIV
AIR SWITCH SOL	CONV	CONV	CONV	CONV	CONV	CONV	CONV
COOLANT (degC)	15	15	15	15	15	15	15
START CLNT (degC)	15	15	15	15	15	15	15
VATS PROBLEM	NO	NO	NO	NO	NO	NO	NO
BATTERY (V)	14.5	14.6	14.5	14.4	14.5	14.6	12.6
VEH SPEED (MPH)	0	0	0	0	0	0	0
SHIFT LIGHT	OFF	OFF	OFF	OFF	OFF	OFF	OFF
A/C REQUEST	NO	NO	NO	NO	NO	NO	NO
A/C CLUTCH	OFF	OFF	OFF	OFF	OFF	OFF	OFF

Figure 7-21 Scan-Gra-Fix software displayed as text and graphics.

Scan-Gra-Fix includes these features:

1. Display, in text or graphics, *live* data on a PC with the vehicle running in the key-on engine off mode.
2. Transfer, display, and save a previously-recorded movie or on-screen live data to a PC.
3. Recall a previously-recorded file to review.
4. Automatically find suspected intermittent sensor or PCM output problems from recorded data using the exclusive Snap-On *Glitch Snitch* feature. This command is available when reviewing a movie or a recalled PC file. When activated, this feature analyzes and re-sorts the data, displaying the most suspect parameters first.

INTELLIGENT BREAKOUT BOX

Since some of the more popular vehicles on the road today do not have a data stream, MPSI has developed a tool called an Intelligent Breakout Box (IBOB). It is designed to be used with the MPSI Pro-Link 9000 scan tool (Figure 7-22) and works on vehicles that are not equipped with a data stream, such as Ford EEC IV, Toyota EFI/TCCS, Acura/Honda, and Mazda.

The IBOB has two primary purposes. First, it collects input and output analog voltage readings (Figure 7-23) entering and exiting the computer and transmits them to the Pro-Link 9000 scan tool for display. Second, the IBOB is capable of displaying frequency/pulse data in terms of frequency and duty-cycle.

The IBOB and Pro-Link are properly connected in parallel (Figure 7-24) with the on-board computer and the vehicle harness.

These additional features are similar to those available on vehicles with a data stream, such as:

1. Freeze-frame data that locks in specific inputs and outputs while looking for additional input/output data.
2. Developing a Custom Data List of inputs/outputs in an order other than the one pre-programmed into the Pro-Link 9000.

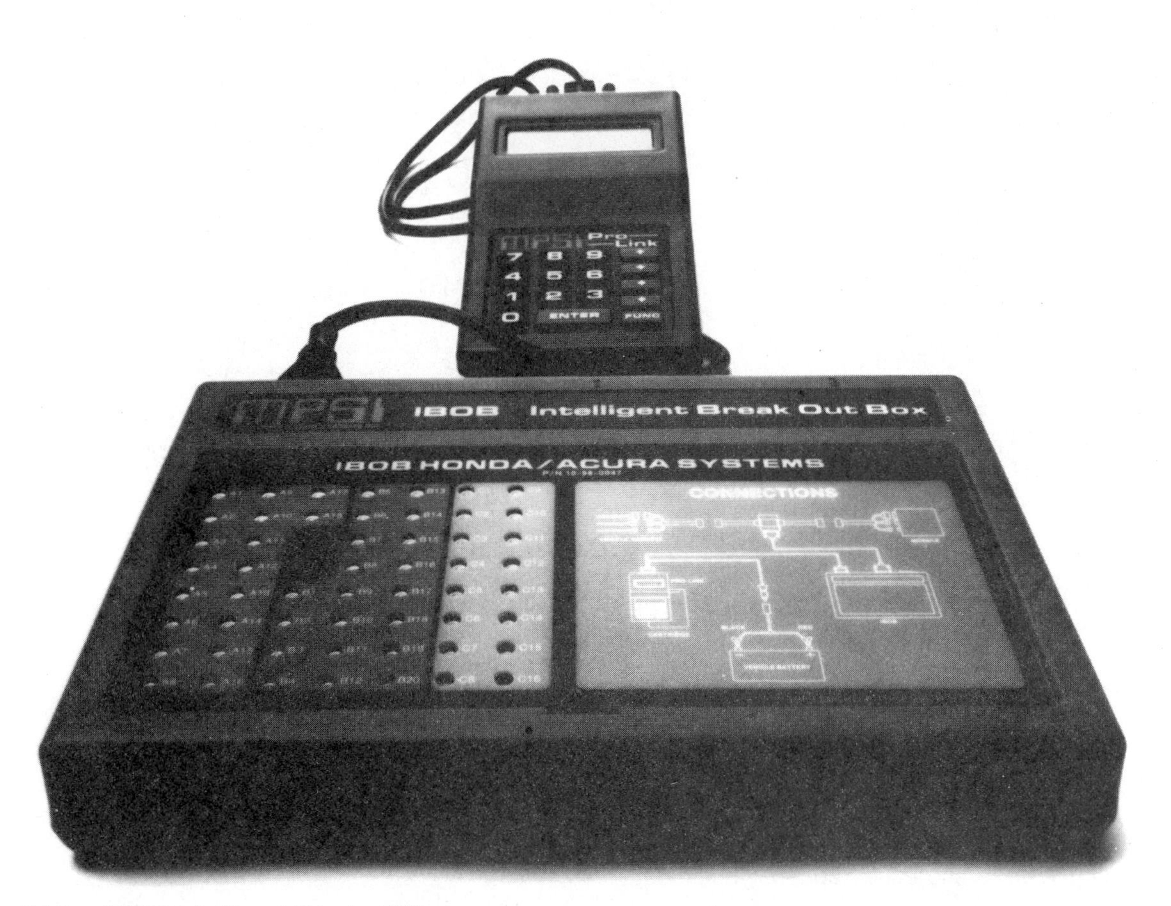

Figure 7-22 MPSI Intelligent Breakout Box.

```
TOYOTA IBOB
1990 3.0L OHC EFI
VIN---V-----L-------
Analog Data List

  E01 ENGINE GND   0.0
  E02 ENGINE GND   0.0
No10 INJ 1,3,5    13.6
No20 INJ 2,4,6    13.9
   E1 ENGINE GND  13.8
  ACV A/C VSV      0.0
   HT O2 HEATER    0.0
   AS AS VSV       0.0
  STJ COLD INJ    13.6
  EGR EGR VSV      0.0
  FPU FUEL PRESS  13.6
  IGT IGNITER      0.0
   S1 ECT SOL     13.6
   L4 T/CASE SW    0.0
   S2 ECT SOL     13.6
    N NEUTRAL SW  12.5
   S3 ECT SOL     13.6
    2 NEUTRAL SW   0.0
   S4 ECT SOL     13.7
    L NEUTRAL SW   0.0
  IGF IGNITER      0.7
  SP2 SPEED SENS   0.0
   G1 DISTR        0.0
   G2 DISTR        0.0
   NE DISTR        0.0
  G - DISTR        0.0
   VF CK CONN      2.1
  TSW COOLNT SW   13.4
 THO1 T/M TEMP     0.0
   T1 CK CONN     13.4
   OX OXYGEN       0.1
 THO2 T/M TEMP     0.0
  KNK KNOCK SENS   0.0
  THG EGR TEMP     0.0
  THW COOLNT TMP   0.2
  IDL THROT SENS  13.1
  THA AIR TEMP     2.1
  VTA THROT SENS   0.5
   VS AIR FLOW     1.5
 OX + OXYGEN       3.3
   VC AIR FLOW     4.8
   E2 SENS GND     0.0
  STA START SW     0.0
  A/C A/C SWITCH   0.0
  OD1 C/C ECU     13.4
  SP1 SPEED SENS  13.1
   DG CK CONN      0.0
  4WD 4WD SW       4.5
    P ECT PWR SW   0.0
  STP BRAKE SW     0.0
    W CHK ENG LT  13.6
  OD2 C/C ECU      0.0
  OIL OIL TMP LT  12.7
  E21 SENS GND     0.0
  +B1 MAIN RELAY  13.4
 BATT BATTERY +B  13.7
   +B MAIN RELAY  13.4
```

```
TOYOTA IBOB
1990 3.0L OHC EFI
VIN---V-----L-------
Frequency/Pulse Data

No10 INJ 1,3,5
FREQ  263.16 Hz
HIGH    0.7mSEC   18 %
LOW     3.1mSEC   82 %

No20 INJ 2,4,6
FREQ  264.61 Hz
HIGH    0.7mSEC   18 %
LOW     3.1mSEC   82 %

 IGT IGNITER
FREQ    0.00 Hz
HIGH    0.0mSEC    0 %
LOW     0.0mSEC    0 %

 IGF IGNITER
FREQ  117.84 Hz
HIGH    1.5mSEC   18 %
LOW     7.1mSEC   82 %

 SP2 SPEED SENS
FREQ    0.00 Hz
HIGH    0.0mSEC    0 %
LOW     0.0mSEC    0 %

  G1 DISTR
FREQ   19.80 Hz
HIGH    2.1mSEC    4 %
LOW    49.0mSEC   96 %

  G2 DISTR
FREQ   19.75 Hz
HIGH    2.2mSEC    4 %
LOW    49.0mSEC   96 %

  NE DISTR
FREQ  480.22 Hz
HIGH    0.8mSEC   36 %
LOW     1.4mSEC   64 %

 SP1 SPEED SENS
FREQ    0.00 Hz
HIGH    0.0mSEC    0 %
LOW     0.0mSEC    0 %
```

Figure 7-23 IBOB analog and frequency/pulse data.

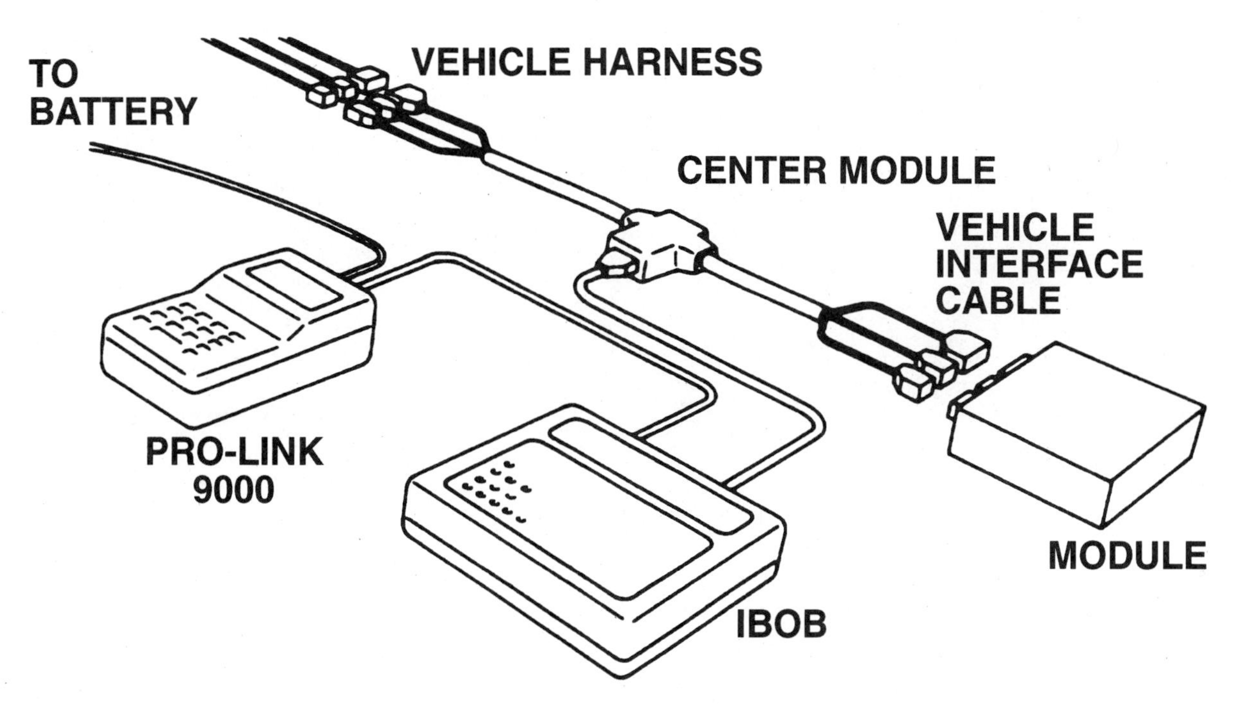

Figure 7-24 Properly-connected IBOB and Pro-Link.

3. Snapshot data recorded with the ignition on and the engine off, engine running stationary, or being driven down the road.
4. Printing analog data, frequency/pulse data, and snapshot data.

The IBOB is another tool that can provide a look at a variety of information. One should keep in mind that the readings reflect raw analog voltages, not processed serial data stream.

Scan tools can be a helpful diagnostic tool, but by no means are they the answer to fixing every driveability complaint. A scan tool is only as good as the computer it is connected to, the information available from the data stream, and the person interpreting the data.

CHAPTER 8

Confirming Air/Fuel Control

IMPORTANCE OF AIR/FUEL CONTROL

In order for the three-way catalytic converter to control HC, CO, and NO_x, the air/fuel ratio must remain in a *window* of approximately 14.7 to 1.

As seen in Figure 8-1, NO_x is reduced when the air/fuel ratio is on the rich side. If the air/fuel ratio moves to the lean side, CO and HC are reduced.

As the air/fuel ratio swings above and below 14.7 to 1, it serves two purposes:

1. As the air/fuel ratio swings to the lean side, NO_x is created, and the excess oxygen is stored in the catalytic converter. A limited amount of HC and CO oxidation occurs at this time.
2. When the air/fuel ratio swings to the rich side, the carbon monoxide produced helps to reduce NO_x because the rich air/fuel has a cooling effect on the exhaust gases. The oxygen previously stored in the catalytic converter is released to oxidize HC and CO.

To convert HC, CO, and NO_x efficiently, the exhaust must be rich and lean at the same time. Since that is not possible, the next best thing is to alternate between rich and lean so the converter can control emissions efficiently.

It's this alternating between rich and lean, called *closed loop*, that allows the three-way converter to control HC, CO, and reduce NO_x.

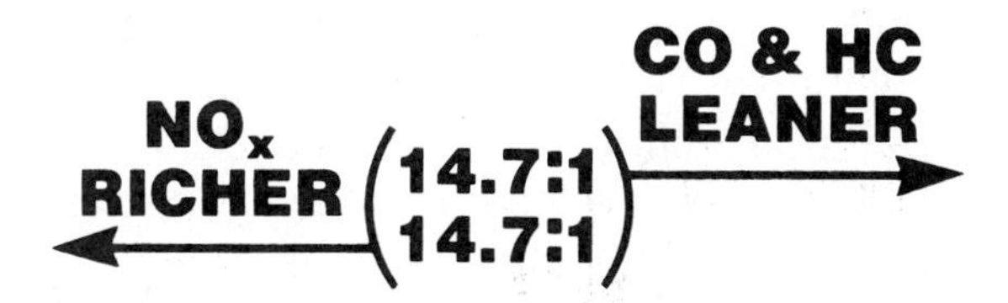

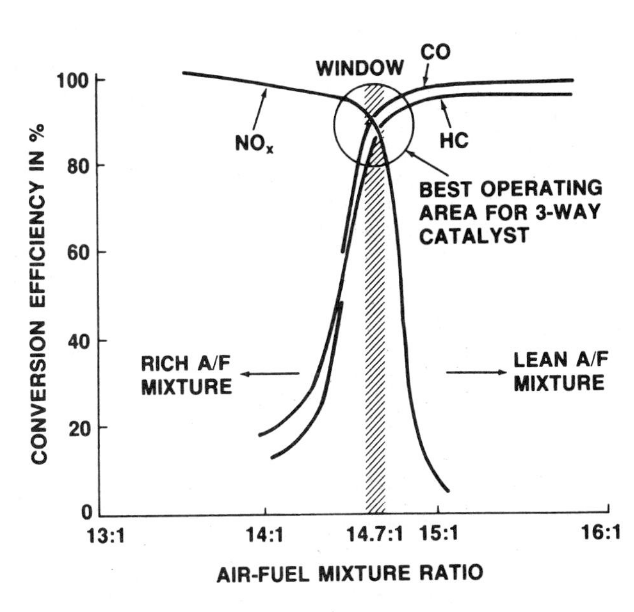

Figure 8-1 Controlling fuel control air/fuel ratios (courtesy GM).

CLOSED-LOOP AIR/FUEL CONTROL

The objective of closed loop is to control the variables that affect combustion and breakdown emissions through a three-way catalyst to render the least possible amount of pollutants into the atmosphere. During closed loop, the computer monitors the oxygen content of the exhaust through the O_2 sensor. It adjusts the air/fuel ratio in response to the oxygen sensor inputs, monitors the adjustments just made, and so on in a continuous cycle.

The heart of a closed-loop system is the oxygen sensor. It is inserted into the exhaust system ahead of the catalytic converter. It generates (zirconium type) or modifies (titania type) a voltage in proportion to the exhaust gas oxygen content. A decrease in oxygen is the result of a rich exhaust, while an increase in oxygen is the result of a lean exhaust. These signals are sent to the computer, which alters the air/fuel ratio by rapidly cycling a fuel metering device, such as a mixture control solenoid, in the carburetor or a fuel injector.

On carbureted engines, the mixture switches from rich to lean 10 times per second to maintain an air/fuel ratio of 14.7 to 1 (stoichiometric). Injector pulse-width (on-time), measured in milliseconds, is increased or decreased to supply the fuel necessary to maintain a 14.7 to 1 ratio.

For the oxygen sensor to produce usable signals, it must be heated to a minimum of 600°F (1-wire or 2-wire non-heated type) or 400°F (3-wire or 4-wire heated type). Until the oxygen sensor can join the feedback system, the engine operates in an *open-loop* mode.

An open-loop mixture is fixed at a rich air/fuel ratio to facilitate easier cold starts and improve warm-up performance and driveability. The initial duration of open loop after starting is determined by inputs from sensors and a timer circuit in the computer.

The computer enters closed loop when some or all of the following conditions have been met:

1. The computer is receiving an RPM signal.
2. The oxygen sensor is providing a signal to the computer.
3. The coolant temperature sensor is above minimum specifications. Some systems enter closed loop at 120°F, while earlier carbureted systems need to reach 180°F before entering closed loop. The minimum temperature requirements will be in the 120°F to 180°F range, depending on the system.
4. The computer's internal clock timer has expired.
   a. Start-up/warm-up mode times will vary (30 seconds to 8 minutes).
   b. Hot restarts can enter closed loop instantly.
5. Vehicle is at cruise, not under load (all vehicles achieve closed loop at light to moderate load at cruise).
6. Vehicle is at curb idle (variable, depending on manufacturer and engine calibration).
7. Gear selector switch is in Drive on some systems.
8. Catalytic converter temperature sensor (if used) is at operating temperature.

Closed-loop operation will be affected by power enrichment and/or wide-open-throttle conditions. These conditions require a rich air/fuel ratio. The computer reverts back to open loop to provide a rich mixture when signals are received from the load sensor and throttle position sensor. Once the enrichment requirements are met and no longer needed, closed-loop operation is resumed.

FEEDBACK CARBURETORS

The installation of an oxygen sensor allowed engineers to redesign the carburetor to meet tough emissions standards. Feedback carburetors allow for rapid corrections to the air/fuel mixture. The computer energizes a solenoid mounted in the carburetor or external of the carburetor (Figure 8-2).

The power circuit of a conventional carburetor is needed to supply additional fuel to the main metering circuit during heavy engine loads or low vacuum conditions. A power valve creates a second path for fuel to leave the float bowl and enter the main metering circuit. The power valve, combined with the fuel from the main jets, provides the additional fuel required for heavy engine loads.

Carburetors that use metering rods (Figure 8-2) instead of a power valve are controlled mechanically or by engine vacuum. As throttle demands increase and vacuum decreases, a tapered metering

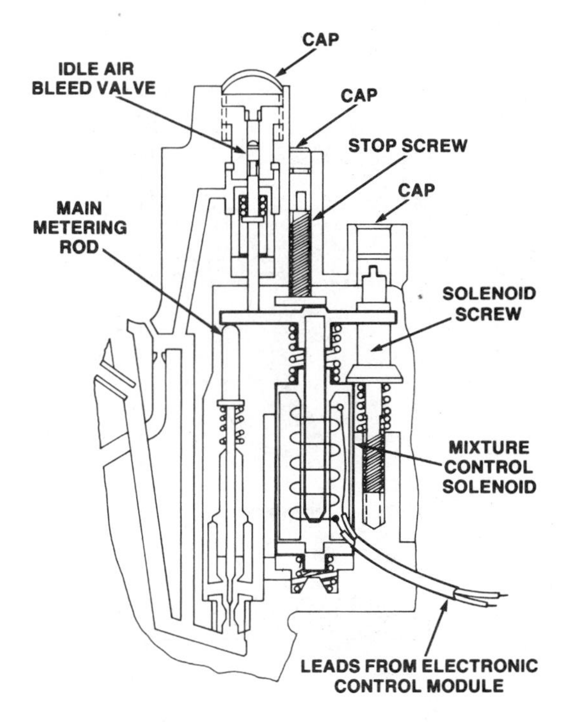

Figure 8-2 Internally-mounted mixture control solenoid (courtesy GM).

rod lifts up, increasing the size of the main jet. The additional fuel fills the main well to supply additional fuel to the main metering circuit (Figure 8-3).

The feedback carburetor uses a mixture control solenoid or fuel control solenoid (Figure 8-3) to regulate the additional fuel leaving the float bowl. The solenoid is energized and de-energized at a rate of 10 times per second. This is called a *duty-cycle.* Since the solenoid is normally open, energizing it creates a lean air/fuel mixture and de-energizing it forces the mixture to go full rich.

All duty-cycle solenoids have two wires (Figure 8-4). Power is supplied to the solenoid through the ignition switch and is protected by either a fuse or fusible link. The ground side of the solenoid is controlled inside the computer by using a transistor (driver). All feedback solenoids will have some type of resistance through their coils. It is important to check the resistance when performing diagnostic procedures. An open solenoid coil will drive the system full rich, and a shorted solenoid coil will drive the system full lean, and possibly damage the computer.

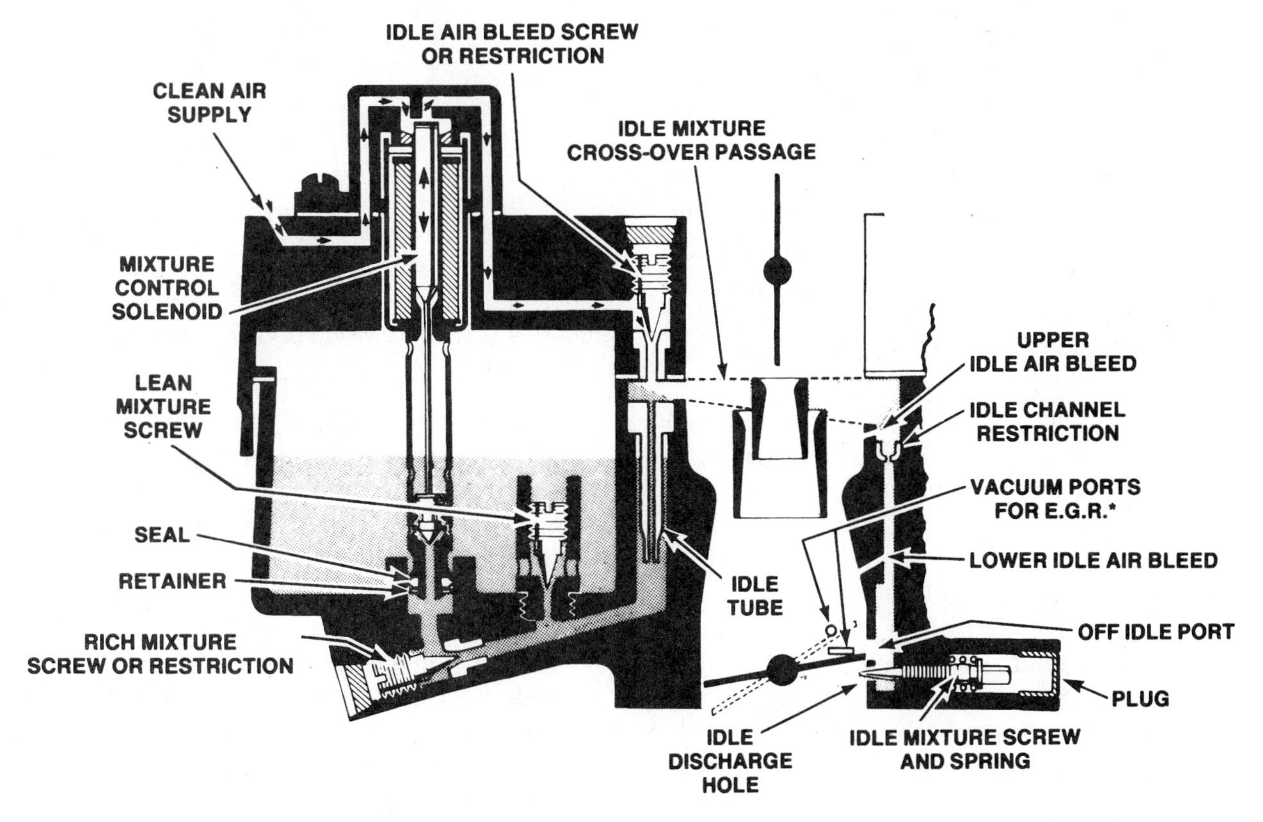

Figure 8-3 The mixture control solenoid replaces the power valve (courtesy GM).

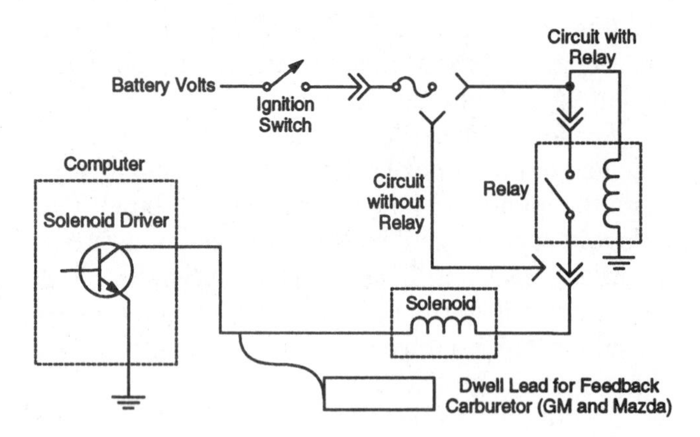

Figure 8-4 Typical feedback carburetor solenoid circuit.

MEASURING "DUTY-CYCLE"

The duty-cycle can be measured in a number of ways:

1. The most common method is to use an ignition dwellmeter. GM and Mazda set the standard for using this method by providing a test connector (Figure 8-4). The connector is located near the carburetor, or around the engine compartment. An analog dwellmeter is recommended because response time is faster and needle movement can be easily monitored. The dwellmeter is connected to the ground side of the circuit, between the solenoid and the computer (this method will work on most other feedback carburetors). The dwellmeter should be placed on the six-cylinder scale because it provides a smooth, gradual needle movement. Other scales can be used, but they may produce very small or very large needle swings.
2. Since a dwellmeter is nothing more than a voltmeter with a recalibrated scale, it is possible to use an analog or digital voltmeter (Figure 8-5). The meter connections are the same as a dwellmeter. Chrysler and some imports recommend this method for testing their systems.
3. On some Ford systems, the duty-cycle can be read using a vacuum gauge teed into the fuel control solenoid vacuum line that goes to the power piston inside the carburetor. On these systems, the solenoid is mounted on the fender or firewall. The system is working correctly if the vacuum gauge is rapidly fluctuating between 2" and 3". A lean exhaust condition will show a 0" to 2" vacuum reading. A rich exhaust condition will show a 4" to 5" vacuum reading.
4. A scan tool can be used on GM vehicles to monitor carburetor dwell. Since the scan tool receives serial data from the computer, it does not show the wide sweeps of the dwell.
5. Some digital multimeters (Figure 8-5) read duty-cycle and dwell. They also include a minimum/maximum record feature (Figure 8-6) that allows the technician to record the minimum and maximum readings for a given time. This can be helpful when diagnosing intermittent problems.

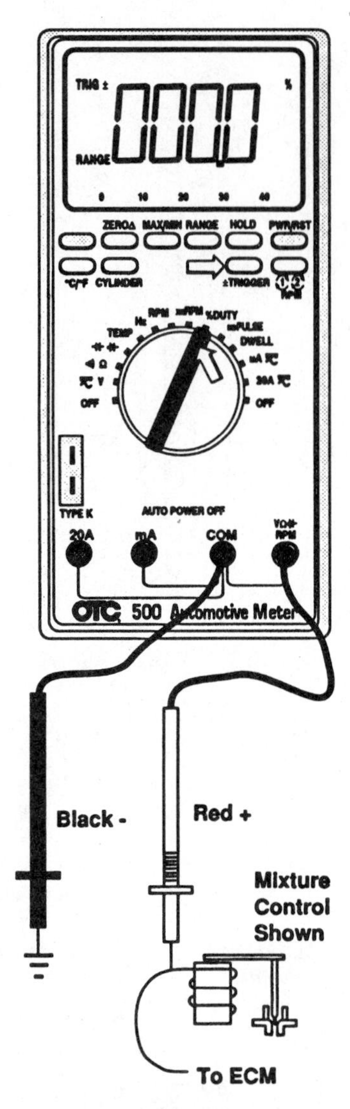

Figure 8-5 Carburetor duty-cycle or dwell can be measured using a digital Volt-Ohm Meter (VOM) (courtesy OTC, a division of SPX Corporation).

VEHICLE PREPARATION

Before attempting to confirm closed-loop operation, make sure that the engine is at operating temperature and has been run for a minimum of two minutes at 2000 RPM. This will insure that the oxygen sensor is at operating temperature.

If the engine is equipped with air injection and a line to the catalytic converter, be sure the computer has switched the air downstream to the converter. This can be done by squeezing the hose and feeling for pulsations of air. If equipped with an air-injection system without a line to the catalytic converter, the air will be diverted to the air cleaner or atmosphere. To verify this, remove the air cleaner lid and listen for air being pumped (dumped) inside. When air is being diverted to the atmosphere, it can be felt at the diverter valve.

CONFIRMING CLOSED-LOOP OPERATION

Methods to confirm closed-loop operation will vary among the various manufacturers. Not every method can be used on all engines, and it also depends on whether the vehicle has a carburetor or fuel-injection system. The three most common methods include the use of a:

1. Digital Multi-Meter (DMM).
2. *Dwellmeter* — Analog or digital.
3. Scan tool.

DMM – Carbureted Systems

To use a DMM, connect a short jumper wire in series at the O_2 sensor's connector to the main harness. Connect the DMM positive lead to the jumper wire and the negative lead to the battery negative terminal and set your meter on DC volts (Figure 8-7).

Data Record (MAX/MIN)

The Data Record feature stores the highest or lowest reading in memory.

- **First, connect the meter probes to the test points. Then, press the MAX/MIN button once to start MIN recording. The minimum reading will be displayed.**
- **Press the MAX/MIN button twice to start the MAX recording. The maximum reading will be displayed.**
- **Press the HOLD button to stop the recording, press again to restart the recording.**

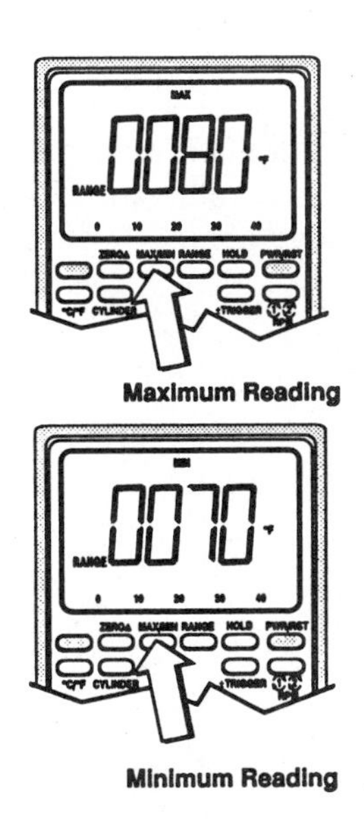

Figure 8-6 Oxygen sensor voltage can be recorded using the Min/Max feature of a DMM (courtesy OTC, a division of SPX Corporation).

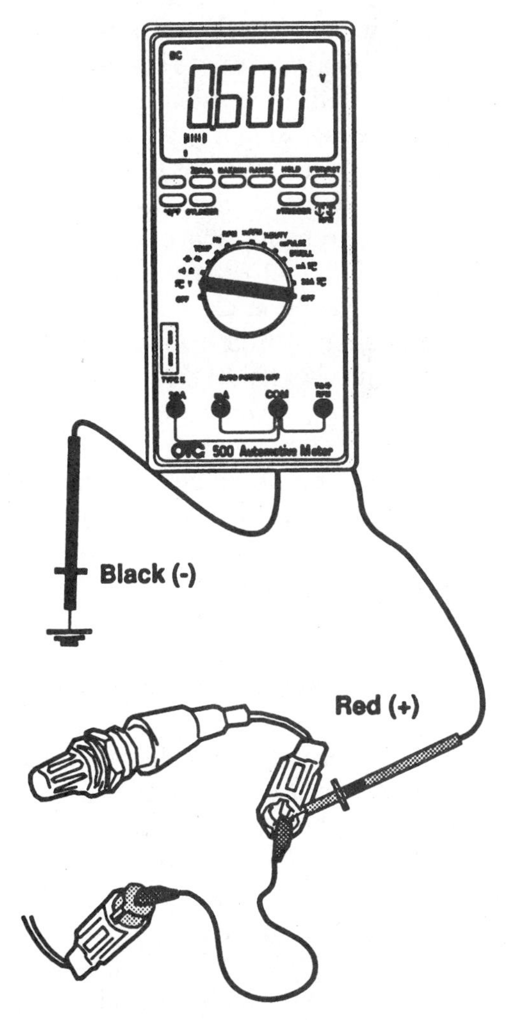

Figure 8-7 Oxygen sensor voltage can be monitored by connecting the multimeter to a jumper wire installed in series (courtesy OTC, a division of SPX Corporation).

If the meter's leads will not reach the battery negative terminal, connect to a good, clean ground under the hood.

Run the engine at 2500 to 3000 RPM. If the system is in closed loop and switching correctly, the voltage should swing rapidly above and below 0.5 volts. Ideal readings would be 0.2v (lean) to 0.8v (rich). Watch the voltage closely—if the oxygen sensor is switching between 100mv and 600mv, the system is running on the lean side. If the oxygen sensor is switching between 400mv and 1.0v, the system is running on the rich side. In both cases, the system is making the necessary corrections, but it is having to overcompensate to achieve the desired air/fuel ratio.

When the voltage remains under 0.5 volts, create a rich exhaust condition. This is done by restricting the air entering the carburetor. Propane enrichment is also a good way to create the required rich condition. With the system in closed loop, a good oxygen sensor signal should rise above 0.5 volts within 100ms. If O_2 passes this test, then look for something else that could cause a lean exhaust.

If the voltage remains above 0.5 volts, create a lean exhaust condition by creating a large vacuum leak. With the system in closed loop, the oxygen sensor signal should drop below 0.5 volts within 100ms. If the O_2 sensor fails this test, the sensor may be contaminated or poisoned. Before replacing the sensor, find the root cause of the problem.

Remember that a properly-functioning system should be switching between lean and rich approximately 8 times in 10 seconds at 2500 to 3000 RPM. If the ratio is less than this, inspect the oxygen sensor and feedback system.

Dwellmeter – Carbureted Systems

A dwellmeter can be used to monitor closed-loop operation if the carburetor is equipped with a two-wire duty-cycle solenoid. Connect your dwellmeter to the ground side of the solenoid. If a connector is not provided, backprobe the ground side of the solenoid and connect your meter.

Using the dwellmeter, select dwell and the six-cylinder scale. This is the industry standard, but any number of cylinders will work. Run the engine at 2500 to 3000 RPM. If the vehicle is in closed loop and switching correctly, the dwell reading will drift back and forth in 2°-10° increments, somewhere between 6° and 54°. A dwellmeter needle floating between 25° and 35° is considered ideal (Figure 8-8).

The farther away the dwell moves from 30° (Figure 8-9), the larger the correction to maintain fuel control. If the dwell reading is below 30°, the computer is adding fuel to compensate for a lean oxygen sensor signal. If the dwell reading is above 30°, the computer is subtracting fuel to compensate for a rich oxygen sensor signal.

If the dwellmeter reads 6° or less (Figure 8-10), the computer is not in control of the fuel because of an excessively lean exhaust condition. If the dwellmeter reads 54° or above (Figure 8-11), the computer is not in control because of an excessively rich exhaust condition.

Run the engine at 2500 to 3000 RPM. If the dwell remains fixed under 30°, create a rich condition. This can be accomplished by restricting the air entering the carburetor. Another technique is to

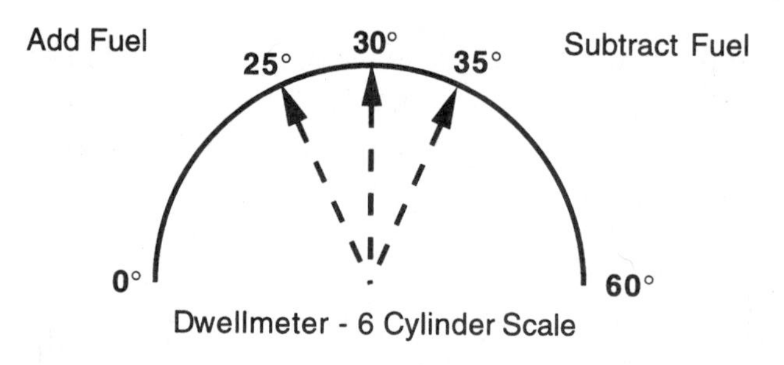

Figure 8-8 In closed loop the dwell should drift in 2°-10° increments. (The dashed lines represent needle movement.)

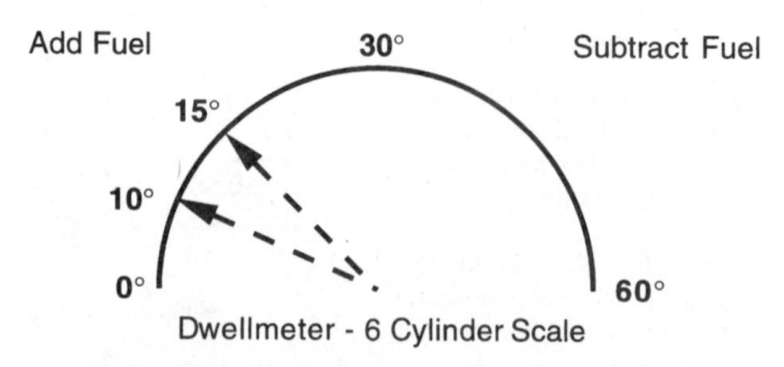

Figure 8-9 This reading shows a closed-loop dwell, but the computer is having to add additional fuel to maintain a 14.7:1 air/fuel ratio.

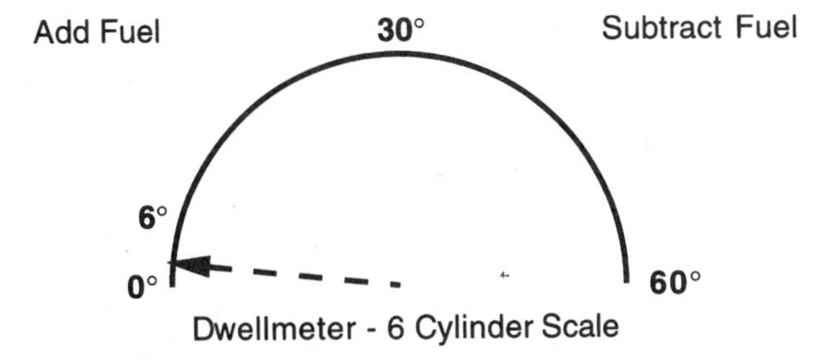

Figure 8-10 A dwell reading under 6° shows the computer is not in fuel control because of an excessively lean condition.

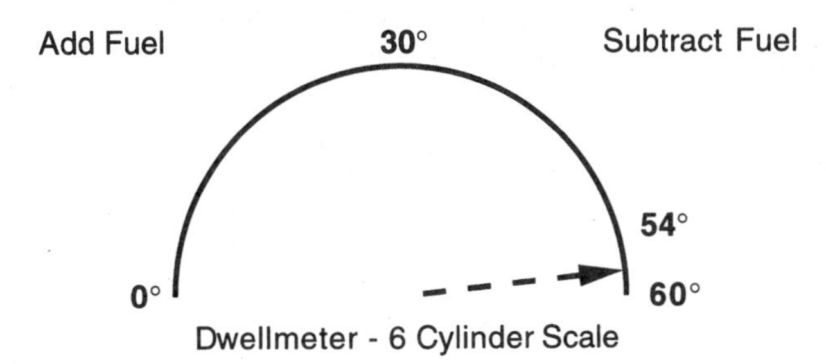

Figure 8-11 A dwell reading above 54° shows the computer is not in fuel control because of an excessively rich condition.

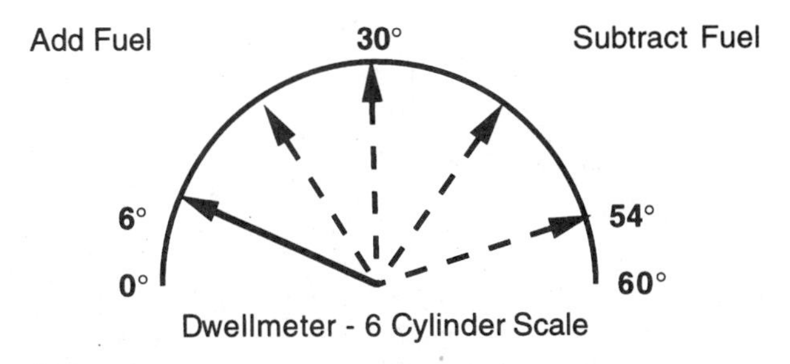

Figure 8-12 Creating a rich condition forces the dwell to increase. This verifies that the feedback system is working correctly.

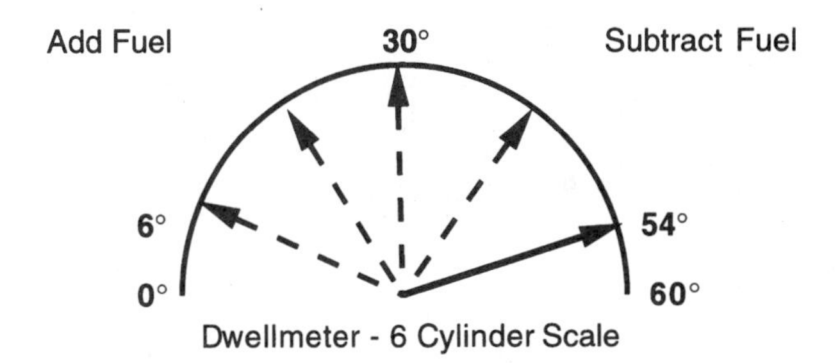

Figure 8-13 Creating a lean condition forces the dwell to decrease. This verifies that the feedback system is working correctly.

slowly inject propane into the carburetor. The dwell should move toward 60° if the feedback system is functioning correctly (Figure 8-12).

Run the engine at 2500-3000 RPM. If the dwell remains fixed above 30°, create a lean exhaust condition. This can be accomplished on most engines by creating a large vacuum leak. The dwell should move toward 0° if the feedback system is functioning correctly (Figure 8-13).

If the vehicle enters closed loop at idle, all of the checks previously discussed can be performed. On a carbureted engine, the response time at idle will be slightly slower than at cruise, but the same results should be achieved.

Scan Tool – Carbureted Systems

There are a few ways to confirm closed-loop operation using a scan tool. A scan tool can be used to read mixture control (MC) dwell, oxygen sensor voltage and cross count, and the rich/lean status of the exhaust. The scan tool will have a slower response time depending on the refresh rate of serial data. This is because the oxygen sensor signal is sent to the computer, then the computer processes the information and sends an output command to the carburetor. At the same that it sends a command to the carburetor, it sends this information through the serial data line to the scan tool. Be aware of this to avoid problems when diagnosing the system using a scan tool (Figure 8-14).

To confirm closed loop, read the oxygen sensor voltage on the scan tool. Perform the same tests as with the DMM or dwellmeter.

Confirming closed loop on a GM system can be accomplished using oxygen sensor cross counts. A cross count is the number of times that the oxygen sensor switches between rich and lean. When using cross counts, the technician should look for activity. If the number is jumping around between 2 and 20, then the oxygen sensor is switching okay (Figure 8-14). A high steady number above 20 is an indication of an engine misfire because the system is having to overcompensate for a problem.

```
84VZ2804                 E-Z EVENT PLUS
G.M. 1984 FULL VIN Z
EVENT 4    TAG --     FRAME -30
DIAGNOSTIC STATE         8 ALCL
TROUBLE CODES: NO
```

	-30	-29	-28	-27	-26	-25	-24	-23	-22	-21	-20	-19	-18	-17	-16	-15
RPM	2525	2550	2550	2625	2925	3100	3150	3175	3150	3100	3050	3050	3025	3050	3050	3050
O2 VOLTS	0.80	0.21	0.29	0.78	0.51	0.75	0.67	0.69	0.71	0.69	0.38	0.50	0.81	0.78	0.81	0.77
O2 CROSS COUNTS	007	004	007	003	008	002	005	005	007	005	003	007	009	004	003	002
O2 FUEL STATE	RICH	LEAN	LEAN	RICH	RICH	RICH	RICH	RICH	RICH	RICH	LEAN	RICH	RICH	RICH	RICH	RICH
M/C SOLENOID DWEL	30 D	28 D	27 D	28 D	26 D	29 D	28 D	29 D	29 D	26 D	27 D	27 D	23 D	25 D	27 D	30 D

	-15	-14	-13	-12	-11	-10	-9	-8	-7	-6	-5	-4	-3	-2	-1	0
RPM	3050	3025	3000	3050	3075	3075	3050	3075	3075	3075	3050	3100	3050	3050	3050	3075
O2 VOLTS	0.77	0.81	0.72	0.73	0.69	0.72	0.69	0.42	0.55	0.73	0.55	0.79	0.77	0.73	0.72	0.55
O2 CROSS COUNTS	002	004	001	004	006	006	008	007	004	002	006	002	002	003	004	011
O2 FUEL STATE	RICH	RICH	RICH	RICH	RICH	RICH	RICH	LEAN	RICH	RICH	RICH	RICH	RICH	RICH	RICH	RICH
M/C SOLENOID DWEL	30 D	28 D	32 D	31 D	32 D	32 D	31 D	26 D	26 D	28 D	28 D	30 D	33 D	37 D	35 D	32 D

	0	1	2	3	4	5	6	7	8	9	10	11	12	13	14	15
RPM	3075	3075	3075	3075	3075	3125	3100	3050	3075	3100	3125	3100	3100	3100	3075	3125
O2 VOLTS	0.55	0.23	0.67	0.78	0.61	0.67	0.70	0.70	0.39	0.58	0.51	0.36	0.79	0.72	0.65	0.60
O2 CROSS COUNTS	011	004	004	006	004	000	003	006	006	009	006	004	002	004	002	004
O2 FUEL STATE	RICH	LEAN	RICH	RICH	RICH	RICH	RICH	RICH	LEAN	RICH	RICH	LEAN	RICH	RICH	RICH	RICH
M/C SOLENOID DWEL	32 D	29 D	28 D	29 D	29 D	33 D	36 D	35 D	33 D	31 D	27 D	26 D	30 D	33 D	33 D	33 D

	15	16	17	18	19	20	21	22	23	24	25	26	27	28	29	30
RPM	3125	3100	3075	3100	3075	3125	3100	3125	3100	3075	3100	3125	3150	3150	3225	3150
O2 VOLTS	0.60	0.33	0.30	0.51	0.74	0.71	0.73	0.69	0.69	0.39	0.58	0.55	0.65	0.77	0.80	0.76
O2 CROSS COUNTS	004	005	006	008	004	003	002	004	000	007	010	005	005	011	000	002
O2 FUEL STATE	RICH	LEAN	LEAN	RICH	RICH	RICH	RICH	RICH	RICH	LEAN	RICH	RICH	RICH	RICH	RICH	RICH
M/C SOLENOID DWEL	33 D	35 D	34 D	29 D	29 D	30 D	33 D	35 D	38 D	39 D	34 D	34 D	33 D	27 D	30 D	32 D

Figure 8-14 Measuring the feedback system using an OTC Monitor 2000 or 4000.

If the cross count is zero at idle, raise the engine RPM to 2500 to 3000 RPM. If the oxygen sensor activity is within the 2 to 20 range, the system is in closed loop and functioning correctly. Return to idle and recheck the counts. If they remain near zero, then the system is in open loop or the oxygen sensor may be contaminated, lazy, or bad. Perform the checks for a lean or rich exhaust condition and watch the response of the oxygen sensor.

Another method is to look at the red Light Emitting Diodes (LEDs) on the front of the scan tool. Some scan tools have a red LED that stays on when the system is in closed loop and another that flashes on/off in response to the exhaust condition. When the light is on, this indicates a high oxygen sensor voltage. If the light is off, this indicates a low oxygen sensor voltage or failed sensor.

Port Fuel-Injection Timing and Pulse-Width

To understand the testing of air/fuel control on fuel-injected systems, two concepts should be understood:

1. Injection timing.
2. Control of injector pulse-width.

Injection timing control determines when each injector will deliver fuel to its corresponding intake port. There are three different methods (Figure 8-15) of injector timing, depending on the application. These methods include:

1. *Simultaneous Injection*—All injectors are pulsed simultaneously by a common driver circuit. Injection occurs once per crankshaft revolution

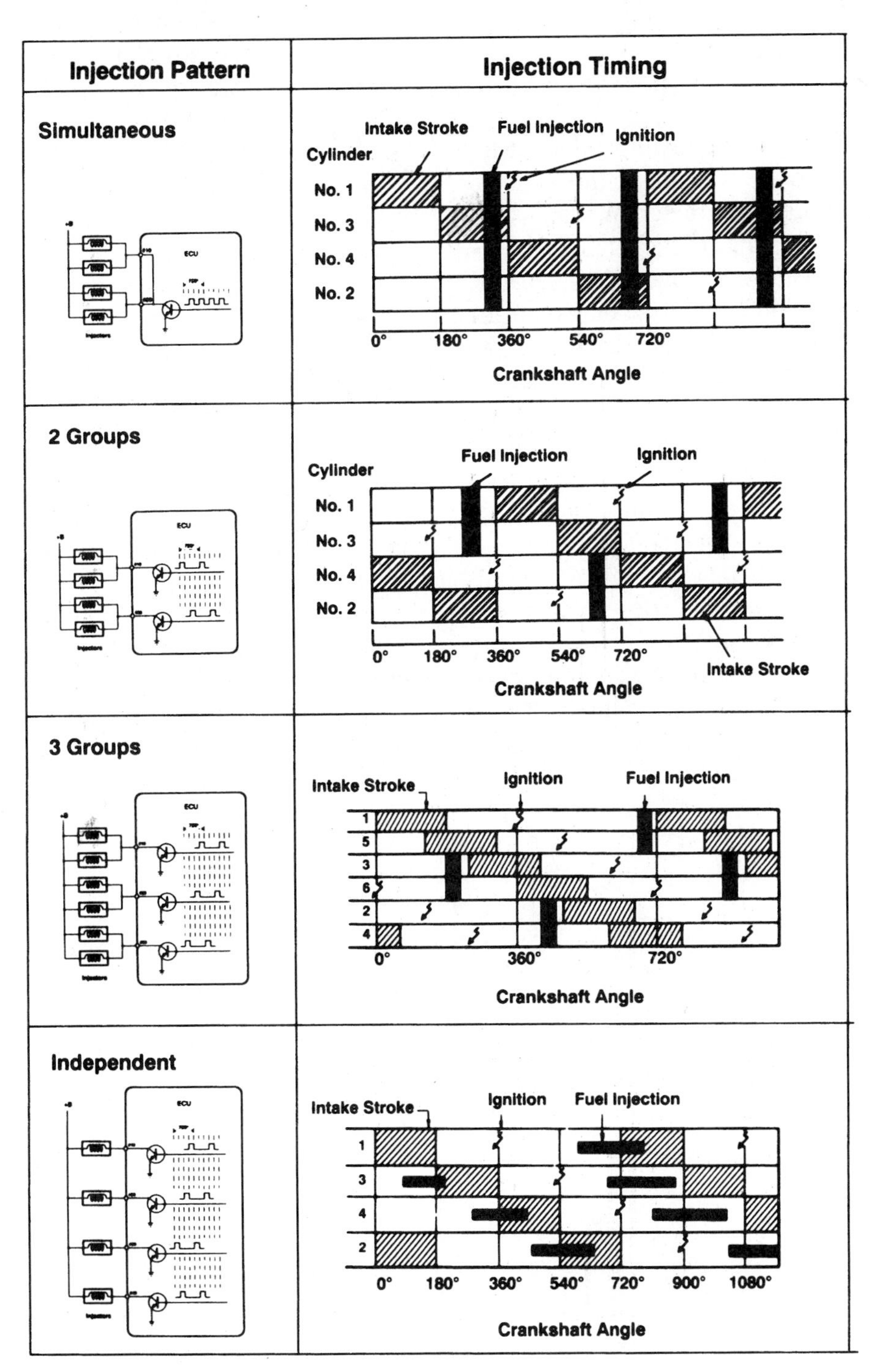

Figure 8-15 Types of injection control.

just before the crankshaft reaches TDC of cylinder #1. This means that twice per engine cycle, one-half of the calculated fuel is delivered by the injectors. This is the simplest method of injection timing in use.

2. *Grouped Injection* — Injectors can be in groups of two, three, or four. The injectors can be grouped:
   a. By pairing two consecutive cylinders in the firing order.
   b. Based on the position of *running mate* cylinders—these are the cylinders in the same position in relationship to the crankshaft.
   c. By left and right banks of the engine.

   Each group of injectors is driven by a separate driver circuit. Injection is timed to deliver fuel immediately preceding the intake stroke for the leading cylinder in the group. The entire group is pulsed once per engine cycle.
3. *Sequential Injection* — Injectors are driven independently and sequentially by a separate driver. Injection is timed to deliver the entire fuel charge just before each intake valve opening.

Injector pulse-width control is determined by a three-step process. The first step involves the calculation of the base pulse-width. The computer calculates the base pulse-width based upon engine speed (RPM) and air flow volume (load). These two inputs together establish the engine load.

The second step involves how corrections are made to the base pulse-width. Once the base pulse-width is calculated, the computer must make corrections to the pulse-width based on changing variables. Variables considered in the correction calculation are the coolant and intake air temperature, throttle position, and oxygen sensor feedback signal when operating in closed loop. The changes that might occur are:

1. As the engine and intake air temperatures move from cold to warm, injector pulse-width is reduced.
2. As the throttle opens, injection pulse-width is momentarily increased or the frequency of injection is increased. On deceleration, the injector pulse-width can be shut off.
3. Fuel injector pulse-width will swing back and forth longer and shorter durations to correct conditions detected by the exhaust oxygen sensor.

On some vehicles, the final step in the process includes battery voltage correction. To determine final injector pulse-width, the computer corrects for injector opening delay using a battery voltage correction factor. The battery correction factor increases injector pulse-width as battery voltage falls because of the injector opening delay that varies with the charging systems voltage as shown in Figure 8-16.

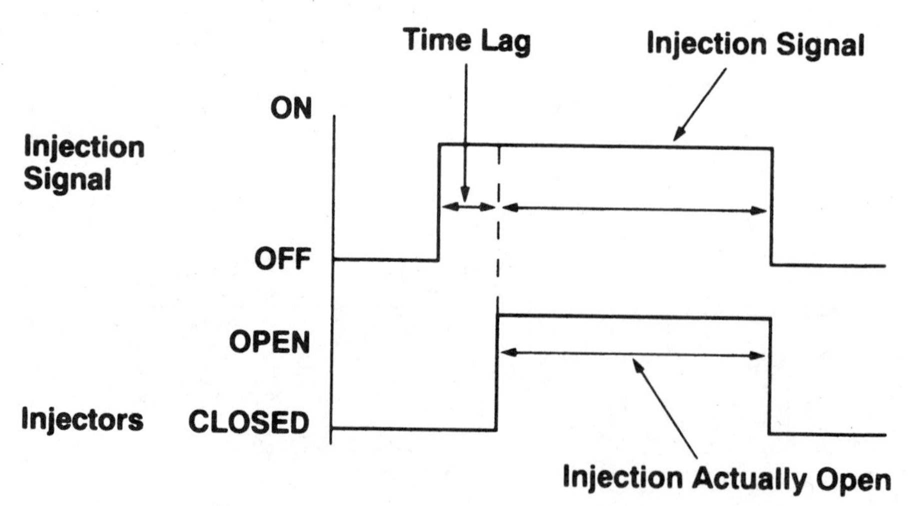

Figure 8-16 Low battery voltage will affect injector opening time.

DMM – Fuel-Injected Systems

To use a DMM, connect a short jumper wire in series at the O_2 sensor's connector to the main harness, or use the diagnostic connector located under the hood on some Japanese vehicles. Connect the DMM positive lead to the jumper wire and the negative lead to the battery negative terminal and set your meter on DC volts. If the meter's leads will not reach the battery negative terminal, connect to a good, clean ground under the hood.

Run the engine at 2500 to 3000 RPM. If the voltage remains under 0.5 volts, create a rich exhaust condition. For a fuel-injected system, propane enrichment is the best method to create the required rich condition. The fuel pressure regulator hose can be removed and plugged to create a rich condition, but the computer will usually compensate for this condition within a few seconds. If the system is in closed loop and functioning correctly, the oxygen sensor signal should rise above 0.5 volts within 100ms (Figure 8-17).

If the voltage remains above 0.5 volts, create a lean exhaust condition. This can be accomplished on most engines by disconnecting the power brake booster hose or PCV hose. Removing a fuel injector's electrical connector will also create a lean condition. If the system is in closed loop and functioning correctly, the oxygen sensor signal should drop below 0.5 volts within 100ms (Figure 8-17).

A properly-functioning system should be switching between lean and rich approximately eight times in ten seconds at 2500 to 3000 RPM. If the ratio is less, inspect the oxygen sensor and fuel-injection system.

Toyota provides another way to verify fuel control on fuel-injected systems. Using the under-hood diagnostic connecter (Figure 8-18) and a voltmeter, voltage feedback (Vf) is used to test the oxygen sensor switching rate. The feedback command from the computer to the fuel injectors can also be monitored.

Connecting a jumper wire across Terminal T or TE1 to E1 allows the voltage signal on Vf to imitate the oxygen sensor signal (Figure 8-18). At 2500 RPM, oxygen sensor switching should occur eight times in ten seconds if the closed-loop system is

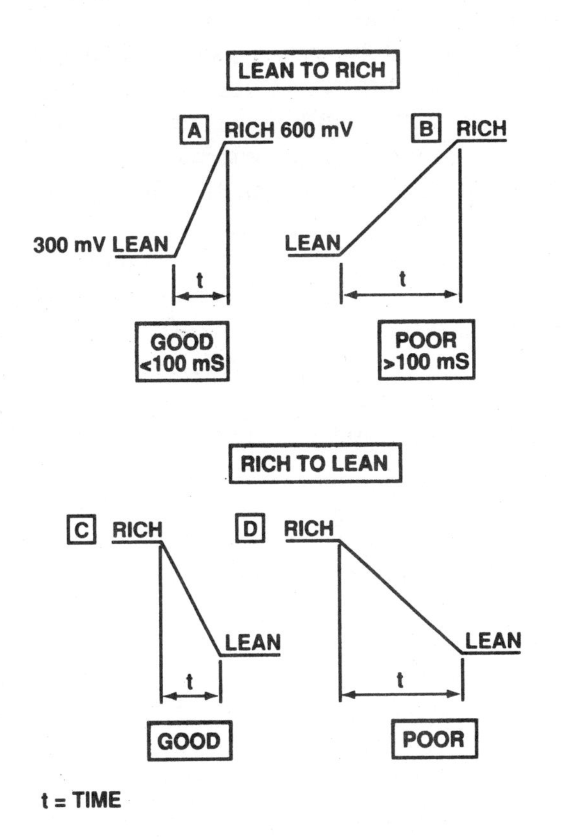

Figure 8-17 A good oxygen sensor is capable of responding rapidly to changing exhaust oxygen conditions (courtesy GM).

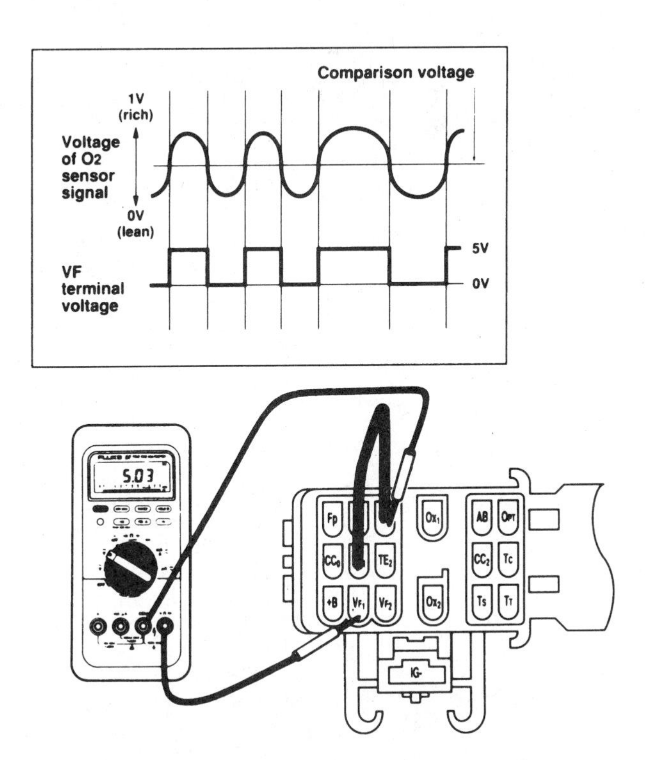

Figure 8-18 Using the under-hood diagnostic connector allows for the testing of the oxygen sensor switching rate.

operating normally. Whenever the oxygen sensor signal is high, indicating a rich exhaust condition, the Vf voltage will be 5 volts. When the oxygen sensor signal is low, indicating a lean exhaust condition, the Vf voltage will be 0 volts.

The Vf terminal also provides information about the amount of fuel correction the computer is making while the engine is running. The computer compares the amount it is correcting the air/fuel ratio to a set of values in memory. If the amount of correction is within its normal range, the computer puts approximately 2.5 volts at the Vf terminal (Figure 8-19).

If the computer is subtracting fuel because of a rich exhaust condition, it will output a lower voltage (either 1.25v or 0v) to the Vf terminal. If the computer is adding fuel because of a lean exhaust condition, it will output a higher voltage (either 3.75v or 5v) to the Vf terminal.

The voltage moves in steps depending on the type of load sensor used, as shown in Figure 8-17. When a system is equipped with a MAP (D-Jet), three different voltage corrections are possible: 5v, 2.5v, and 0v. When a system is equipped with an air flow meter (L-Jet), five different voltage corrections are possible: 5v, 3.75v, 2.5v, 1.25v, and 0v.

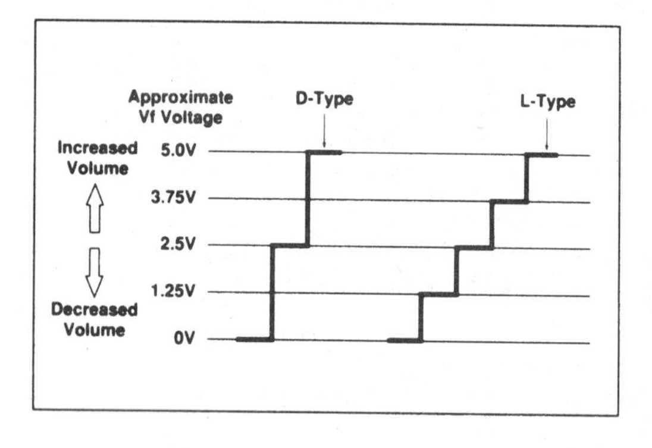

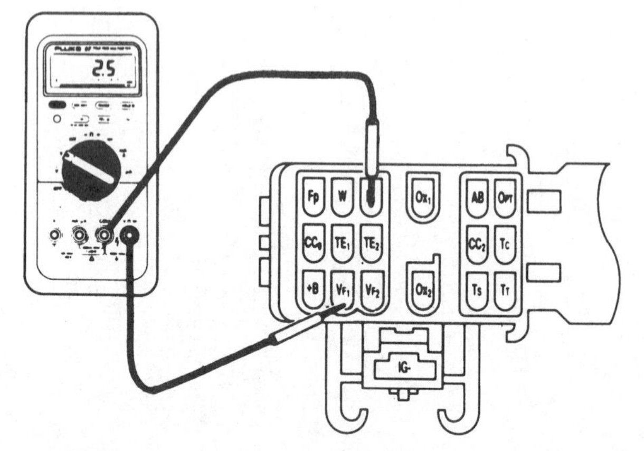

Figure 8-19 Corrections to the air/fuel ratio may be monitored at the Vf terminal of the diagnostic connector.

Scan Tool – Fuel-Injected Systems

The dwellmeter used for carbureted systems is of no use on a fuel-injected system. A scan tool is an excellent way to monitor the amount of fuel correction being made to a fuel-injected engine.

Currently, GM, Ford, and Chrysler display fuel correction data on a scan tool. They all use different terms for fuel correction, such as: Block Learn and Block Integrator; Short-Term Fuel Trim and Long-Term Fuel Trim; and Adaptive Memory and Additive Fuel Factor. Whatever the name, they all indicate the amount of correction being made to the air/fuel ratio.

Block Integrator (INT) has been used on General Motors fuel-injected engines since 1982. Block Integrator represents a short-term correction to the amount of fuel delivered to the engine during closed-loop operation. It responds to the amount of time the oxygen sensor spends above or below the 450mv threshold. If the oxygen sensor voltage remains mostly below 450mv, indicating a lean exhaust condition, the Block Integrator will increase to add fuel. If the oxygen sensor voltage remains mostly above 450mv, indicating a rich exhaust condition, the Block Integrator will decrease to subtract fuel.

The data is displayed on the scan tool as a digital count from 0 to 255. The average value is 128. The farther the number moves away from 128, the larger the correction to the air/fuel ratio. When the number is above 128, the computer is making a short-term correction to the air/fuel ratio by adding fuel. When the number is below 128, the computer is making a short-term correction to the air/fuel ratio by subtracting fuel (Figure 8-20). The computer makes these adjustments by increasing or decreasing injector pulse-width.

Block Integrator is part of the volatile RAM in the CPU. Turning off the ignition switch will cause these learned values to be lost and the Block Integrator reverts to a neutral setting of 128. Each time the vehicle enters closed loop, the Block Integrator must relearn from the oxygen sensor.

Although the Block Integrator can correct the air/fuel ratio over a wide range, it is only for a short-term correction. Therefore, *Block Learn (BLM)* was added to the system to make long-term corrections.

```
90CL3802                 E-Z EVENT PLUS
G.M. 1990 3800 VIN C                                                    2  of  2  pages
EVENT 2   TAG --     FRAME -25
DIAGNOSTIC STATE        9 ROAD
TROUBLE CODES: NO
CURRENT CODES: NO

                    5    6    7    8    9    10   11   12   13   14   15   16   17   18   19
RPM                0950 0950 0975 1000 0975 0950 0950 0975 0950 0950 0975 0950 0950 0975 0975
INTEGRATOR         129  129  128  131  127  126  130  130  126  126  127  127  128  132  128
BLOCK LEARN MULT   136  136  136  136  136  136  136  136  136  136  136  136  136  136  136
BLM CELL           01   01   01   01   01   01   01   01   01   01   01   01   01   01   01
INJ PULSE WIDTH    06.0 05.6 05.5 05.7 05.2 05.1 05.4 05.2 05.2 05.3 05.2 05.3 05.2 05.3 05.1


                    20   21   22   23   24   25

RPM                0950 1000 0975 0950 1000 0975
INTEGRATOR         128  132  132  128  132  132
BLOCK LEARN MULT   136  136  136  136  136  136
BLM CELL           01   01   01   01   01   01
INJ PULSE WIDTH    05.2 05.4 05.2 05.2 05.2 05.3
```

Figure 8-20 Typical Block Integrator and Block Learn readings captured by an OTC Monitor 2000 or 4000.

Block Learn is part of the non-volatile RAM in the CPU, which means that these values remain in memory when the ignition switch is turned off. Once the engine is restarted and enters closed loop, fuel delivery is based on the learned values stored in memory.

The operating range of the engine for any combination of RPM and load is divided into 16 cells or blocks (Figure 8-21). When the engine operates under any combination of RPM and load, the system will operate in one of these blocks. Stored in each cell are the learned values the computer registered the last time that RPM and load were experienced. Once a particular RPM/load combination is achieved, the computer will automatically add or subtract fuel depending on the value stored in memory.

If the engine always operated at a stoichiometric air/fuel ratio, then the Block Learn and Block Integrator would read 128. Unfortunately, as mentioned in Chapter 1, the ideal engine is only a dream, and the real-world engine is reality. Constant corrections are necessary to maintain a smooth, clean running engine.

Block Learn and Block Integrator should be checked during closed-loop operation. They should be checked in 500 RPM increments from idle up to 3000 RPM. This will help to identify the difference between a small vacuum leak at idle that does not affect off-idle. It can also help to identify a dirty fuel injector or ignition misfire that might affect all driving ranges. If the technician follows a few simple rules, Block Learn and Block Integrator can become an excellent diagnostic tool. These rules are as follows:

1. If the vehicle is operating in closed loop and both readings are in the 128 to 132 range, then the system is working okay.

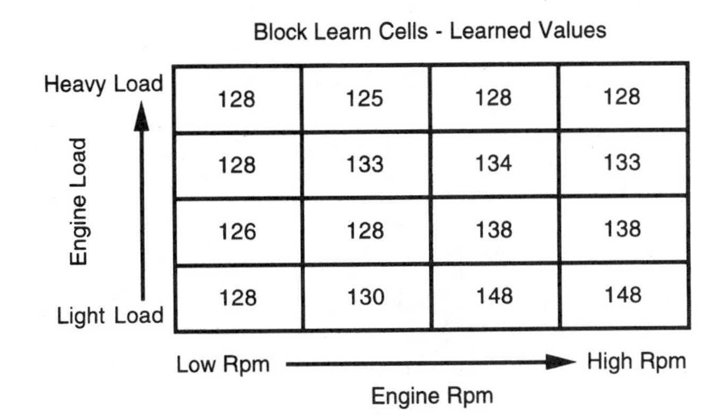

Figure 8-21 Long-term fuel corrections are retained in the Block Learn cells.

2. If the Block Integrator is between 128 and 132, but the Block Learn is higher (above 132) or lower (below 128) than the integrator, the computer is effectively correcting for a small problem.
3. If the Block Learn and Block Integrator are both above 132, then the computer is attempting to correct for a lean exhaust condition. The farther away from 132, the more severe the problem may be.
4. If the Block Learn and Block Integrator are both below 128, then the computer is attempting to correct for a rich exhaust condition. The farther away from 128, the more severe the problem may be.
5. If the Block Learn and Block Integrator are at opposite ends of the scale (the commands are going in opposite directions), the computer and circuit grounds should be checked.

Both Block Integrator and Block Learn have limits to the amount of correction they can make. If the mixture is off enough that Block Learn reaches its limit of control, and still can not correct the condition, the integrator will also go to its limit in the same direction and the engine would run poorly. The high/low limits for each one will vary, but a good range to use is 100 to 150. The following are examples of Block Learn and Block Integrator readings:

- *Vehicle #1* — These are normal readings in closed loop at idle and at cruise.

  Engine RPM: 550
  Engine RPM: 2550
  Block Integrator: 129
  Block Integrator: 128
  Block Learn: 130
  Block Learn: 127

- *Vehicle #2* — A vacuum leak at idle forces the Block Learn to add fuel to bring the Block Integrator back to 129. Testing the vehicle at higher RPM shows that the problem is not major because the readings move back to normal.

  Engine RPM: 650
  Engine RPM: 2450
  Block Integrator: 130
  Block Integrator: 127
  Block Learn: 140
  Block Learn: 128

- *Vehicle #3* — The readings show that the Block Learn has reached its lower limit at idle and only partially corrects for a rich exhaust condition. High fuel pressure, a leaking fuel pressure regulator, or leaking injectors might cause this problem because the condition exists at low and high speeds.

  Engine RPM: 575
  Engine RPM: 3050
  Block Integrator: 100
  Block Integrator: 119
  Block Learn: 100
  Block Learn: 115

- *Vehicle #4* — The readings show that the Block Learn has approached its upper limit and is unable to correct for a lean exhaust condition. Restricted or non-operation fuel injectors, engine misfire, or large vacuum leaks might cause this problem because the condition exists at low and high speeds.

  Engine RPM: 675
  Engine RPM: 3250
  Block Integrator: 145
  Block Integrator: 140
  Block Learn: 145
  Block Learn: 143

- *Vehicle #5* — In this case, the Block Integrator shows a lean condition and the Block Learn is commanding the system to subtract fuel. A bad computer or system ground might cause this condition.

  Engine RPM: 675
  Engine RPM: 1765
  Block Integrator: 145
  Block Integrator: 140
  Block Learn: 115
  Block Learn: 118

Short-Term Fuel Trim (SFTRIM) and *Long-Term Fuel Trim (LFTRIM)* used by Ford perform the same jobs as Block Integrator and Block Learn on a GM. The difference between them is that Short-Term Fuel Trim and Long-Term Fuel Trim are displayed in percentage (%) of correction. The neutral (ideal) percentage is 0%. If the oxygen sensor indicates a lean exhaust condition, the computer will attempt to make short-term correction by increasing the percentage of fuel. This is indicated by a SFTRIM reading of a positive percentage (0 to

+50%). The opposite is true of a rich exhaust condition; in that case, the correction percentage will be a minus percentage (0 to -50%). If the SFTRIM is unable to correct for the rich condition, then the computer will make a LFTRIM correction in an attempt to bring the SFTRIM back to 0%.

Figure 8-22 shows an example of a problem that had just occurred when this data was captured. A high O_2S1 voltage indicates a rich exhaust condition. The SFTRIM begins subtracting fuel but the LFTRIM has not yet begun to make a long-term correction.

On Chrysler systems, the computer has two oxygen feedback memory functions that affect injector pulse-width. They are called Adaptive Memory and Additive Adaptive Memory. The computer looks at its own memory of feedback operation to determine the adaptive memory factor. Adaptive memory is a correction factor or an updating of the memory cells based on feedback information. It is factored into the injector pulse-width calculation.

Each memory cell is based on RPM and load. Each cell stores feedback information about the system when the engine operates within the parameters of a given cell. If the cell stores any information beyond the neutral feedback state, the correction is figured into the pulse-width calculations. The correction will be a positive (lean condition) or negative (rich condition) percentage depending on the condition. It will appear on the scan as a percentage (Figure 8-23).

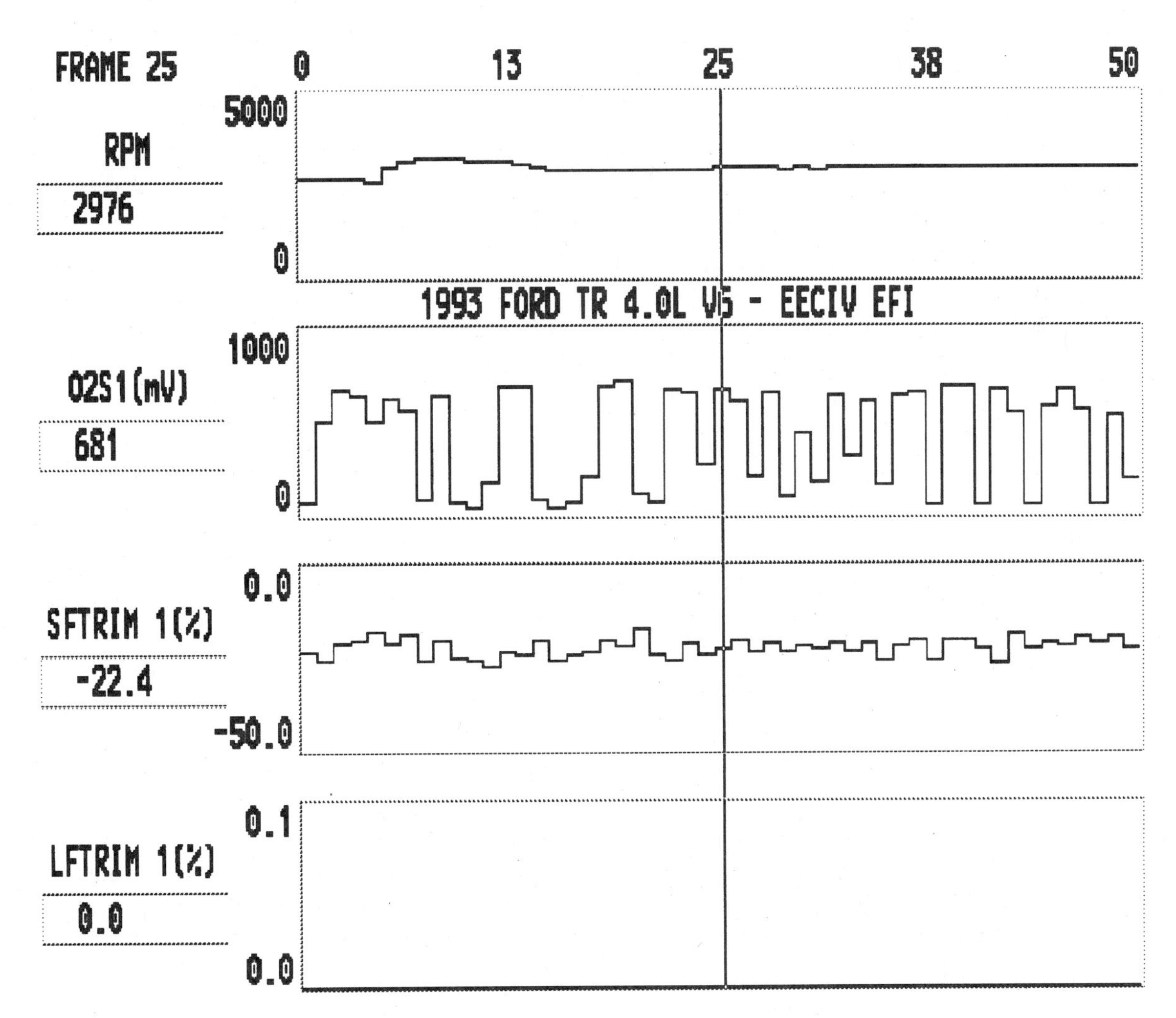

Figure 8-22 Short-Term Fuel Trim corrections are made before Long-Term Fuel Trim corrections.

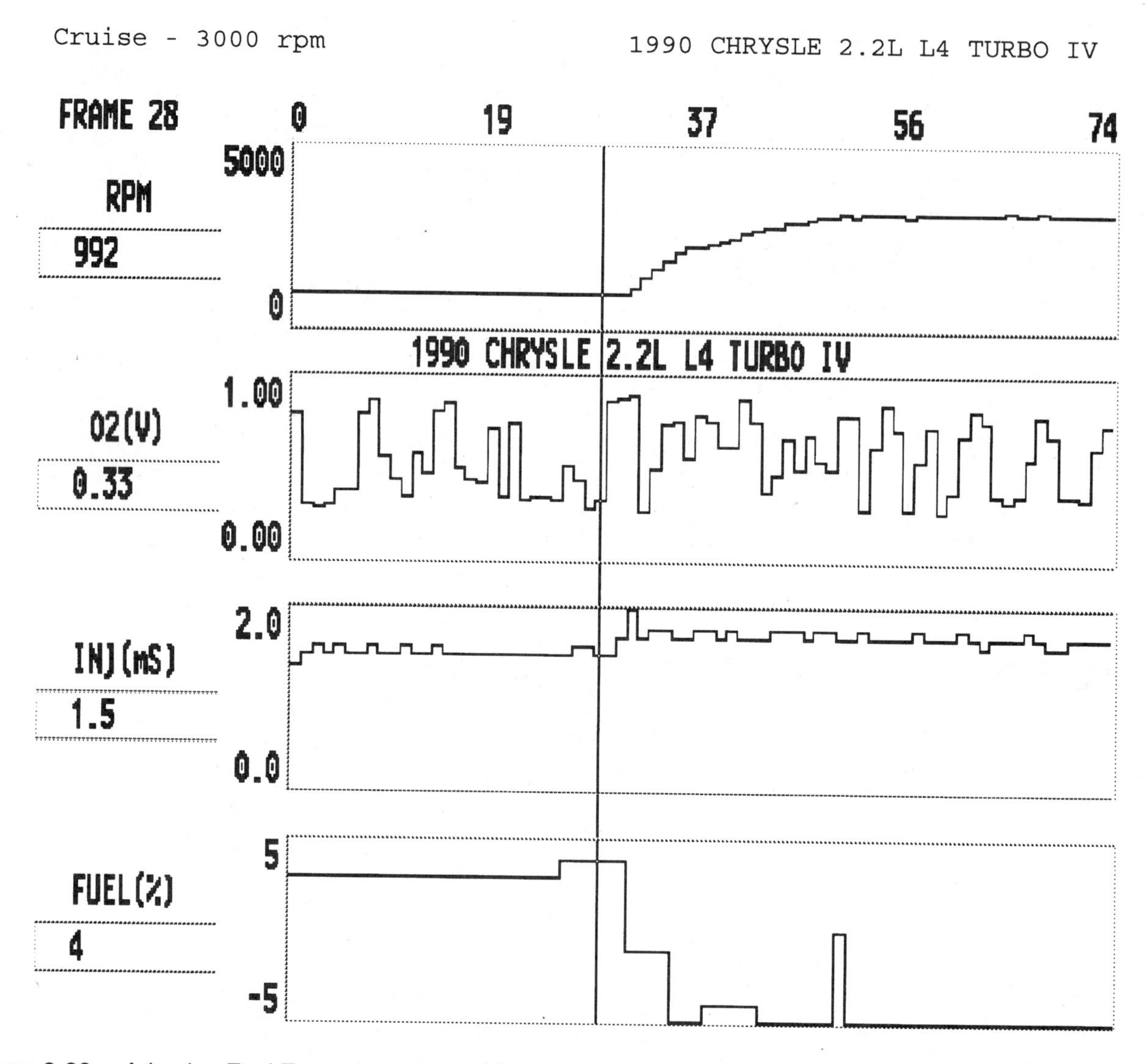

Figure 8-23 Adaptive Fuel Factor is measured in percentage and changes with engine RPM and load.

An example of how adaptive memory is factored into the corrected pulse-width is shown in this chart:

Idle Pulse-Width: 1.2 milliseconds
W.O.T. Pulse-Width: 6.0 milliseconds
Adaptive Memory: 20%
Adaptive Memory: 20%
Corrected Pulse-Width: **1.44 milliseconds**
Corrected Pulse-Width: **7.2 milliseconds**

There is a second adaptive memory factor called *Additive Adaptive Memory.* The additive adaptive factor adds or subtracts a specific amount of milliseconds to the correction pulse-width. It can compensate for problems like rich injector drift or a MAP tolerance that has exceeded the ability of the adaptive memory to correct the air/fuel ratio.

This factor is figured into the corrected pulse-width. For example, if the additive adaptive function is at its maximum additive valve, say 400 microseconds, the following would result:

Idle Pulse-Width: 1.2 milliseconds
W.O.T. Pulse-Width: 6.0 milliseconds
Additive Adaptive: 400 microseconds
Additive Adaptive: 400 microseconds
Corrected Pulse-Width: **1.60 milliseconds**
Corrected Pulse-Width: **6.4 milliseconds**

The scan tool will show the additive adaptive factor in microseconds. It is only updated at certain idle conditions and has the most effect at lower pulse-width conditions and less effect at higher pulse-width conditions.

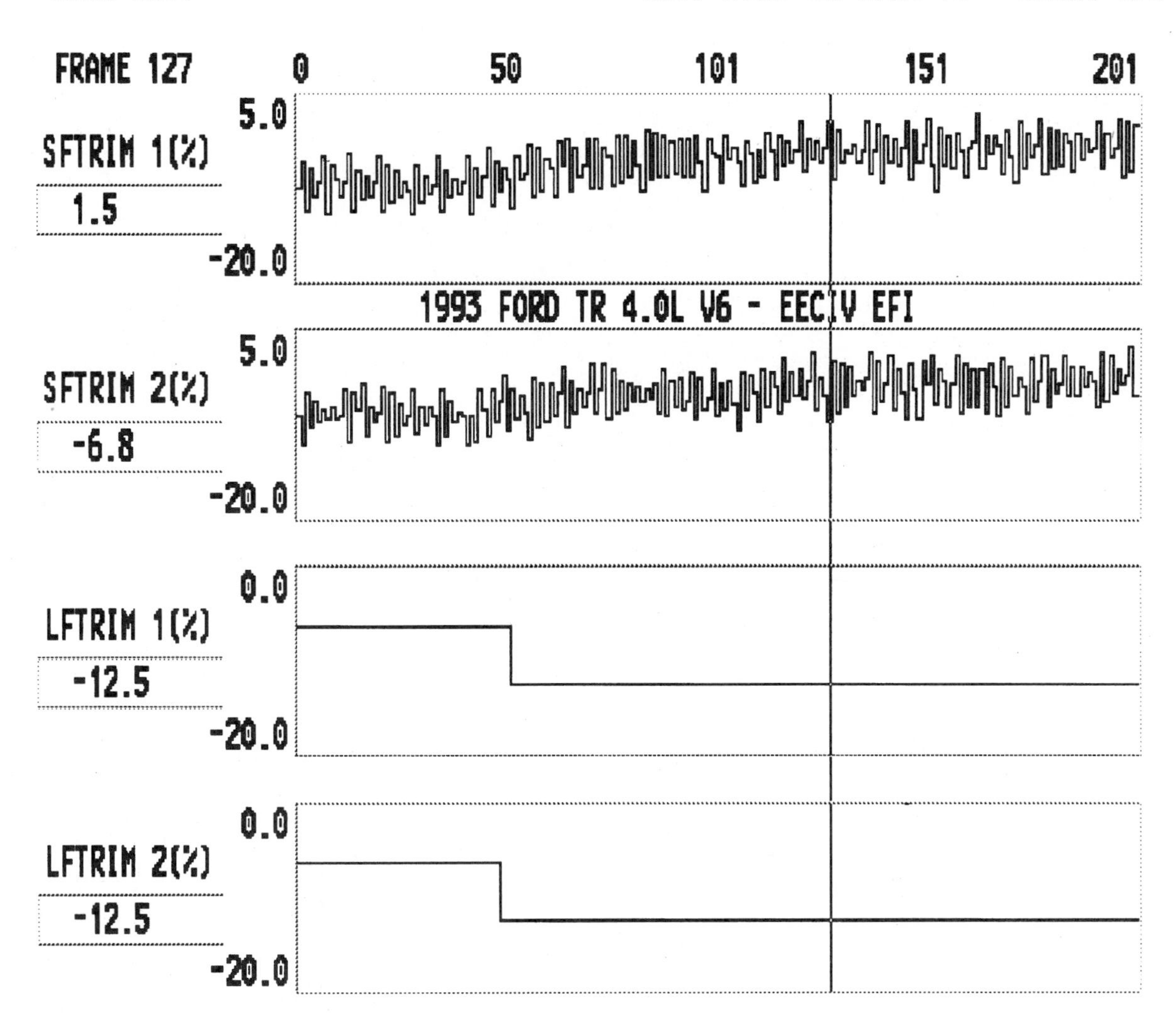

Figure 8-24 Systems with two oxygen sensors have Short-Term Fuel Trim and Long-Term Fuel Trim for both banks of the engine.

Dual Oxygen Sensor Systems

Vehicles equipped with left bank and right bank oxygen sensors use the same terminology. Block Learn and Block Integrator, Short-Term and Long-Term Fuel Trim, Adaptive Memory and Additive Fuel Factor will have left and right correction readings and will be displayed on the scan tool as shown in Figure 8-24.

When diagnosing systems with dual oxygen sensors, see if the problem is common to both banks or one bank. If both sides are showing abnormal readings, look at the various inputs used to calculate pulse-width, check fuel pressure, battery voltage, circuit grounds, and the computer. If the problem is isolated to one side, use an engine analyzer to scope the engine, perform a power balance, check the exhaust manifold for leaks, etc. This does not cover all possibilities—it is meant only to give the technician a starting point. Always check thoroughly before making a system diagnosis or repair.

CHAPTER 9

Automotive Lab Scopes

With the increased use of electronic components in today's automobiles, there is a demand for equipment capable of helping the technician solve elusive electronic problems. Powerful, yet easy-to-use tools are required which can show more than a simple overview of ignition difficulties.

For years, the automotive repair technician most likely has had an oscilloscope in the shop to help with troubleshooting problems. Originally designed to detect problems in the primary and secondary ignition systems, it also displayed waveforms indicating problems with injector systems, alternators, and other electrical components. These scopes, often large and seemingly complicated to use, only revealed surface information about the different electrical systems.

In recent years, the electronic *Lab (Laboratory) Scope* has been incorporated into larger engine analyzers. Also, equipment manufacturers have redesigned the *Lab Scope* as a standalone unit for use in the automotive repair shop.

How does the Lab Scope compare to other handheld equipment used in most shops? Analog voltmeters only display the average values of an input signal, and digital voltmeters merely sample the input several times a second and update the display at some specific rate. If the voltage is a steady DC or AC signal, good readings can be obtained from either type of voltmeter, providing the signal is free of RFI and glitches. A scan tool allows the technician to see momentary values of what the computer thinks is happening. In contrast, the Lab Scope shows both the quantity and quality — how good or bad the signal is.

WHAT IS A LAB SCOPE?

A Lab Scope could be called a visual voltmeter. It allows the technician to see the rapidly-changing voltages by displaying an electrical signal as a waveform. The need to observe signals from sensors and actuators as a waveform, instead of just a number, is very important for diagnosing intermittent problems.

Unlike most ignition scopes, the values of these divisions can be adjusted to best display the incoming signal. For example, if the vertical scale is adjusted so that each division represents two volts, any signal taking up an entire division at this setting is a 2-volt signal. This allows more precision than an ignition scope, which draws graphs of voltage on the vertical scale and degrees of crankshaft rotation on the horizontal scale.

Since the Lab Scope can display a waveform complete with noise (Figure 9-1) and glitches riding on input signals, it is ideal for catching problems easily missed by other test equipment. RFI from the ignition system is an unwanted signal that rides on a steady AC or DC signal back to the computer. This causes intermittent and unpredictable results from

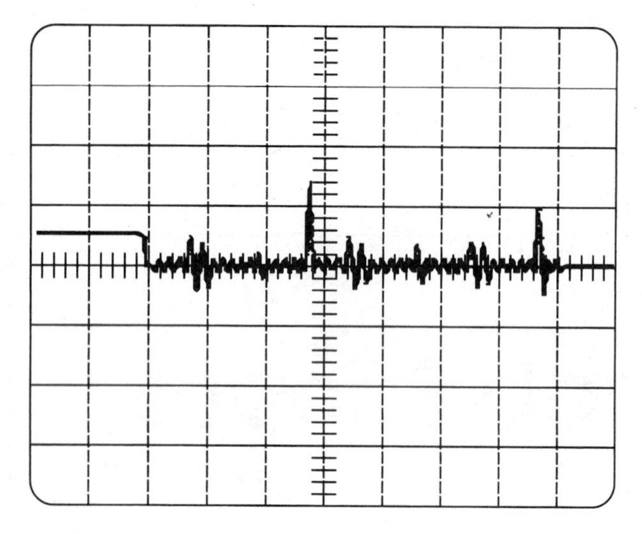

Figure 9-1 Unwanted noise identified by Lab Scope.

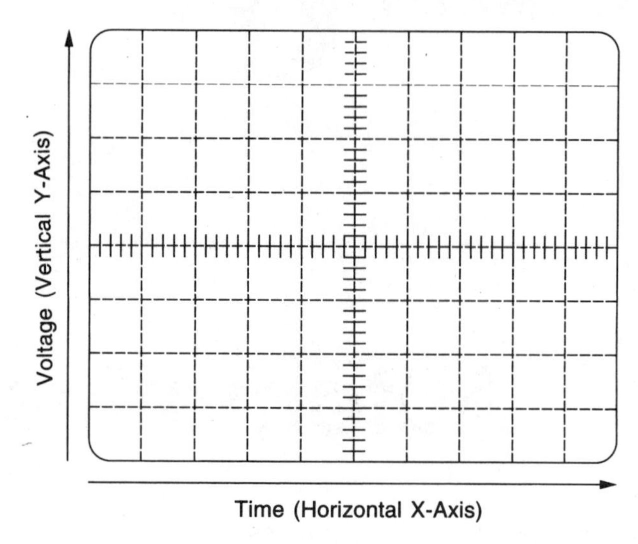

Figure 9-2 A Lab Scope displays voltage versus time.

the computer. Information, good or bad, can be seen and examined with a Lab Scope. A Lab Scope can also find glitches that might occur during tapping, wiggling, or pulling on a wire, harness, or component. This may uncover faulty internal or external connections that could be responsible for intermittent problems.

The technician can measure almost any electrical signal with the two-dimensional graph drawn by an oscilloscope. In most applications, the scope shows a graph of voltage (on the vertical axis) versus time (on the horizontal axis) (Figure 9-2). This general-purpose display presents far more information than is available from other tests and measurement instruments such as frequency counters, multimeters, and scan tools.

For example, a Lab Scope can show how much of a signal is direct (DC), how much is alternating (AC), how much is noise (and if the noise is changing with time), what the frequency of the signal is, and more. With a scope, the technician can see everything at once rather than having to perform many separate tests.

HOW DOES A SCOPE WORK?

An oscilloscope is very similar to a TV screen. The oscilloscope draws a graph by moving an electron beam across a phosphor coating inside the CRT (Cathode Ray Tube). The result is a glow that for a short time afterwards traces the path of the beam. A simple trace is shown in Figure 9-3 as a solid line.

There is a grid pattern or graticule covering the face of the CRT that serves as the reference for measurements (Figure 9-4). It is partitioned into squares called *major divisions,* which are broken down further into *minor divisions.* These divisions are used as reference points when evaluating the frequency or timing and voltage of an incoming signal. The vertical scale represents voltage and the horizontal scale represents time.

Analog scopes are also called *real-time scopes.* This means that what is at the probe tip is what is on the screen. They offer the fastest screen update rate available. The update rate is the time it takes the beam to turn off and return to the left side of the screen for the next sweep.

The Digital Storage Oscilloscope (DSO) converts a signal at its input to digital information and stores it in its computer memory. A captured signal (stored in memory) can be frozen on the scope display for analysis. Some DSOs send the signal data directly to a computer or a printer, or save it to a disk. Another important advantage is the ability to capture a one-shot event and very low-frequency signals because analog scopes flicker when viewing low-frequency signals. The DSO allows the technician to capture an intermittent problem that might be missed with a real-time analog scope. Digital scopes also eliminate the flicker by holding the waveform on-screen until the next update.

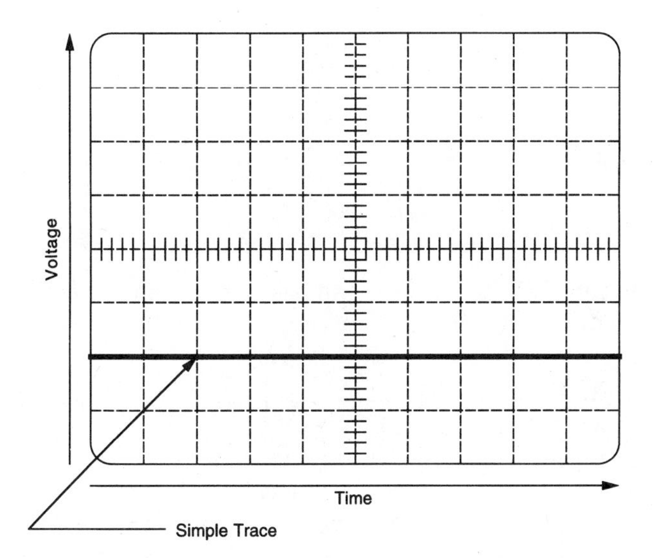

Figure 9-3 An electron beam is displayed as a trace.

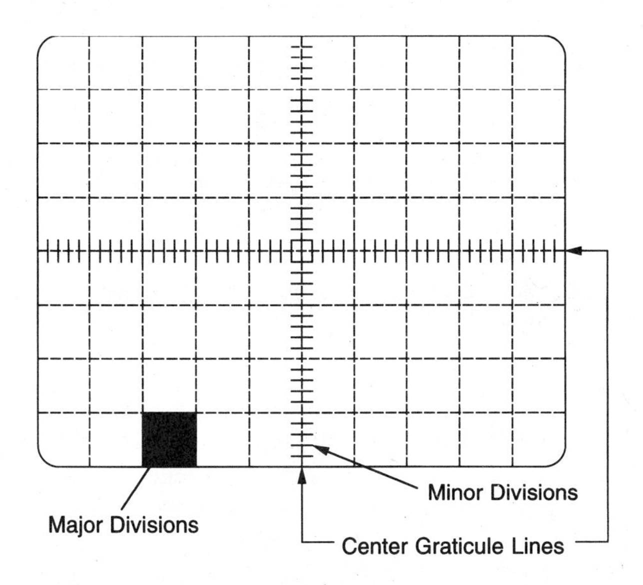

Figure 9-4 A graticule is divided into major/minor divisions.

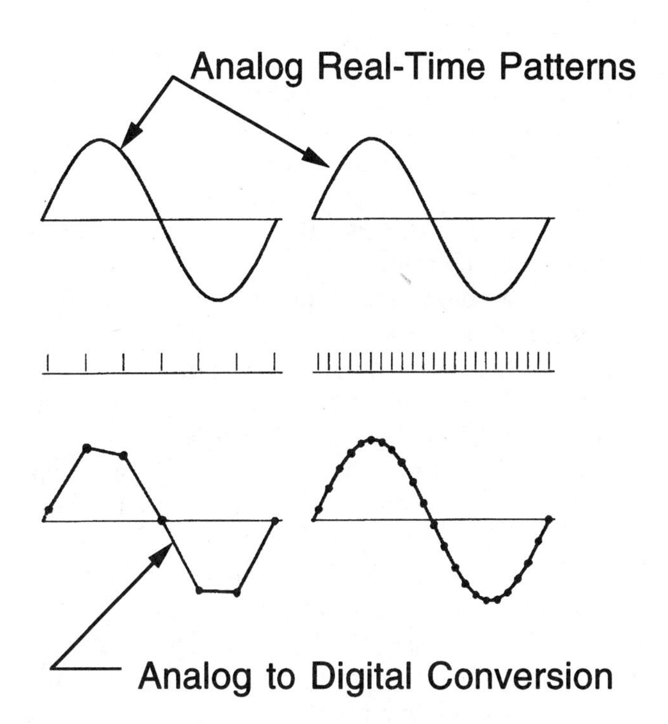

Figure 9-5 Analog to digital conversion, showing a trace drawn at two different sampling rates.

The main difference between an analog and a digital oscilloscope is the way in which signals are acquired and displayed. With the analog oscilloscope, the signal must be repetitive and it must be occurring in real-time (live). With the digital oscilloscope, the signal is not quite real-time. Instead, a DSO displays a drawing of a signal that took place just a split second earlier (Figure 9-5).

There may be some concerns about the accuracy of a DSO. Actually, this strategy of delayed timing is a major benefit. The sampling rate is so high (typically one million samples per second for automotive applications) that all the detail a technician could ever want is displayed. In addition, this fast sampling rate insures that momentary irregularities that can be the cause of intermittent driveability problems can be captured. Also, since captured waveforms can be stored in memory, they can be recalled and used for comparison to a known standard or for other types of analysis.

The screen of an analog oscilloscope is a CRT (Cathode Ray Tube), similar to that of an ordinary TV. The screen is the anode. It has a positive charge and attracts negatively-charged particles from the cathode, an electron gun. The beam of electrons is called a cathode ray. The inside of the screen is coated with phosphorous, which glows or fluoresces wherever the beam of electrons strikes it.

An analog oscilloscope produces a waveform by sweeping the beam across the screen from left to right. As the beam travels across the screen, it is deflected up and down in relation to the instantaneous voltage present at the input probe, representing signal amplitude. It then repeats or updates this sweep at either a fixed or a variable rate. To remain visible, the waveform must be redrawn on the screen repeatedly.

Deflection coils produce a varying magnetic field that bends the electron beam. The vertical deflection coils cause the beam to move up or down to change the height of the trace or pattern. These coils are controlled by the input voltage applied to the scope's probe tip. As voltage increases, so does the height of the trace.

The horizontal deflection coils steer the electron beam from side to side. The speed at which the beam moves is determined by the frequency of the input signal, or an internal signal of a user-variable frequency, and is called the *sweep rate.*

A digital storage oscilloscope appears to sweep a beam across the screen. The DSO uses an analog to digital (AD) converter to digitize the input waveform. This means it will sample an incoming waveform signal at many points in time, and then convert the instantaneous amplitudes at each point to a binary number (Figure 9-6) proportional to the amplitude. These binary numbers are then stored in memory.

Typically, the memory is scanned many times per second. The CRT screen on the digital monitor is constantly being *refreshed* at a fixed rate by the data stored in memory, or until a new waveform overwrites the old data (Figure 9-7).

Both scopes have the capability of displaying *Dual Traces* on the screen. This means two traces are displayed on the CRT at once. The traces displayed can be unrelated signals or related signals that need to be compared for time or voltage. For example, a fuel injector's on-time is triggered by a primary ignition signal from the ignition module (Figure 9-8). Another example is when an O_2 sensor is driven full lean; the expected response would be an increase in the injector pulse-width or on-time. The key to comparing two signals is setting the correct trigger. This will be covered in the next section.

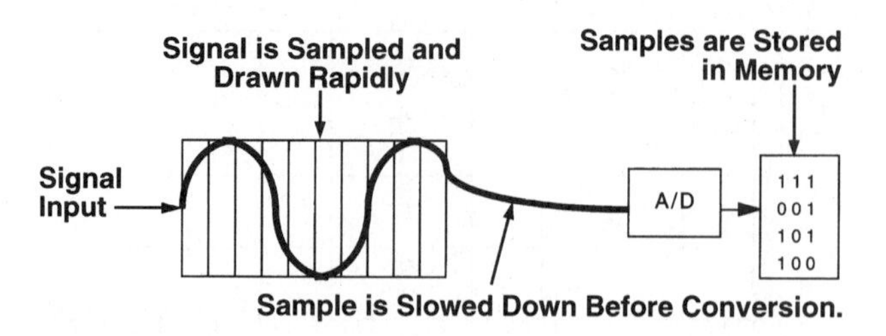

Figure 9-7 The trace is sampled, drawn, and stored in memory.

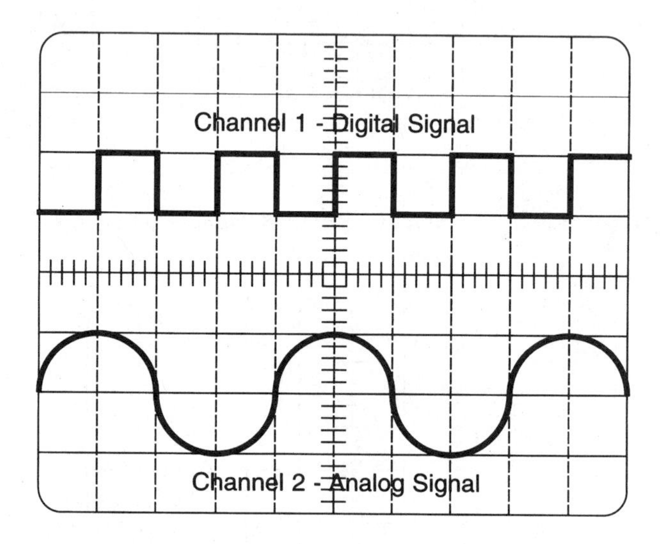

Figure 9-8 Dual trace allows two patterns to be shown at once.

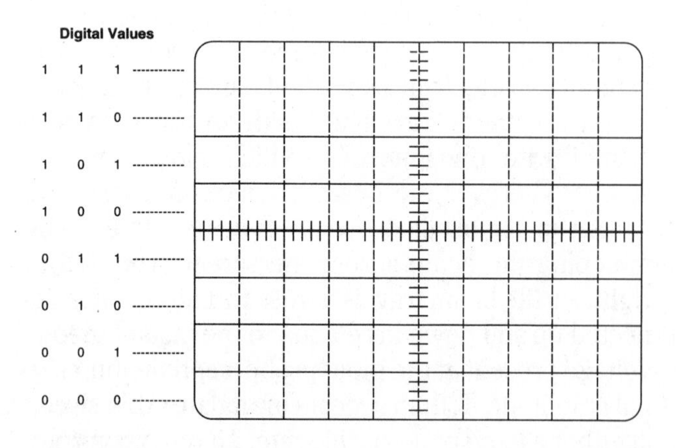

Figure 9-6 Analog voltages are converted to binary numbers.

Dual Tracing capabilities are included as a feature of most Lab Scopes sold today. Sometimes, a Lab Scope can show as many as four traces. While this might sound like a good feature, it really is not needed and makes for a cluttered screen.

STANDARD CONTROLS

Common controls found on most Lab Scope displays include:

1. Intensity.
2. Beam finder.
3. Trace rotation.

4. Vertical (voltage) system.
5. Horizontal (time) system.
6. Trigger system.

Intensity

The *Intensity* control adjusts the brightness of the trace. It is necessary because the scope will be used in different ambient-light conditions and with many kinds of signals. For instance, on square waves, because the slower horizontal segments look brighter than the faster vertical segments, the intensity can be turned up to make the fainter parts of the waveform easier to see.

The Intensity control is also useful because the intensity of a trace depends on two factors: how bright the beam is and how long it is on screen. As different sweep speeds are selected with the TIME/DIV switch, the beam on-time and beam off-time changes, and the beam then has more or less time to excite the phosphor.

Beam Finder

For convenience, the *Beam Finder* lets the technician locate the electron beam when it is off-screen. When the Beam Finder button is pushed, it will reduce the vertical and horizontal deflection voltages and override the intensity control so the beam always appears within the screen. The beam should now appear in a quadrant of the screen. The *Horizontal* and *Vertical Position* controls are used to position the trace. If the technician inadvertently turns down the display intensity, the beam can still be located because the intensity is automatically increased while the Beam Finder is pressed.

Trace Rotation

Another display control on the front panel of a Lab Scope is *Trace Rotation*. This adjustment lets the technician electrically align the horizontal deflection of the trace with the fixed CRT graticule. To avoid accidental misalignments when the scope is in use, the control is recessed and must be adjusted with an adjustment tool or a screwdriver.

If this seems like a calibration item that should be adjusted once and then forgotten, it is for most oscilloscope applications. Unfortunately, the earth's magnetic field affects trace alignment, and when a Lab Scope is used in many different positions, as an automotive Lab Scope will be, it's very handy to have a front-panel trace-rotation adjustment. Figure 9-9 shows a trace that is out of alignment and needs to be adjusted.

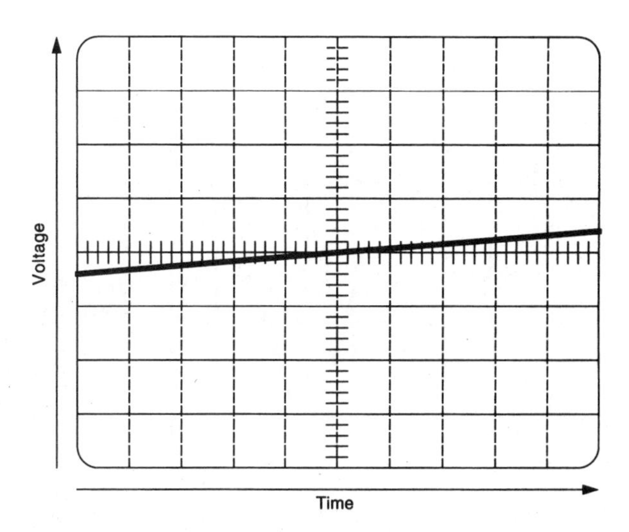

Figure 9-9 A trace is altered by the Earth's magnetic fields.

The Vertical System

The scope's vertical system supplies the display system with the vertical, or Y-axis, information for the graph on the CRT screen. The vertical system takes the input signals, develops deflection voltages, then uses the deflection voltages to control the CRT electron beam.

The vertical system controls the voltage displayed on vertical axis of the graph. Any time the electron beam that draws the graph moves up or down, it does so under control of the vertical system.

The voltage setting on the oscilloscope for displaying a signal is the voltage per division (VOLTS/DIV) on the grid (Figure 9-10). For most automotive low-voltage circuits, the technician will work with voltage levels between 0 and 5 volts and 0 and 12 volts.

By assigning one volt per division, a five-volt trace will cover five divisions. At one-half volts per division, or 500 millivolts, the same five-volt signal will go right off the screen. Selecting too many volts per division might cause the trace to be so flat that it is unreadable.

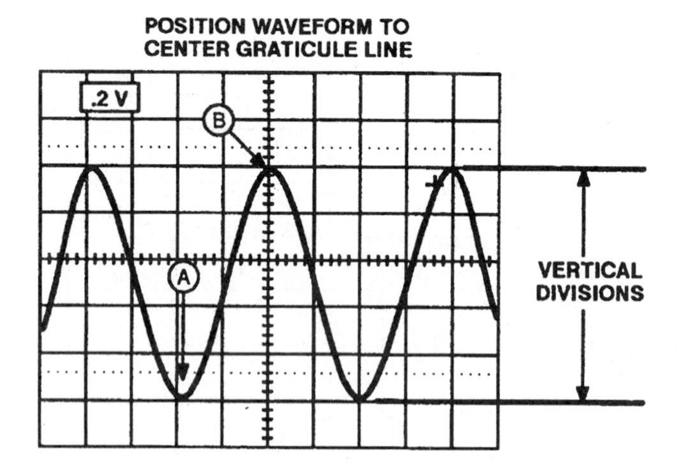

Figure 9-10 Vertical divisions represent voltage: Point A = Low volts, Point B = High volts.

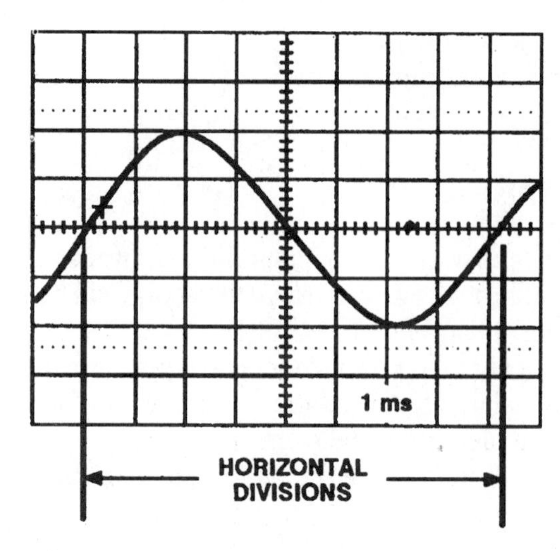

Figure 9-11 Horizontal divisions represent time.

The *Vertical Position* control allows the trace to be placed anywhere on the screen. The position control for each channel changes the vertical placement of its respective trace.

The Horizontal System

To draw a graph, your scope needs both vertical and horizontal data. The horizontal system supplies the second dimension by providing deflection voltages that move the electron beam horizontally. It also contains a sweep generator that produces a waveform or ramp. The ramp is used to control the scope's sweep rates.

Because the sweep generator is calibrated in time, it is usually called the *time base.* The time base lets the technician observe the signal for a unit of time, from very short times measured in nanoseconds or microseconds, to longer times measured in seconds.

The horizontal system controls the left-to-right movement of the beam and is the time per division on the grid (Figure 9-11). The time setting on the oscilloscope for displaying a signal is represented by the TIME/DIV switch. Because the same time base will not work for all signals, it must be adjusted whenever a different signal needs to be analyzed.

If the TIME/DIV is set too low (slow), a series of square waves may appear as a straight line. If the TIME/DIV is set too high (fast), a series of square waves may appear as an extremely close mass of vertical lines. When making setup decisions for time base, use the highest TIME/VOLTS speed suitable for the signal or for the detail to be measured, to obtain the best possible image on the screen.

For example, if a fuel injector takes 10 milliseconds to fire, a time base of one millisecond per division will cause the trace to occupy the full width of the screen. A better choice would be a TIME/VOLTS setting of two milliseconds per division or more. When setting up the voltage and time base scales, adjust the horizontal and vertical controls until the best possible image is obtained.

Trigger

At this point, the technician should know that the display system draws the waveforms on the screen, the vertical system supplies the voltage for the graph, and the horizontal system provides the time base for the graph. The only thing missing is the *when.* How does the scope know when to start drawing the signal?

The *trigger* is the *when,* because it determines the instant that the scope is to start drawing the signal. Triggering is important when acquiring time-related information from signals such as fuel injectors, RPM, camshaft, Mass Air Flow sensor, etc. Equally important is that the graph of each signal should start predictably at the same point on the waveform so that the display remains stable (Figure 9-12).

The graph seen on the screen is not a static one, although it appears to be. It's always changing and being updated with the input signal. If the technician

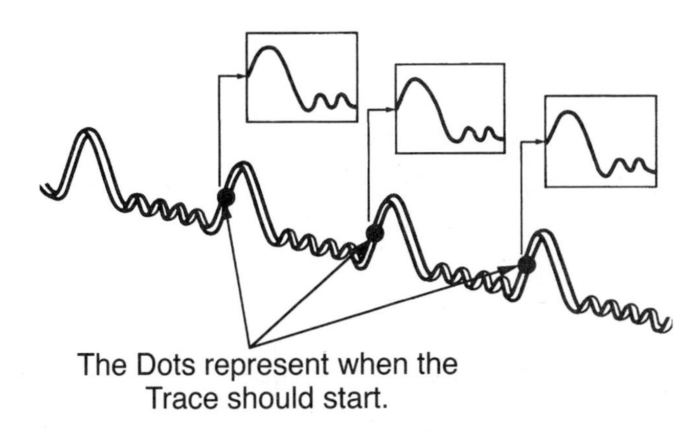

Figure 9-12 The trigger is critical to time-based signals.

is using the 0.05 µs TIME/DIV setting, the scope is drawing one graph every 0.5 µs (0.05 µs per division times 10 screen divisions). That is 2,000,000 graphs every minute (not counting retraces and hold-off times). Imagine the jumble on the screen if each sweep started at a different place for each signal.

The trigger system determines when the oscilloscope begins drawing by starting the horizontal sweep across the screen. Each sweep will start at the right time, if the right trigger system control settings are selected. Triggering the Lab Scope is often the most difficult operation to perform because of the many options available and the exacting requirements of some signals. It should be performed as recommended in the following paragraphs.

When the Normal *(NORM)* trigger mode is selected, the CRT beam is not swept horizontally across the face of the CRT until a sample of the signal being observed, or another signal related to it, triggers the time base. However, this trigger mode is inconvenient because no trace appears on the CRT screen without a signal, or if the trigger controls are improperly set. Since the absence of a trace can also be due to an improperly-set VOLTS/DIV switch, time can be wasted in determining the cause. The *AUTO* trigger mode solves this problem by causing the time base to automatically free-run a trace on the CRT when not triggered. This gives a single horizontal line with no signal, and a vertically-deflected but non-synchronized display when the vertical signal is present but the trigger controls are improperly set. The only hitch with AUTO operation is that signals below 25 Hz cannot reliably trigger the time base. Therefore, the usual practice is to leave the trigger MODE switch set for AUTO, but reset it to NORM if any signal fails to produce a stable display.

Next, set the *slope* and *level* controls to tell the trigger circuit to recognize a particular voltage level on the trigger signal. The *slope* switch determines whether the sweep will begin on the positive (rising) or negative (falling) transition of the trigger signal (Figure 9-13). Always select the most stable slope or edge. The *level* control determines the point on the selected slope where the time base will be triggered (Figure 9-14). Every time that level occurs, the sweep generator is turned on.

Trigger Source switches tell the trigger circuit which trigger signal to select: Channel 1, Channel 2, line voltage, or an external signal. To use an

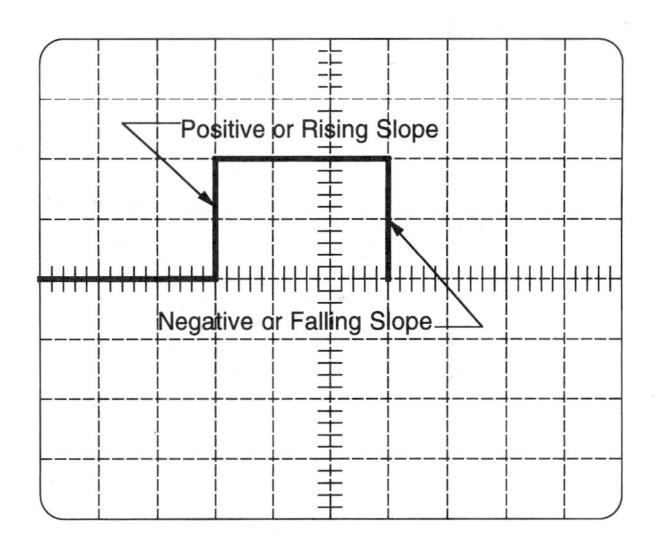

Figure 9-13 The trigger can be set to start on the rising or falling slope.

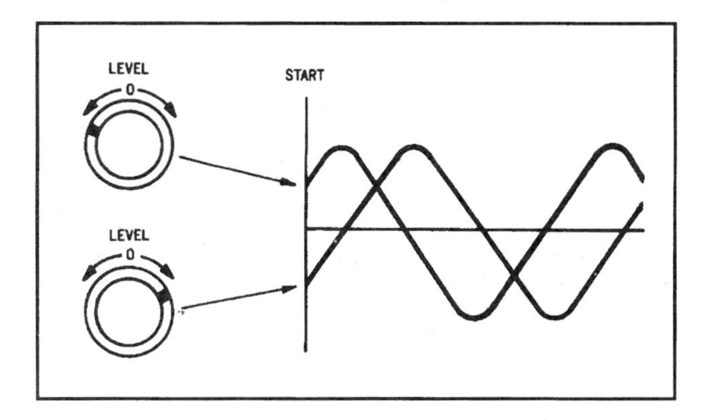

Figure 9-14 The trigger level control is for selection of the voltage starting point.

external trigger signal, connect it to the trigger system circuit using the external input (EXT INPUT) connector. Use the COUPLING switch to determine how the selected signal is to be coupled to the trigger circuits.

Trigger Sources are grouped into two categories, depending on whether the trigger signal is provided internally or externally. The source makes no difference in how the trigger circuit operates, but the internal triggering usually uses the signal that is being displayed. That has the advantage of letting the technician see where the trigger is.

Most scopes can only trigger on either one channel or the other when the two input signals are not synchronized. But triggering on the displayed signal is not always what is needed, so *External Triggering* is also provided. This can provide more control over the display. To use an external trigger, set the left SOURCE switch to EXT and the right SOURCE switch to either EXT/10 or EXT. Connect the external triggering signal to the EXT INPUT connector on the front panel. This increases the usefulness of the scope. The external trigger allows the display to be synchronized to an external source. Triggering from the #1 plug wire is very useful.

The two Trigger Source switches determine the source signal for the trigger. Internal triggering sources are enabled when the left SOURCE switch is set to the appropriate channel setting (CH 1 or CH 2), which allows triggering the scope on the signal coming from the selected channel. When triggering on one channel, set the scope to trigger the sweep on some part of the waveform present on that channel.

When *Trigger Coupling* is selected, the scope's Vertical Mode switches determine what signal is used for triggering. If the Vertical Mode switch is set to CH 1, then the signal on CH 1 triggers the scope. If the switch is set to CH 2, then the CH 2 signal triggers the scope.

If the switch is set to BOTH/ALT, then both signals alternately trigger the scope. This is accomplished by alternating the CH 1 and CH 2 trigger signals synchronously with the display. In the ADD mode, the algebraic sum of CH 1 and CH 2 is the triggering signal. In the CHOP mode, the scope triggers as it does in ADD mode, which prevents the instrument from triggering on the chop frequency instead of on the signals. Vertical Mode triggering is a kind of automatic source selection that allows switching back and forth between vertical modes to look at different signals. This mode makes alternate triggering possible, with the scope triggering first on one channel, then on the other channel. That means you can look at two completely unrelated signals at the same time.

PROBES

Connecting measurement test points to the inputs of an oscilloscope is best done with a probe like the one illustrated in Figure 9-15. Scope probes are the most popular method of connecting the oscilloscope to circuitry. These probes are available with 1X attenuation (direct connection) and 10X attenuation. The 10X attenuator probes increase the effective input impedance of the probe/scope combination to 10 megohms.

The reduction in input capacitance is the most important reason for using attenuator probes at high frequencies, where capacitance is the major factor in loading down a circuit and distorting the signal. When 10X attenuator probes are used, the scale factor (VOLTS/DIV switch setting) must be multiplied by ten.

Using a probe instead of a bare wire reduces stray signals, but circuit loading is still an undesirable side effect. Depending on how great the loading is, the circuit loading effect modifies the circuit environment and changes the signals in the circuit being tested, either a little or a lot.

When a probe is connected to the scope, it should be checked for proper calibration. This is called *probe compensation*. Figure 9-16 shows the effects of improper compensation. If not properly compensated, the waveform will be distorted, possibly leading the technician down the wrong path.

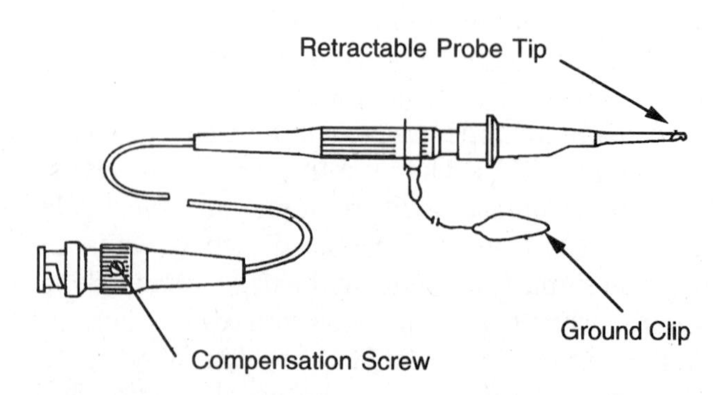

Figure 9-15 Typical probe and lead.

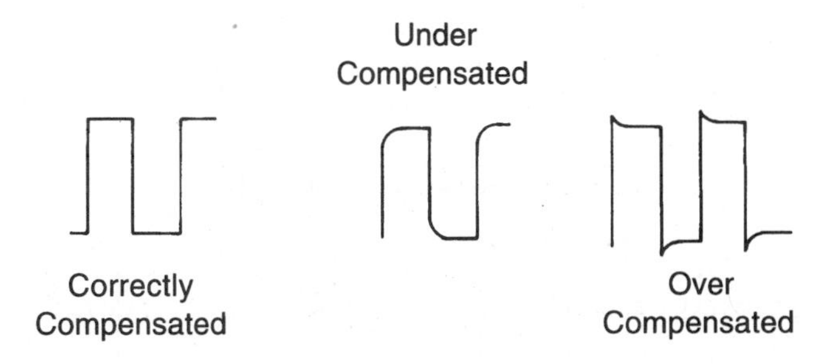

Figure 9-16 Probe compensation is critical to proper operation.

Compensation of the probe can be done by connecting the probe to the CAL connector on the front of most scopes. This port provides a fast-rise square wave of precise amplitude for probe adjustment. Probe compensation is accomplished by turning the screw shown in Figure 9-15.

The probe's ruggedness, flexibility, and the length of its cable are also important. The longer the cable is, the greater the capacitance at the probe tip. Most modern probes feature interchangeable tips and adapters for many applications. Retractable hook tips allow the probe to be connected to most circuit components. Alligator clips for contacting large-diameter test points is another possibility.

For the reasons already mentioned, the best way to guarantee that your scope and probe measurement system has the least distortion on your measurements is to use the probe recommended for your scope. Always make sure the probe is properly compensated.

WAVEFORMS

A wave is a disturbance traveling through a medium, and a waveform is a graphic representation of a wave. Like a wave, a waveform depends on movement and time. The ripple on the surface of a pond is a movement of water in time. The waveform on the oscilloscope screen is a movement of an electron beam over time.

Changes over time form the wave shape, the most readily-identifiable characteristic of a waveform. Waveform shapes tell a great deal about the signal. Any change in the vertical dimension of a signal represents a change in voltage. Any time there is a flat horizontal line, there is no change for that length of time. Straight diagonal lines mean a linear change; that is, a rise or fall in voltage at a steady rate over time. Sharp angles on a waveform mean sudden change.

Wave shapes alone do not tell the whole story. To completely describe a waveform, the technician will need to find its particular parameters. Depending on the signal, these parameters might be amplitude, time, frequency, width, rise time, etc.

The simplest of all waveforms is just a straight line, either at some specific voltage level or at zero voltage. If a circuit stays on or off for a long time and then has one change of state (voltage goes from low to high, or from high to low), the waveform can be called a *step waveform.* For example, as the transmission is shifted into or out of gear, the Park/Neutral switch changes the voltage signal up or down (Figure 9-17).

There are basically four variations of waveform types:

1. *Alternating Current (AC)* — This waveform changes amplitude (voltage) and polarity over time. This is commonly called a *sine wave.* An AC alternating polarity signal may be a square wave or a sinusoidal wave. It is sinusoidal if the rise and fall from peak to peak are identical or symmetrical. If the rise and fall are not symmetrical, the waveform is non-sinusoidal.

   When testing an AC signal, always check it for amplitude, time, and shape. The amplitude of an AC voltage signal is measured from peak to peak as shown in Figure 9-18. The time is usually

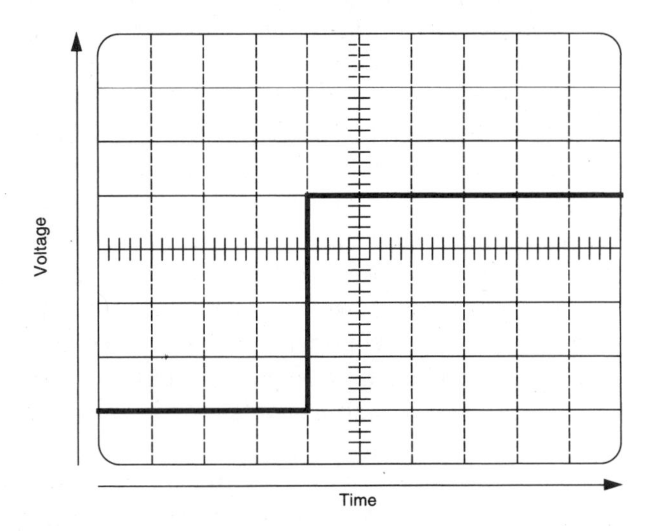

Figure 9-17 Step waveform of an On or Off switch.

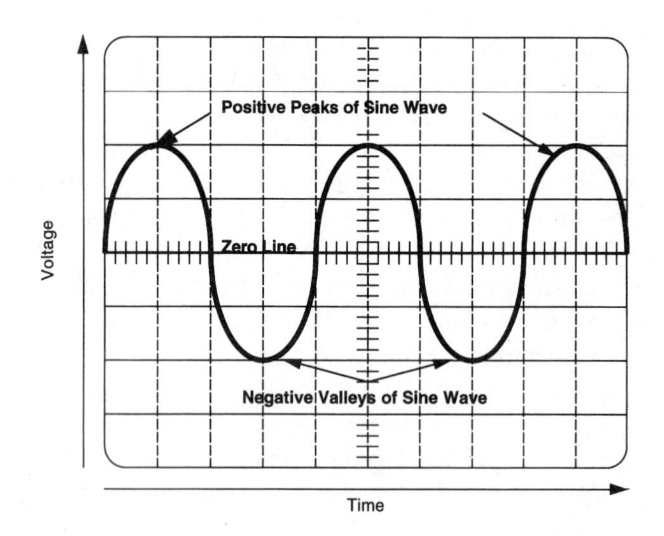

Figure 9-18 AC sine wave shows positive and negative voltages.

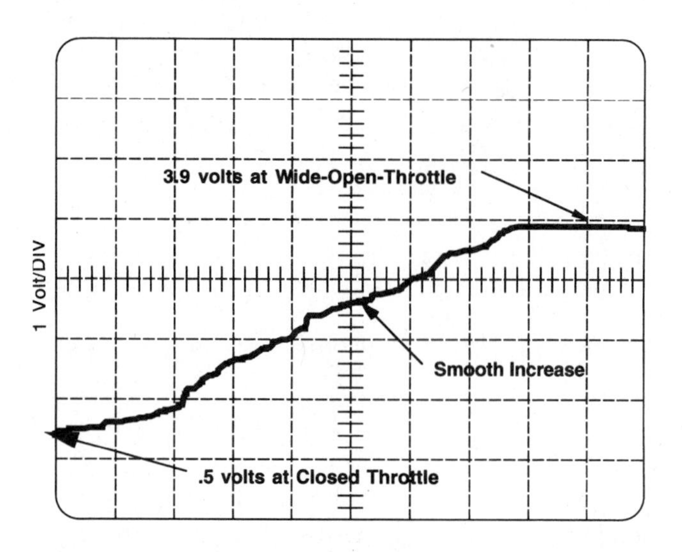

Figure 9-19 DC signal of a good TPS, measured from closed throttle to wide-open-throttle.

measured by the number of cycles per second, called the *frequency.* The shape should be checked for unwanted noise or glitches that could send false trigger signals to the vehicle's on-board computer.

AC waveforms may be produced by:

a. Permanent magnet-type generators (ignition pickup coils, vehicle speed sensors, ABS wheel speed sensors, etc.).
b. Cam and crankshaft position sensors.
c. Detonation (knock) sensors.

2. *Direct Current (DC)* — Direct current waveforms may be a straight line, a varying voltage waveform, square waves, pulse-width modulated waves, frequency modulated waves, or step patterns.

For DC waves, the amplitude is the vertical positive voltage rising above the reference point. The reference point is the baseline on the scope's display and is assigned a value of zero. A DC voltage is constantly applied and should change with respect to time. The voltage will generally appear on the scope as a straight line (Figure 9-19). This signal should be checked for unwanted spikes, both upward and downward, which might indicate an open circuit or one with a high resistance.

DC waveforms may be produced by:

a. Thermistors, such as coolant and intake air temperature sensors, etc.
b. A potentiometer, such as throttle position, EGR valve position sensor, Vane Air Flow meters, etc.
c. Resistive network sensors, such as non-Ford MAP sensors.
d. Oxygen (O_2) sensors.

3. *Direct Current Fixed Digital Pulse Signals* — The term *square wave* is often applied to any waveform that has vertical sides and a flat top. However, a fixed digital square wave alternates between two voltage levels, giving *equal time* to each level. Fixed digital square waves are said to be symmetrical.

Computers *think* only in terms of zero or one. Square waves are signals with two voltage levels that represent these values. In most automotive systems, 0v and 5v or 12v represent voltages going into the computer from such things as Hall-Effect switches or MAF (Mass Air Flow) sensors.

When digital signals are controlled by the computer, this is called *ground side switching.* The on/off time ratio remains fixed and can be measured in duty-cycle or dwell (Figure 9-20). The waveform should be checked for amplitude, time, and shape. A defective pattern might have unwanted spikes, rounded corners, or chopped-off tops of the waveform.

Fixed Digital Pulse Signal waveforms may be produced by:

a. Hall-Effect switches used for distributor reference signals and crankshaft position, etc.
b. Digital-type Mass Air Flow sensors, Ford MAP and Baro sensors, etc.

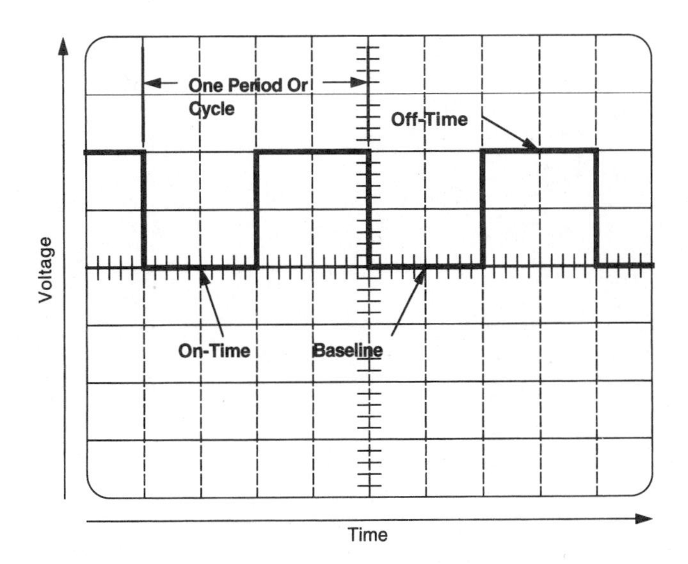

Figure 9-20 Fixed digital pulse-width signal waveform.

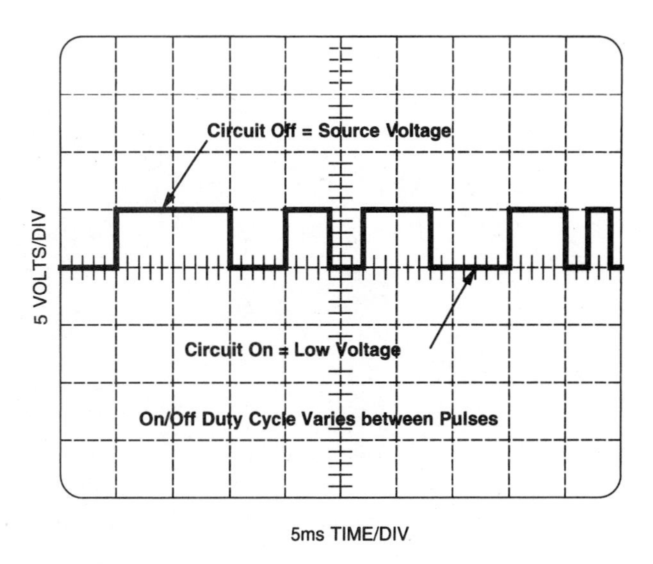

Figure 9-21 PWM solenoid's ground side controlled by the computer.

4. *Pulse-Width Modulated Signals* — This type of signal spends more time at one voltage level than another. Changing the pulse-width over time to control a device is called a *Pulse-Width Modulation (PWM)*. Pulse trains are said to be asymmetrical (Figure 9-21).

   Voltage levels are typically *on* and *off*. This type of signal also changes its ratio of on and off time and is measured in duty-cycle or dwell (Figure 9-21). The waveform should be checked for amplitude, time, and shape. A defective pattern might have unwanted spikes, rounded corners, or chopped-off tops of the waveform.

   Pulse-width modulated waveforms may be produced by:

   a. The computer.
   b. Fuel injectors (Figure 9-22).
   c. Ignition system control modules.
   d. Emission control system solenoids, such as: EGR, purge canister, etc.
   e. Some transmission torque converter clutch solenoids.

In summary, the best way to become proficient with a Lab Scope is to use it frequently. Knowing what good patterns should consist of and look like helps the technician to quickly identify bad patterns.

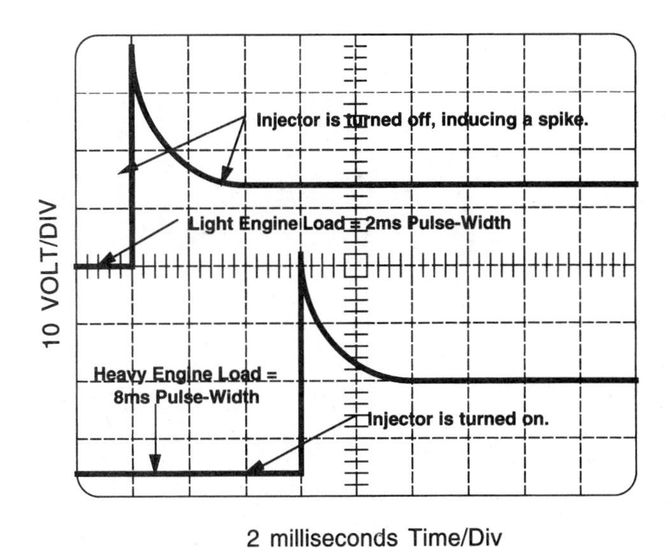

Figure 9-22 PWM fuel injector under different engine loads.

For this to become reality, the technician must become familiar with the setup and operation of various controls on a Lab Scope. There are many buttons and knobs incorporated into a Lab Scope; the technician should concentrate on the controls covered in this chapter. By doing this, the Lab Scope will be used frequently and correctly, and intermittent electrical problems will be accurately diagnosed.

CHAPTER 10

Diagnosing Emissions-Related Failures

Most customers bring a vehicle into the shop for a tune-up because the vehicle is hard to start or just plain runs poorly. The technician usually performs a visual inspection, and then uses some type of equipment to diagnose the problem. The technician's primary objective is to find the problem and fix it.

When a customer brings a vehicle into the shop because of an emissions-related failure, the technician has one distinct advantage. The customer will give the technician a Vehicle Inspection Report (VIR) of the emissions test that will provide a starting point. If the I/M test performed was a two-speed idle test, then the VIR will show the results of a non-loaded idle and a 2500 RPM test (Figure 10-1). If the vehicle has failed an IM240 test, the VIR will show the composite (Figure 10-2, pages 159 and 160) and second-by-second emissions results. Both tests provide the technician with a "clue" or starting point as to why the vehicle has been brought to the shop.

Whether the customer is tired of the way the vehicle runs or because the vehicle failed an emissions test, the technician must find the problem and fix it. The main problems that arise are:

1. Not all shops have the same equipment and tools.
2. If the vehicle failed a centralized IM240 test, the repair shop must attempt to recreate the speed and load at the time of failure without a dynamometer.
3. If the vehicle failed a centralized IM240 test, the VIR will display the results in grams per mile. The repair shop will have to verify the repairs in concentrations of parts per million (ppm) and percentages (%).
4. The technician must now interpret NO_x readings that have not been used in prior emissions testing.

DEVELOPING A STARTING POINT

For years, technicians have been successfully repairing vehicles using the available resources in their shops. Whether the vehicle enters the shop with or without a VIR, the technician must start somewhere. They might start with a visual inspection, test drive, or by connecting an engine analyzer to the vehicle. Some businesses have even installed a dynamometer so they can recreate the complaint without leaving the shop (Figure 10-3, page 161). In some cases, technicians believe they can do the job without the high-tech equipment.

Regardless of belief, each technician must establish a starting point for each vehicle in question. By doing this, the technician is developing a *baseline* that can provide the necessary data to make an accurate diagnosis.

Whatever the method, the technician must use the equipment at hand to collect the data. If a gas analyzer is available, the technician will usually check

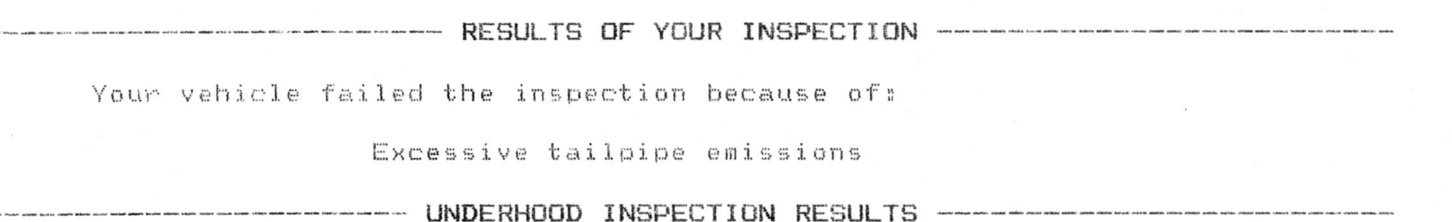

RESULTS OF YOUR INSPECTION

Your vehicle failed the inspection because of:

Excessive tailpipe emissions

UNDERHOOD INSPECTION RESULTS

**** passed ****

PCV system
Fuel evaporative controls
Catalyst
Exhaust gas recirculation
Ignition spark controls
Oxygen sensor and connectors
Wiring of other sensors/switches/computers
Vacuum line connections to sensors/switches
Fuel injection
Other emission related components

EMISSION TEST RESULTS

TRAINING MODE: NOT A VALID REPORT

2500 RPM TEST RESULTS

	MEASURED	MAXIMUM ALLOWED EMISSIONS	AMOUNT OVER LIMIT	RESULT
HC(ppm)	227	220	7	FAILED
CO(%)	8.77	1.2	7.57	FAILED
CO2(%):	8.5			
O2(%):	1.3			
Engine RPM:	2342			

IDLE TEST RESULTS

	MEASURED	MAXIMUM ALLOWED EMISSIONS	AMOUNT OVER LIMIT	RESULT
HC(ppm)	517	100	417	FAILED
CO(%)	8.20	1.0	7.20	FAILED
CO2(%):	8.3			
O2(%):	2.1			
Engine RPM:	758			

Figure 10-1 VIR for an idle/2500 RPM smog check.

the exhaust emissions at idle and 2500 RPM. Some technicians prefer to connect an engine analyzer to perform a series of automatic diagnostic tests. A computer-related problem might warrant the use of a scan tool to pull codes or to look at computer data. Technicians might even use a DMM to monitor the O_2 sensor, the carburetor duty-cycle, or the fuel-injector pulse-width.

TAILPIPE INSPECTION

An important part of the visual inspection which is often overlooked is the tailpipe. The color of the exhaust leaving the tailpipe (light blue, white, or black) can be a baseline as to what the problem may be.

RG-240 per Second TEST RESULTS #1

Plate:	ESP1	Date:	09/27/94
Year:	1994	Time:	11:41
Make:	FORD	Test ID:	00000015
Model:	TBIRD		
Cylinders:	8	Fuel Economy (MPG):	23.30
Class:	1	Trace Distance:	1.981
Trans:	A	Trace Time:	240
Fuel:	G	Bad Trace(s):	1
		Trace Result:	FAIL

EPA Pressure Test

	Pres.Start	Pres.End	Time (sec)	Results
Gas Tank:				N/A
Gas Cap:				N/A

STEADY	HC ppm	NOx ppm	CO %	CO2 %	O2 %	hp	mph
ASM-5015:							
ASM-2525:							
Custom:							

RG-240	HC g/mi	NOx g/mi	CO g/mi	CO2 g/mi	Purge 1	hp	lbs
Limits:	1.20	2.50	20.0	N/A			
Readings:	0.02	1.11	1.9	378.2		11.20	3500
Results:	PASS	PASS	PASS	N/A	N/A		

Sec	HC g	NOx g	CO g	CO2 g	Purge	Sec	HC g	NOx g	CO g	CO2 g	Purge	Sec	HC g	NOx g	CO g	CO2 g	Purge
1	0.000	0.000	0.04	0.66	--.---	31	0.001	0.003	0.12	2.64	--.---	61	0.000	0.010	0.00	3.39	--.---
2	0.000	0.000	0.04	1.03	--.---	32	0.001	0.011	0.10	2.73	--.---	62	0.000	0.008	0.00	3.42	--.---
3	0.000	0.000	0.03	1.47	--.---	33	0.001	0.025	0.13	3.78	--.---	63	0.000	0.006	0.00	3.31	--.---
4	0.000	0.000	0.03	1.64	--.---	34	0.001	0.034	0.13	4.50	--.---	64	0.000	0.004	0.00	2.54	--.---
5	0.000	0.000	0.02	1.75	--.---	35	0.001	0.026	0.07	3.82	--.---	65	0.000	0.003	0.00	2.13	--.---
6	0.000	0.000	0.03	1.81	--.---	36	0.000	0.020	0.04	3.22	--.---	66	0.000	0.005	0.00	2.55	--.---
7	0.000	0.000	0.03	1.82	--.---	37	0.000	0.014	0.04	3.03	--.---	67	0.000	0.016	0.00	3.02	--.---
8	0.000	0.000	0.05	1.89	--.---	38	0.000	0.008	0.02	2.30	--.---	68	0.000	0.017	0.00	2.84	--.---
9	0.001	0.000	0.06	1.88	--.---	39	0.000	0.006	0.02	1.88	--.---	69	0.000	0.013	0.00	2.41	--.---
10	0.001	0.000	0.08	1.88	--.---	40	0.000	0.005	0.01	1.62	--.---	70	0.000	0.008	0.00	1.97	--.---
11	0.001	0.002	0.11	2.36	--.---	41	0.000	0.004	0.01	1.55	--.---	71	0.000	0.006	0.00	1.99	--.---
12	0.001	0.017	0.13	3.08	--.---	42	0.000	0.004	0.02	1.77	--.---	72	0.000	0.012	0.00	2.62	--.---
13	0.001	0.043	0.12	3.99	--.---	43	0.000	0.005	0.07	2.42	--.---	73	0.000	0.027	0.00	3.26	--.---
14	0.001	0.091	0.08	5.33	--.---	44	0.000	0.012	0.06	2.66	--.---	74	0.000	0.023	0.00	3.49	--.---
15	0.001	0.078	0.06	5.21	--.---	45	0.000	0.023	0.06	3.07	--.---	75	0.000	0.014	0.00	3.38	--.---
16	0.001	0.062	0.05	4.51	--.---	46	0.000	0.052	0.05	4.08	--.---	76	0.000	0.006	0.00	2.80	--.---
17	0.001	0.048	0.11	4.47	--.---	47	0.000	0.061	0.04	4.87	--.---	77	0.000	0.003	0.00	1.93	--.---
18	0.001	0.043	0.11	4.44	--.---	48	0.000	0.042	0.02	4.09	--.---	78	0.000	0.002	0.00	1.81	--.---
19	0.001	0.037	0.10	4.04	--.---	49	0.000	0.032	0.01	3.59	--.---	79	0.000	0.002	0.00	2.08	--.---
20	0.001	0.027	0.09	3.24	--.---	50	0.000	0.028	0.02	3.66	--.---	80	0.000	0.003	0.00	2.31	--.---
21	0.001	0.016	0.10	2.99	--.---	51	0.000	0.031	0.03	3.66	--.---	81	0.000	0.008	0.00	2.77	--.---
22	0.001	0.013	0.08	2.87	--.---	52	0.000	0.029	0.02	3.65	--.---	82	0.000	0.009	0.00	2.94	--.---
23	0.000	0.009	0.08	2.52	--.---	53	0.000	0.025	0.01	4.18	--.---	83	0.000	0.006	0.00	3.20	--.---
24	0.000	0.007	0.06	2.27	--.---	54	0.000	0.029	0.01	4.27	--.---	84	0.000	0.004	0.00	3.10	--.---
25	0.000	0.006	0.06	2.43	--.---	55	0.000	0.035	0.00	4.34	--.---	85	0.000	0.004	0.00	2.74	--.---
26	0.000	0.004	0.06	2.32	--.---	56	0.000	0.034	0.00	4.05	--.---	86	0.000	0.003	0.00	2.37	--.---
27	0.000	0.002	0.08	2.17	--.---	57	0.000	0.030	0.00	3.84	--.---	87	0.000	0.003	0.00	2.19	--.---
28	0.001	0.002	0.09	2.06	--.---	58	0.000	0.020	0.00	3.60	--.---	88	0.000	0.003	0.00	1.97	--.---
29	0.001	0.001	0.11	2.19	--.---	59	0.000	0.013	0.00	3.51	--.---	89	0.000	0.003	0.00	1.82	--.---
30	0.001	0.002	0.14	2.64	--.---	60	0.000	0.011	0.00	3.40	--.---	90	0.000	0.002	0.00	1.56	--.---

Figure 10-2A Second-by-second iM240 emissions readings (courtesy Environmental Systems Products, Inc.).

RG-240 per Second TEST RESULTS #2

Plate: ESP1
Year: 1994
Make: FORD
Model: TBIRD

Date: 09/27/94
Time: 11:41
Test ID: 00000015

Sec	HC g	NOx g	CO g	CO2 g	Purge	Sec	HC g	NOx g	CO g	CO2 g	Purge	Sec	HC g	NOx g	CO g	CO2 g	Purge
91	0.000	0.001	0.00	1.41	--.---	141	0.000	0.028	0.01	4.71	--.---	191	0.000	0.001	0.00	4.85	--.---
92	0.000	0.001	0.00	1.43	--.---	142	0.000	0.020	0.00	4.24	--.---	192	0.000	0.000	0.00	4.40	--.---
93	0.000	0.001	0.00	1.14	--.---	143	0.000	0.010	0.00	4.27	--.---	193	0.000	0.000	0.00	4.41	--.---
94	0.000	0.001	0.00	1.08	--.---	144	0.000	0.004	0.00	3.74	--.---	194	0.000	0.000	0.00	4.21	--.---
95	0.000	0.001	0.00	0.97	--.---	145	0.000	0.002	0.00	2.95	--.---	195	0.000	0.000	0.00	4.31	--.---
96	0.000	0.001	0.00	1.04	--.---	146	0.000	0.002	0.00	2.60	--.---	196	0.000	0.000	0.00	4.26	--.---
97	0.000	0.000	0.00	0.95	--.---	147	0.000	0.004	0.00	3.12	--.---	197	0.000	0.000	0.00	4.31	--.---
98	0.000	0.000	0.00	0.84	--.---	148	0.000	0.013	0.00	3.61	--.---	198	0.000	0.000	0.00	4.21	--.---
99	0.000	0.000	0.00	0.97	--.---	149	0.000	0.011	0.00	2.88	--.---	199	0.000	0.000	0.00	3.89	--.---
100	0.000	0.000	0.00	1.09	--.---	150	0.000	0.009	0.00	2.50	--.---	200	0.000	0.000	0.00	3.81	--.---
101	0.000	0.000	0.00	1.10	--.---	151	0.000	0.008	0.00	2.83	--.---	201	0.000	0.000	0.00	3.88	--.---
102	0.000	0.000	0.00	1.25	--.---	152	0.000	0.007	0.00	3.17	--.---	202	0.000	0.000	0.00	3.81	--.---
103	0.000	0.003	0.00	2.12	--.---	153	0.000	0.009	0.00	2.89	--.---	203	0.000	0.000	0.00	3.77	--.---
104	0.000	0.007	0.00	2.44	--.---	154	0.000	0.008	0.00	3.00	--.---	204	0.000	0.000	0.00	3.81	--.---
105	0.000	0.022	0.00	3.41	--.---	155	0.000	0.007	0.00	2.73	--.---	205	0.000	0.000	0.00	3.71	--.---
106	0.000	0.032	0.01	4.67	--.---	156	0.000	0.005	0.00	2.45	--.---	206	0.000	0.000	0.00	3.43	--.---
107	0.000	0.029	0.01	5.67	--.---	157	0.000	0.004	0.00	2.36	--.---	207	0.000	0.000	0.00	2.73	--.---
108	0.000	0.027	0.00	5.98	--.---	158	0.000	0.004	0.00	2.60	--.---	208	0.000	0.000	0.00	2.23	--.---
109	0.000	0.018	0.00	5.23	--.---	159	0.000	0.006	0.00	3.13	--.---	209	0.000	0.000	0.00	1.96	--.---
110	0.000	0.011	0.00	4.71	--.---	160	0.000	0.012	0.00	3.65	--.---	210	0.000	0.000	0.00	1.57	--.---
111	0.000	0.005	0.00	3.98	--.---	161	0.000	0.020	0.00	4.62	--.---	211	0.000	0.000	0.00	1.83	--.---
112	0.000	0.005	0.00	4.11	--.---	162	0.000	0.021	0.00	5.25	--.---	212	0.000	0.011	0.00	2.51	--.---
113	0.000	0.007	0.00	4.19	--.---	163	0.000	0.014	0.00	5.76	--.---	213	0.000	0.031	0.00	4.29	--.---
114	0.000	0.006	0.00	3.19	--.---	164	0.000	0.006	0.00	5.79	--.---	214	0.000	0.018	0.00	5.47	--.---
115	0.000	0.005	0.00	2.84	--.---	165	0.000	0.005	0.00	5.95	--.---	215	0.000	0.027	0.00	5.78	--.---
116	0.000	0.004	0.00	3.15	--.---	166	0.000	0.004	0.00	6.40	--.---	216	0.000	0.026	0.00	6.41	--.---
117	0.000	0.008	0.00	3.89	--.---	167	0.000	0.006	0.00	6.89	--.---	217	0.000	0.016	0.00	6.64	--.---
118	0.000	0.010	0.00	3.79	--.---	168	0.000	0.007	0.01	7.17	--.---	218	0.000	0.010	0.00	6.24	--.---
119	0.000	0.006	0.00	2.84	--.---	169	0.000	0.007	0.00	7.33	--.---	219	0.000	0.004	0.00	5.02	--.---
120	0.000	0.003	0.00	1.95	--.---	170	0.000	0.006	0.01	7.59	--.---	220	0.000	0.002	0.00	3.96	--.---
121	0.000	0.002	0.00	1.64	--.---	171	0.000	0.004	0.00	7.47	--.---	221	0.000	0.002	0.00	3.19	--.---
122	0.000	0.001	0.00	1.37	--.---	172	0.000	0.003	0.01	6.75	--.---	222	0.000	0.001	0.00	1.80	--.---
123	0.000	0.001	0.00	1.32	--.---	173	0.000	0.004	0.00	5.85	--.---	223	0.000	0.001	0.00	1.39	--.---
124	0.000	0.001	0.00	1.31	--.---	174	0.000	0.006	0.01	5.43	--.---	224	0.000	0.000	0.00	1.28	--.---
125	0.000	0.012	0.00	2.76	--.---	175	0.000	0.005	0.00	5.37	--.---	225	0.000	0.000	0.00	1.24	--.---
126	0.000	0.024	0.00	3.37	--.---	176	0.000	0.002	0.01	4.89	--.---	226	0.000	0.000	0.00	1.40	--.---
127	0.000	0.018	0.00	3.08	--.---	177	0.000	0.001	0.02	4.16	--.---	227	0.000	0.000	0.00	1.39	--.---
128	0.000	0.010	0.00	2.42	--.---	178	0.000	0.001	0.06	3.60	--.---	228	0.000	0.000	0.00	1.33	--.---
129	0.000	0.006	0.01	2.27	--.---	179	0.000	0.000	0.07	3.39	--.---	229	0.000	0.000	0.01	1.32	--.---
130	0.000	0.004	0.00	2.40	--.---	180	0.000	0.000	0.07	3.31	--.---	230	0.000	0.000	0.00	1.23	--.---
131	0.000	0.002	0.00	2.16	--.---	181	0.000	0.000	0.07	3.60	--.---	231	0.000	0.000	0.00	1.30	--.---
132	0.000	0.001	0.00	1.63	--.---	182	0.000	0.000	0.05	3.98	--.---	232	0.000	0.000	0.00	1.41	--.---
133	0.000	0.001	0.00	1.42	--.---	183	0.000	0.001	0.03	4.20	--.---	233	0.000	0.000	0.00	1.29	--.---
134	0.000	0.001	0.00	1.24	--.---	184	0.000	0.005	0.02	4.36	--.---	234	0.000	0.000	0.00	1.27	--.---
135	0.000	0.000	0.00	1.18	--.---	185	0.000	0.004	0.01	4.19	--.---	235	0.000	0.001	0.00	1.20	--.---
136	0.000	0.000	0.00	1.14	--.---	186	0.000	0.003	0.01	4.37	--.---	236	0.000	0.000	0.00	1.32	--.---
137	0.000	0.000	0.00	1.38	--.---	187	0.000	0.004	0.00	4.66	--.---	237	0.000	0.000	0.00	1.34	--.---
138	0.000	0.000	0.00	1.55	--.---	188	0.000	0.004	0.01	5.32	--.---	238	0.000	0.000	0.00	1.30	--.---
139	0.000	0.000	0.00	1.98	--.---	189	0.000	0.002	0.00	5.49	--.---	239	0.000	0.000	0.00	1.20	--.---
140	0.000	0.017	0.00	3.57	--.---	190	0.000	0.001	0.00	5.54	--.---	240	0.000	0.001	0.00	1.13	--.---

Figure 10-2B Second-by-second IM240 emissions readings (courtesy Environmental Systems Products, Inc.).

Figure 10-3 A dynamometer can simulate a road test to baseline the vehicle (courtesy Clayton Industries).

Light Blue Smoke

Light blue smoke is an indication that the engine is burning oil; this would result in higher hydrocarbons. It is important to note when the smoke appears and how much is exhausted into the atmosphere:

1. *Start-Up* — This generally indicates that the valve guide seals may be leaking into the combustion chamber. It could also indicate that the valve cover gaskets are leaking and oil has seeped into the exhaust manifold.
2. *Acceleration* — This generally indicates that the oil control rings are not performing up to specifications, because they cannot control the oil when put under load.
3. *Deceleration* — This generally indicates defective valve guide seals. This is more evident on deceleration because of the high manifold vacuum present in the intake manifold during deceleration, which in turns allows oil to be drawn down the valve guide stem.
4. A leaking automatic transmission vacuum modulator can cause the above conditions. Check the modulator's vacuum hose at the intake manifold for signs of fluid; if present, the diaphragm in the modulator is ruptured and must be replaced.

White Color Smoke

White color smoke is a sign of moisture being present in the exhaust system. Excessive moisture can occur in a couple of ways, and might be normal or abnormal:

1. *Cold Warm-Up* — Condensation occurs in the engine, catalytic converter, and exhaust manifold after a hot engine is shut down. During the first few minutes of warm-up, droplets of water and/or steam appear at the tailpipe. This condition should last only a few minutes.
2. Hot Running Engine — A slight amount of moisture (steam) might be visible, but droplets of water should not be present with a hot engine. If there is a considerable amount of water, it is an indication that an internal problem exists with the cooling system, cylinder block, or head gasket(s). This could result in higher hydrocarbons.

Black Color Smoke

Black color smoke is a sign of a rich air/fuel ratio. Check the tailpipe for a dry, sooty black residue. If present, check any component or adjustment that could create a rich air/fuel ratio. Also, see if the black smoke is present at idle, part-throttle, under a snap acceleration, or during deceleration. This could result in higher carbon monoxide and a lower oxygen level.

Spending just a few minutes to check the exhaust smoke at the tailpipe can provide the technician with an excellent starting point, and it doesn't even require any tools or equipment.

EXHAUST GAS ANALYZER

The most basic piece of equipment a shop must have is an exhaust gas analyzer. Both four-gas and five-gas analyzers are now available. The fifth gas is NO_x and it is measured in parts per million. The prominent units currently in use are sold by Snap-On, MPSI, and OTC.

Besides the addition of NO_x, the biggest change with these units is that they are now portable. This means that a shop without a dynamometer can take the vehicle on a test drive with the exhaust probe in the tailpipe (Figure 10-4).

The advantage of this is the fact that these units are designed to *record* the exhaust emissions readings when triggered by the technician. Once the data is collected, the technician can return to the shop and review the data and print a copy of the results (Figure 10-5). If this sounds familiar, it is, because it works similar to a scan tool snapshot. In some cases, the portable exhaust analyzer uses the scan tool as its monitor.

An additional feature of the OTC Gas Analyzers is the diagnostic program designed into the cartridge. The test procedure will be directed by instructions displayed on the Monitor 4000 Enhanced screen. The primary information displays and the paths followed are shown in Figure 10-6, page 164. Additional message displays not shown in Figure 10-6 can also be displayed. These messages guide the technician by requesting information or requiring specific steps to be performed.

Before determining if emissions readings are in the correct range, the technician must understand each of the five gases and how they relate to each other. The following paragraphs explain the basics of each exhaust gas. All readings described in the following sections are for after the catalytic converter.

Three things should be checked when monitoring exhaust gas readings:

1. Be sure the air injection system is disabled to prevent a false oxygen reading or diluted CO_2 reading.

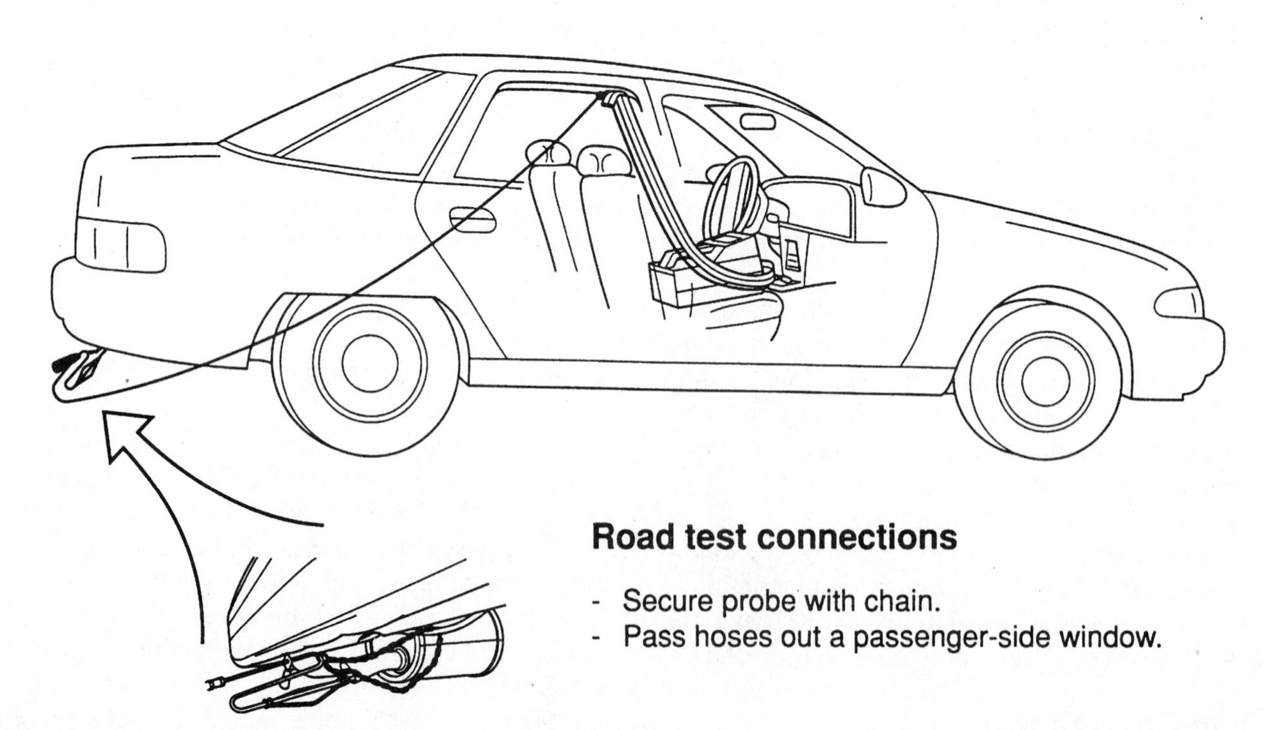

Figure 10-4 A portable exhaust gas analyzer allows the technician to take it along on the road test (courtesy Micro Processor Systems, Inc. – MPSI).

The data will be printed as shown.

Time of Day Recorded
Seconds per Frame
Gas Event
Date Recorded
Frame Number
Data

10-10-91 | 13:38 | Metric

0.72 seconds per frame

* - over limit

FRM	CO	HC	CO2	O2	NOx (5-gas only)	RPM	OIL
11	0.0	0	0.0	00.0	000	000	00
12	0.0	0	0.0	00.0	000	000	00
13	0.0	0	0.0	00.0	000	000	00
14	0.0	0	0.0	00.0	000	000	00
15	0.0	0	0.0	00.0	000	000	00
16	0.0	0	0.0	00.0	000	000	00
17	0.0	0	0.0	00.0	000	000	00
18	0.0	0	0.0	00.0	000	000	00
19	0.0	0	0.0	00.0	000	000	00
20	0.0	0	0.0	00.0	000	000	00
21	0.0	0	0.0	00.0	000	000	00
22	0.0	0	0.0	00.0	000	000	00

-----PRINTING COMPLETE-----

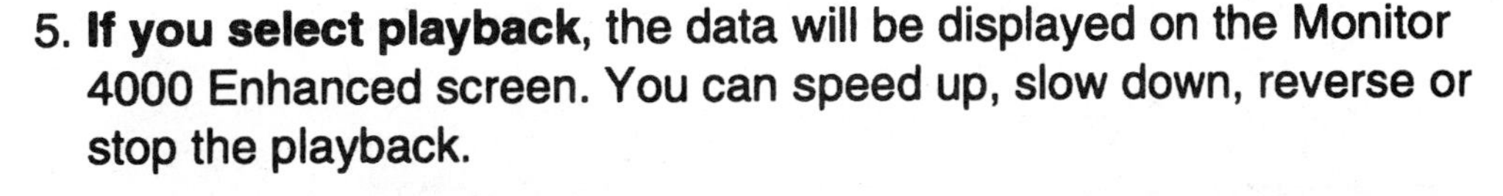

5. **If you select playback**, the data will be displayed on the Monitor 4000 Enhanced screen. You can speed up, slow down, reverse or stop the playback.

- Speed up: press the → key.
- Slow down: press the ← key.
- Stop: press the → or ← key until STP is displayed.
- Reverse: press the ← key until the negative speeds are displayed. Press the → key to return to forward playback.

PLAYBACK SELECTION RANGE

- 8, - 4, - 2, - 1, STOP, 1, 2, 4, 8

Each selection multiplies the playback speed by the number selected i.e. 8 times normal speed.

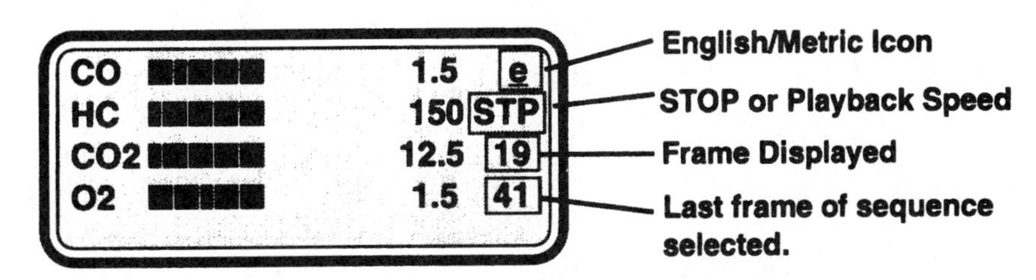

Figure 10-5 Portable exhaust gas analyzers can be used to take a snapshot of the emissions (courtesy OTC, a division of SPX Corporation).

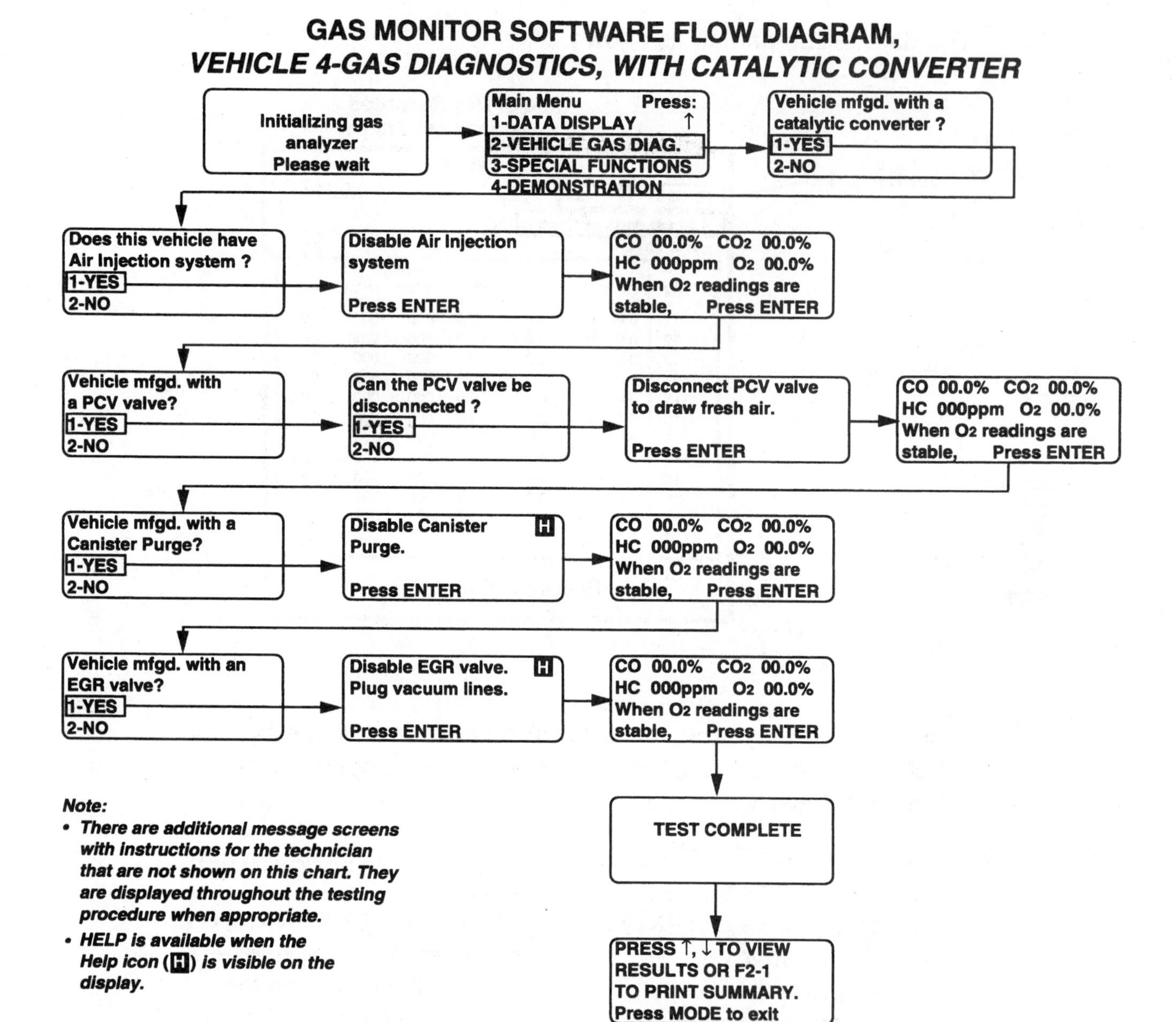

Figure 10-6 Diagnostic software is designed to help the technician (courtesy OTC, a division of SPX Corporation).

2. See if there is a two-way or three-way catalytic converter installed.
3. Bring the engine up to operating temperature.

Hydrocarbons (HC)

Hydrocarbons (HC) are complex organic molecular chains consisting of hydrogen and carbon atoms. Hydrocarbons are a generic term for any fuel, including gasoline, propane, kerosene, oil, etc. When HC is released into the atmosphere it contributes to photochemical smog.

All engines emit a certain amount of HC into the exhaust. No real-world engine burns all of its fuel. Even a tuned engine in good mechanical and electrical condition produces some HC. When the flame front within the combustion chamber reaches the cool cylinder wall, it quenches, leaving some unburned gas (Figure 10-7).

Figure 10-7 Quench area of the combustion chamber, where HC is created.

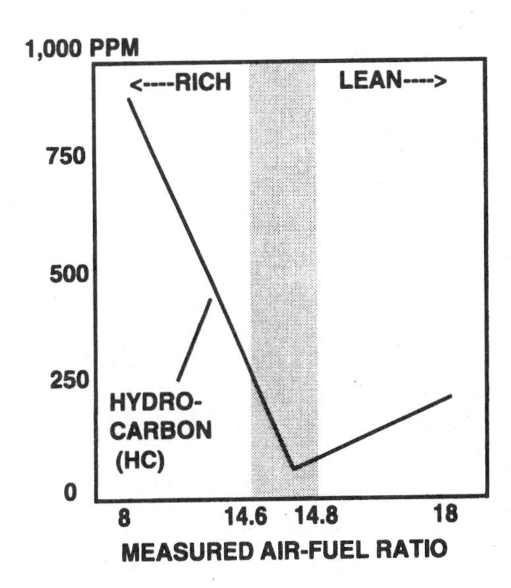

Figure 10-8 How HC relates to the stoichiometric air/fuel ratio (courtesy OTC, a division of SPX Corporation).

Hydrocarbons are measured in concentrations of parts per million (ppm). This measures the ratio of HC molecules to the total molecules sampled. For example, 500 ppm is equal to 500 HC molecules out of the total 1,000,000 molecules sampled by the exhaust analyzer.

When there is excessive HC in the exhaust, the technician should realize that something is wrong. Excessive HC indicates that the fuel entering the combustion to produce power went out the exhaust unburned. This is generally the result of an engine misfire. There are three different kinds of engine misfires:

1. *Total Misfire* — The cylinder is completely dead (i.e., open plug wire, zero compression, etc.).
2. *Partial Misfire* — The cylinder is working but is not 100 percent (i.e., plug with a small amount of carbon buildup, small gap, wire deteriorating, air/fuel ratio excessively lean or rich, etc.).
3. *Intermittent Misfire* — The cylinder sometimes works, and sometimes does not (i.e., coil breaks up at certain times, module acts up, etc.).

Keep in mind that HC in the exhaust only indicates that there is a misfire somewhere under the hood (Figure 10-8). It cannot by itself show the nature or location of the misfire (plug, plug wire, etc.).

Do not make the mistake of automatically assuming the ignition is at fault when a misfire is uncovered. The other possibilities are compression, fuel delivery, timing, too much fuel, too little fuel, oil in the combustion chamber, vacuum leaks, etc. In other words, anything that would cause the burning process to stop sooner than expected will produce higher HC.

The amount of acceptable HC in the exhaust will vary from engine to engine. The technician's goal should be to achieve the lowest possible readings below factory specifications and/or applicable emissions standards.

Carbon Monoxide (CO)

Carbon monoxide (CO) consists of one carbon molecule and one oxygen molecule. It is a poisonous, odorless, and tasteless gas. It can kill a person by attaching itself to the red blood cells, replacing the oxygen in the blood stream and causing suffocation.

CO is the result of burning fuel (HC) in the combustion chamber. When air and fuel are ignited in the cylinder and partial combustion occurs, CO is produced. The amount of CO that is produced is in direct proportion to the air/fuel ratio (Figure 10-9). CO, therefore, is a byproduct of combustion.

Carbon monoxide is measured as a percentage (%) of the total emissions sampled. It could be measured in parts per million, but the numbers would

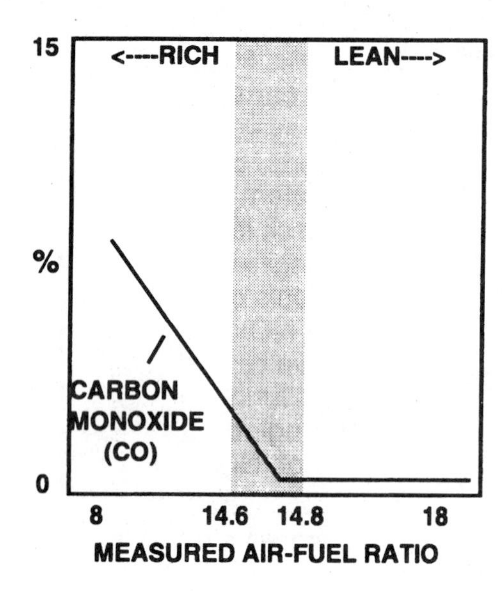

Figure 10-9 How CO relates to the stoichiometric air/fuel ratio (courtesy OTC, a division of SPX Corporation).

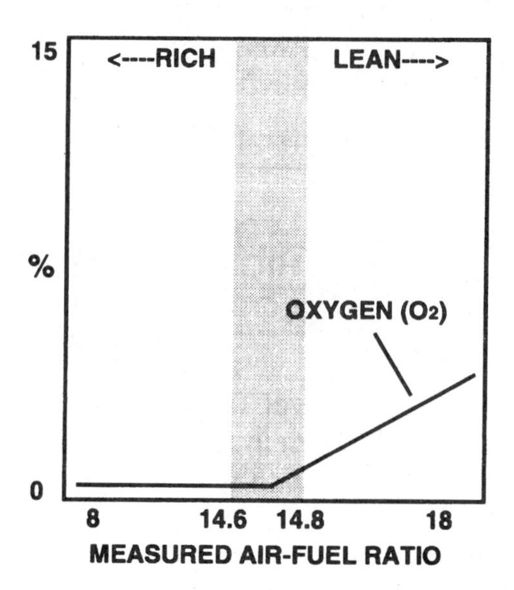

Figure 10-10 How O_2 relates to the stoichiometric air/fuel ratio (courtesy OTC, a division of SPX Corporation).

be in the thousands. For example, 10% would be 100,000 ppm of the total 1,000,000 molecules sampled by the exhaust analyzer.

Carbon monoxide is the best indicator of a rich air/fuel ratio. High CO means that there is either not enough oxygen or there is too much fuel in the combustion chamber to burn the fuel completely. Lack of oxygen or too much fuel creates a rich mixture.

CO levels should be at or below factory specifications. The phrase, "as low as possible" does not apply here. This is because the lower the level, the leaner the air/fuel ratio. If there is too much oxygen present during the combustion process, a lean misfire will occur. This in turn will increase the HC level.

Oxygen (O_2)

Approximately twenty-one percent of the air we breathe is oxygen, seventy-eight percent is nitrogen, and the remaining one percent is made up of other particulates. An oxygen molecule consists of two oxygen atoms (O_2). Oxygen will be produced when normal combustion occurs in the cylinder, a lean misfire is occurring, from the air injection system, leaks in the exhaust system, or from a catalyst that is not working properly.

O_2 is measured as percentage of the exhaust gas sampled (Figure 10-10). It is compared to an ambient oxygen reading of 20.7%.

The use of O_2 in automotive diagnostics can lead the technician to answers more often than any of the other gases. While CO can indicate a rich condition, it cannot indicate a lean one. Once CO drops to approximately 0.5% or lower, it levels out and is virtually useless as a diagnostic aid. O_2, on the other hand, is an excellent indicator of a lean-running engine. When an engine is running lean, O_2 will increase proportionately as the air/fuel ratio becomes leaner. This increase will continue through the lean misfire point. As the engine goes to the lean misfire condition, the O_2 readings will rise dramatically as the misfire increases.

If the air/fuel ratio is adjusted properly, the catalytic converter is working properly, and the exhaust system is good, the O_2 readings will be somewhere between 0.5 and 2.5%. An ideal oxygen level would be 1-2%. The closer the reading is to 0%, the richer the air/fuel ratio, and the higher the readings are above 2.5%, the leaner the air/fuel ratio.

Oxides of Nitrogen (NO_x)

Oxides of nitrogen (NO_x) consist of one nitrogen molecule and one or more oxygen molecules. Oxides of nitrogen irritate the lungs, lower resistance to respiratory infections, and contribute to the development of emphysema, bronchitis, and pneumonia. NO_x

contributes to ozone formation and can also react chemically in the air to form nitric acid.

NO_x is the result of combustion chamber temperatures that exceed 2500°F. Nitrogen in the air combines with oxygen in the combustion chamber to form NO_x. A condition must exist in the combustion chamber that prevents the control of combustion chamber temperatures. The inability to control combustion chamber temperatures can result in NO_x emissions or possibly damage the pistons and valves.

Oxides of nitrogen are measured in concentrations of parts per million (ppm) (Figure 10-11). This measures the ratio of NO_x molecules to the total number of molecules sampled. For example, 1000 ppm is equal to 1000 HC molecules of the total 1,000,000 molecules sampled by the exhaust analyzer.

At this time, no standards for parts per million have been established for NO_x. Obviously, the lower the reading the better, but the maximum emissions in parts per million is not known at this time.

Carbon Dioxide (CO_2)

Like CO, carbon dioxide (CO_2) is also a byproduct of combustion. However, CO_2 is the result of one carbon molecule combining with two oxygen molecules.

CO_2 is the result of burning fuel (HC) in the combustion chamber. If sufficient oxygen is present, then the leftover carbon can combine with two oxygen molecules to create CO_2.

Carbon dioxide is measured as a percentage (%) of the total emissions sampled (Figure 10-12). It could be measured in parts per million, but the numbers would be in the thousands. For example, 15% would be 150,000 ppm of the total 1,000,000 molecules sampled by the exhaust analyzer.

CO_2 can show what HC and CO by themselves cannot and how efficient an engine is running. Any problem under the hood that has to do with the burning process will affect CO_2 levels. Since we have to burn the HC to produce CO_2, the technician cannot expect to see good CO_2 levels with a misfire occurring. Air/fuel ratios also affect CO_2 because the air/fuel has to be right on the money or close to it to have the best possible (highest) CO_2 levels.

CO_2 levels will vary from engine to engine. A reading of 12% and above is an indication of an efficient engine and a tight exhaust system. The higher the level is above 12%, the more efficient the system. If the CO_2 level drops below 12%, then one or more of the other gases should indicate a problem. If the problem is only a leaking exhaust system and not an engine-related problem, then the oxygen readings will increase.

3000 PPM
<----RICH
LEAN---->
2000
1000
0
OXIDES of NITROGEN (NOx)
8
14.6
14.8
18
MEASURED AIR-FUEL RATIO

Figure 10-11 How NO_x relates to the stoichiometric air/fuel ratio (courtesy OTC, a division of SPX Corporation).

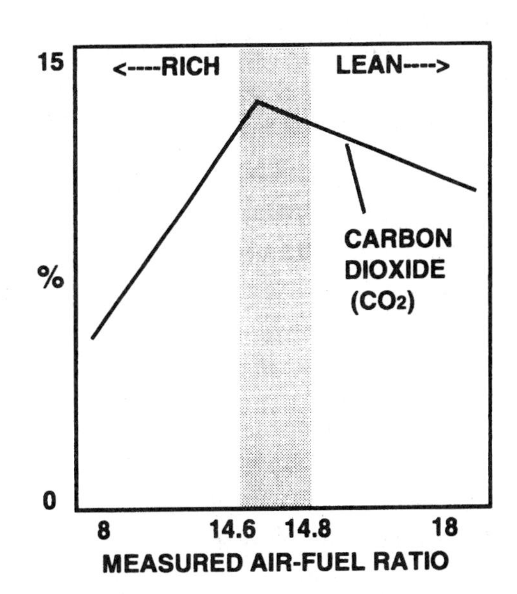

Figure 10-12 How CO_2 relates to the stoichiometric air/fuel ratio (courtesy OTC, a division of SPX Corporation).

ANALYZING EXHAUST GAS READINGS

While numerous possibilities exist, certain procedures should be followed to diagnose exhaust gas readings. The procedures will vary slightly when a catalytic converter is installed. Always start out by looking at the CO readings first. This should be done at idle and 2500 RPM.

For non-catalyst vehicles, a high CO level at either speed is caused by a rich air/fuel ratio. This must be adjusted first, because it will affect the HC readings if severe enough.

Since most non-catalyst engines are equipped with a carburetor, start there. If the CO reading was high during cruise only, begin by checking for:

1. A dirty air filter.
2. A fully-open choke butterfly.
3. An evaporative canister loaded with fuel.
4. An EGR valve that does not open.
5. A PCV valve that is stuck closed or a crankcase that is loaded with fuel.

If the CO reading was high at idle only, check for a carburetor main nozzle discharging fuel because of a high float level or a stuck open/leaking power valve. Next, check the mixture screw(s) settings. This can be done by turning the mixture screw(s) in and counting the turns to bottom the screw(s). Back out the screw(s) 2 to 3 full turns. Set the final air/fuel ratio to factory specifications or slightly below. At the same time, be sure that the curb idle speed is set correctly, because it will affect the CO readings.

For catalyst vehicles, carbureted or fuel-injected, always start out by looking at the CO readings first. If CO is above 0.5% and O_2 is at or near 0%, then the air/fuel ratio is too rich. When the CO is below 0.5% and O_2 is between 0.5 to 2.5% at idle and cruise, then the air/fuel is okay. If O_2 is above CO, and CO is above 0.5%, then the possibility exists that the catalyst may be bad.

When the HC reading is high and CO is low, first make sure that a lean misfire does not exist. This would be indicated by high HC, a CO reading well under specifications, and O_2 over 3%. Begin by looking for anything that could cause a lean misfire, such as vacuum leaks, low fuel pressure, or an air/fuel ratio set too lean, etc.

If the CO is at or slightly below specifications, check for an ignition or mechanical engine problem that could cause high HC. An engine misfire causing high HC can also drive O_2 up, because excess O_2 is left when a cylinder misfires.

This can be checked with an engine analyzer that should include a vacuum gauge, ignition scope, and a cylinder power balance unit. A vacuum gauge reading that intermittently drops or rapidly fluctuates can help identify a mechanical problem; a floating needle can indicate an air/fuel problem; and a needle reading lower than normal can indicate retarded ignition timing or a vacuum leak. The ignition scope is best for detecting ignition problems and some air/fuel-related problems. A mechanical problem can show up in a scope pattern and can be isolated with a cylinder balance test. Performing a cylinder balance test while monitoring the vacuum gauge, ignition scope, and exhaust gas analyzer can help pinpoint a mechanical problem to a specific cylinder. When performing a cylinder balance, the cylinder(s) causing the high HC readings will show little or no change when it is canceled out.

For high HC and O_2 readings at 2500 RPM, look for low fuel pressure, low carburetor float level, intermittent misfire, or over-advanced ignition timing. Use a fuel pressure gauge, vacuum gauge, ignition scope, or a timing light with an advance meter to locate the problem. If the problem is a total lean misfire, try slowly injecting propane into the engine to confirm a lean air/fuel ratio.

When HC and CO are slightly higher than normal, check for a catalytic converter that is not reaching operating temperature or one that is not working at all. Eliminate all other possibilities before condemning a catalytic converter.

To verify catalyst operation on a closed-loop feedback system, perform the following procedures:

1. Hold the engine at 2000 RPM, and watch the exhaust readings. If the converter's cold, the readings should continue to drop until the converter reaches "light-off" temperature.
2. When the numbers stop dropping, check the oxygen levels. Oxygen should be approximately 0.5 to 1.0%. This shows that the converter is using most of the available oxygen. There's one exception to this: if there is no CO left, there can be more oxygen in the exhaust, but it should be less than 2.5%.

3. If there is too much oxygen left, and no CO in the exhaust, stop the test, and make sure the system is in air/fuel control as explained in Chapter 8. If not, take care of this problem first. If the system is in control, use your propane enrichment tool to bring the CO level up to approximately 0.5%; now the oxygen level should drop to zero.
4. Once you have a solid oxygen reading, snap the throttle open, then let it drop back to idle. Check the rise in oxygen level; it should not rise above approximately 1.2%.

If the converter passes these tests, it is working properly. If the converter fails either test, chances are that it is working improperly or not at all.

On non-feedback vehicles, it is possible for the vehicle to be running lean. With the lean mixture, the converter has nothing to do, so the oxygen level at 2000 RPM will be higher than normal. This doesn't mean the converter is bad. On non-feedback vehicles, there is a different procedure:

1. Disable the AIR system.
2. Check the exhaust oxygen level at 2000 RPM. If the oxygen drops to zero, skip the snap throttle part of the test. If there is still oxygen in the exhaust at 2000 RPM, and CO drops to zero, the mixture is too lean to test the converter.
3. If the fuel mixture is lean during the test, the converter has nothing to do, so the oxygen level at 2000 RPM will be higher than normal. This does not mean the converter is bad.
4. Put the hose from the propane enrichment tool into the air cleaner inlet and slowly add propane to the mixture until the CO levels at 2000 RPM are about 0.5%. The oxygen level should drop as the catalyst starts using the oxygen to convert CO to CO_2. Keep the CO level steady. The converter should now use all of its oxygen, so the exhaust oxygen level should drop to zero.
5. Once there is a solid oxygen reading, snap the throttle open, then let it drop back to idle. Watch the oxygen level; it should not rise above approximately 1.2%.

If the converter passes these tests, it is working properly. If the converter fails either test, chances are that it is working improperly or not at all.

When reading CO_2, the main thing to look for is that it is as high as possible and that the other readings are within specifications. If there is a problem with any of the other readings, then the CO_2 will decrease. If CO_2 and O_2 are both low, look to see if the AIR is still hooked up or the vehicle is in open loop. If O_2 is higher than CO_2, look for exhaust leaks somewhere in the system. Remember, CO_2 is a byproduct of combustion and represents the overall efficiency of the engine and exhaust system.

The final gas is NO_x. Currently, there are no standards for parts per million. The technician will need to baseline the vehicle when testing begins. This way, the baseline readings can be compared to the after-repair readings to see if a significant reduction was achieved. If the vehicle is in for high NO_x readings, remember to cover the basics of nonloaded testing:

1. Be sure that the EGR valve is working correctly and that the passages are free of carbon, allowing sufficient flow.
2. Check that the air/fuel ratio is correct, because a lean condition or misfire will cause NO_x to rise.
3. Make sure that the base ignition timing is set correctly and that any applicable advance mechanisms, including the knock sensor, are working correctly. Over-advanced timing will increase compression pressures and raise cylinder temperatures.
4. Check to be sure that the engine is not overheating—the correct thermostat and engine operating temperature is critical to the control of NO_x.
5. If the engine's compression ratio has been modified, chances are that NO_x will be affected.
6. A high-mileage engine or one that is regularly operated without being sufficiently brought up to operating temperature can cause carbon buildup in the cylinder. If the engine has been burning oil or has been run excessively rich for long periods of time, this will cause carbon buildup. Use a quality decarbonizing chemical to reduce the amount of carbon. Carbon does two things: it raises the compression ratio, and it absorbs and releases heat that can cause pre-ignition.
7. Last but not least, if all else checks out okay, then the NO_x portion of the catalyst might be bad. To date, there is no specific test for diagnosing the reduction bed of the catalytic converter. One test that may work is to probe the exhaust before and after the catalytic converter to see if there is

reduction occurring. Presently, there has not been sufficient time to validate this test by testing a large number of vehicles.

ANALYZING IM240 TEST RESULTS

To effectively diagnose vehicles that fail an IM240 test, technicians will need to use the skills that most of them already have. In other cases, the material presented to this point has enhanced those skills. Three important procedures are necessary to diagnose and repair IM240 failures:

1. The technician must relate the failed emission(s) to the average speed and load at which the vehicle failed the drive cycle. It is important to know that the test results are the composite (overall average) for each emission. Intermittent emissions spikes are included in the composite, but they must be of sufficient time length and quantity to cause the vehicle to fail the test. Deceleration modes are not included in the composite readings.
2. The technician will need to baseline a vehicle in order to compare the repaired vehicle to the failed VIR. This baseline allows the technician to know if significant reductions have been achieved. The technician will be working in concentrations of parts per million and percentage and not in grams per mile.
3. The technician must attempt to recreate the driving conditions under which the vehicle failed. In most cases, this will be done on a road test using a portable gas analyzer.

Figure 10-13 shows a second-by-second trace of a good HC signal. The vertical scale of the trace represents the grams per mile. The horizontal scale represents the time in seconds. The HC standard for this vehicle was 2.0 grams per mile. Notice that there is one spike late in drive cycle, but the vehicle still passed. Each emission will have its own second-by-second trace reading.

For the technician to understand and diagnose an emissions failure, the duration of the high emissions and the speed and load must be looked at. Figure 10-14 shows the duration of time that the emissions level was above the cutpoints. Emissions

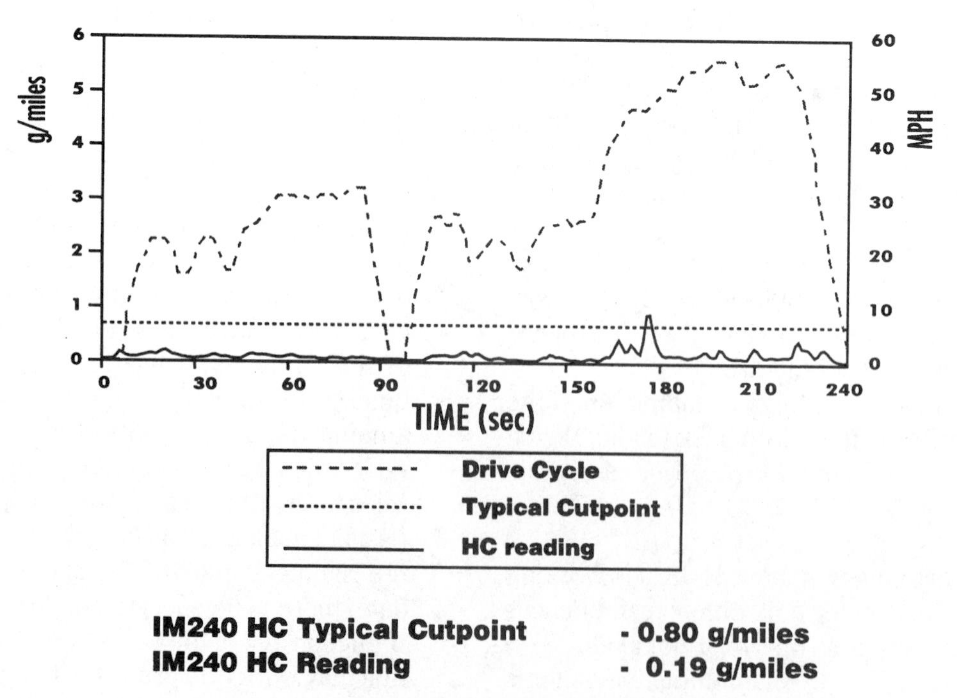

Figure 10-13 Hydrocarbon readings from a vehicle that passed the IM240 drive cycle test (courtesy Aspire, Inc.).

standards for each model year are called cutpoints. If the average emissions level is below the cutpoint during the test, the emission will pass.

High emissions levels that occur for a long duration of time indicate that the emissions were excessive, which could result in a failed test. It is to the advantage of the technician to look at the total area under the cutpoint.

The second-by-second trace allows the technician to see the actual emissions levels at different points in the IM240 drive cycle. Figure 10-15 (top) shows the second-by-second trace for an emissions level that passed. Notice that the overall highlighted area spent more time below the cutpoint line. Figure 10-15 (bottom) shows the trace for an emissions level that failed. Notice that the overall highlighted area spent more time above the cutpoint line. The emission shown failed one acceleration mode in Phase One and two different acceleration modes in Phase Two. This information provides the technician with a place to start.

The next step is to recreate the problem during a test drive so the technician can develop a baseline before starting diagnosis and repairs. The technician can now relate the high emissions to a specific time and speed.

During the IM240 drive cycle, six modes of operation are simulated:

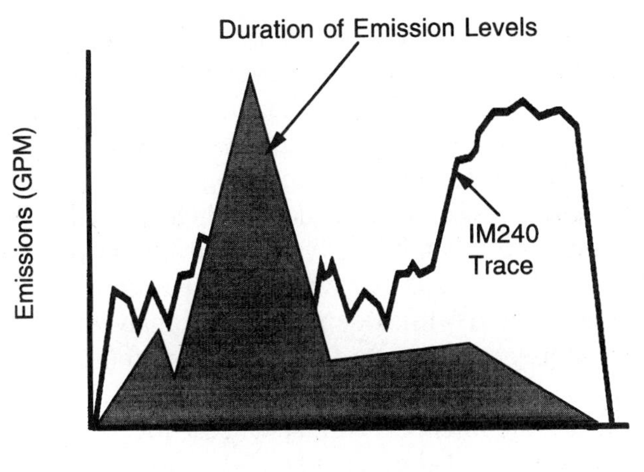

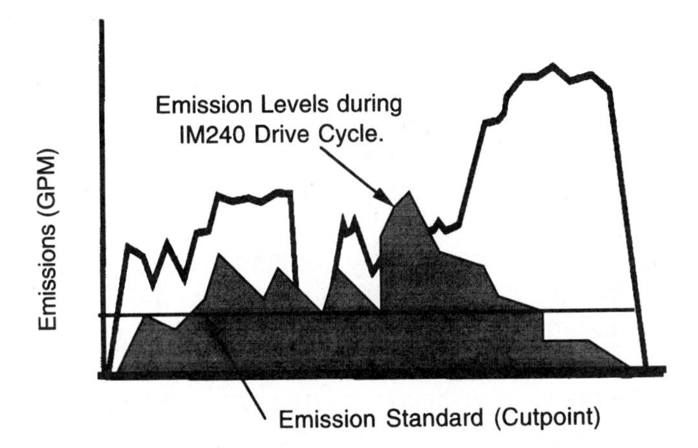

Figure 10-14 Emissions levels duration and cutpoint.

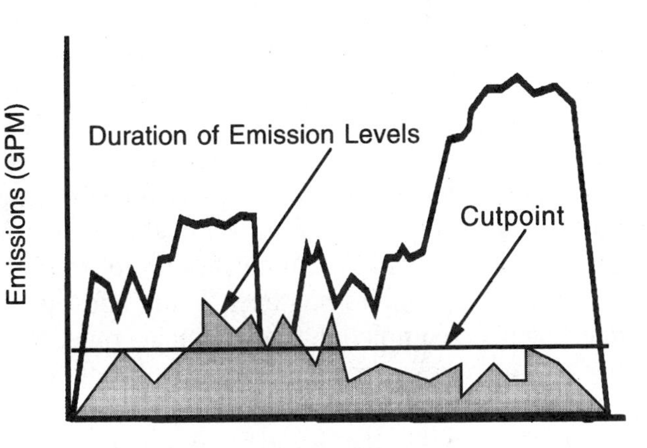

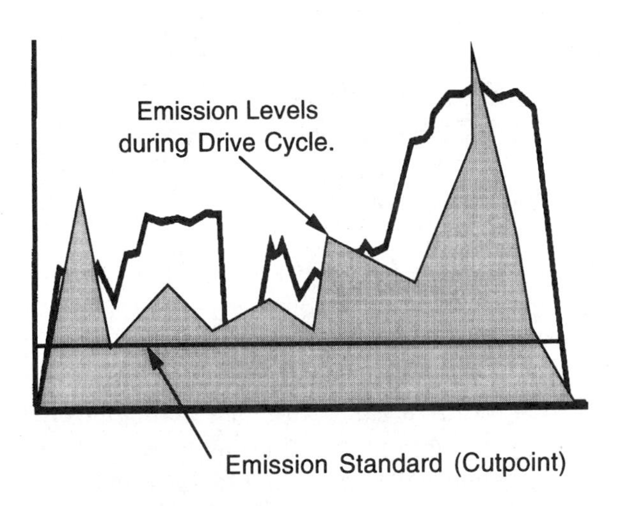

Figure 10-15 Top: Emission passes test. Bottom: Emission fails test.

1. *Mode 1* — Idle, no load at 0 mph.
2. *Mode 2* — Acceleration from 0-35 mph.
3. *Mode 3* — Acceleration from 35-55 mph.
4. *Mode 4* — Steady-state low cruise at 35 mph.
5. *Mode 5* — Steady-state high cruise at 55 mph.
6. *Mode 6* — Two decelerations, one from 35 to 0 mph and one from 55-0 mph.

At no time does the maximum acceleration speed change exceed 3.3 mph/sec. By carefully evaluating the second-by-second trace, the technician can recreate the trace by simulating one or more of the driving modes.

To recreate the failure shown on the bottom of Figure 10-15, the technician should rapidly accelerate the vehicle. If a portable gas analyzer is available, then the readings can be recorded during the baseline test and then compared to the final readings after the repairs are made.

The ability of the technician to recreate the emissions failure is critical to the success of the IM240 program. Repairing a vehicle the first time will prevent the ping-pong effect of a customer bouncing back and forth between a centralized station and the repair shop. This will lead to customer satisfaction that will help the program to succeed.

Once the problem has been found and the repair has been completed, the technician must refer back to the baseline reading. Since the baseline and after-repair readings are in parts per million, the amount of reduction achieved can be compared. Unfortunately, the IM240 VIR will not be of help because it is in grams per mile.

CAUSES AND REPAIR OF PURGE AND PRESSURE FAILURES

Purge Failures

The evaporative purge test is conducted during an IM240 transient dynamometer test to detect vehicles with inoperative evaporative canister purge systems. The test procedure includes disconnecting the test vehicle's vapor purge line running from the canister to the engine and installing a gas flowmeter in the line. After installing the flowmeter in the evaporative purge system, the vehicle is operated over the IM240 transient cycle, and the cumulative vapor purge flow in units of liters are recorded. The vehicle is recorded as a failure if its cumulative vapor purge is less than 1.0 liter.

There are a number of typical causes of purge failures. First, the canister purge solenoid or vacuum-operated purge valve can be missing, disconnected, bypassed, or otherwise inoperative. In addition, vacuum or vent lines can be disconnected, or missing, plugged, damaged, or incorrectly routed. Moreover, the purge hose can be disconnected, missing, split, or not sealed. Also, the canister purge thermal vacuum switch can be stuck, or there may be no emissions control module signal to the purge solenoid.

Pressure Failures

The pressure test of the evaporative system is used to determine the integrity of a vehicle's evaporative system and fuel tank. In order to perform evaporative system pressure testing, the following equipment is needed: a helium or nitrogen gas bottle, a standard regulator, hoses connecting the tank to a pressure meter and to the vehicle's evaporative system, and computer hardware to interface the metering system with a computerized analyzer. The test sequence consists of the following steps:

1. Test equipment is connected to the fuel tank canister hose at the canister end. The gas cap is checked to ensure that it is properly, but not excessively, tightened.
2. The system is pressurized to 14 ±0.5 inches of water without exceeding 26 inches of water system pressure. Fuel tank pressurization is done by modulating the nitrogen flow into the fuel system by successive opening and closing of the control valve by the operator. Modulating the nitrogen flow into the system allows a higher pressure nitrogen flow to be safely used to pressurize the system.
3. The gas cap is loosened, then retightened, and the vehicle is allowed to stand for up to two minutes to determine if it can continue to hold pressure.
4. If pressure in the system remains above 8 inches of water after two minutes, the vehicle passes the test.

Pressure failures indicate the potential existence of leakage on vehicles' evaporative emission control system components, which include the gas cap, filler neck, sending unit, rollover valve, and vent hoses.

The HC channel of the emissions analyzer is extremely sensitive to fuel vapor and can detect fuel leaks where visible signs are not present. To check for evaporative pressure loss or leaks:

1. Observe the HC readings.
2. The engine should be off and the fuel gas cap should be on tight.
3. Use the exhaust probe to "sniff" for leaks from the fuel pump, tanks, lines, and connections as shown in Figure 10-16.
4. A sudden rise in HC will indicate a leak in the system.
5. Allow the HC level to clear, and move the probe slowly back to the approximate place where the high HC was detected.
6. To pinpoint the leak, move the probe slowly back and forth over the suspected area and watch the HC levels. Remember that there is a slight delay between sample entry at the probe and the HC displayed on the analyzer.

REFERENCE MATERIALS

For any technician to be successful at repairing emissions failures and/or other problems, they will need to build a library of reference materials. With the tremendous amount of information available today, no one person can be expected to memorize it all.

A variety of manuals, books, video tapes, and technical bulletins are available. They vary in cost, so obviously the technician must select what best suits their needs and budget. They are a valuable diagnostic tool, so don't skimp when selecting reference materials or other resources such as the ones listed below.

1. *Factory manuals* provide preventive maintenance schedules, specifications, component testing, disassembly/reassembly procedures, component locations, wiring diagrams, and vacuum diagrams.

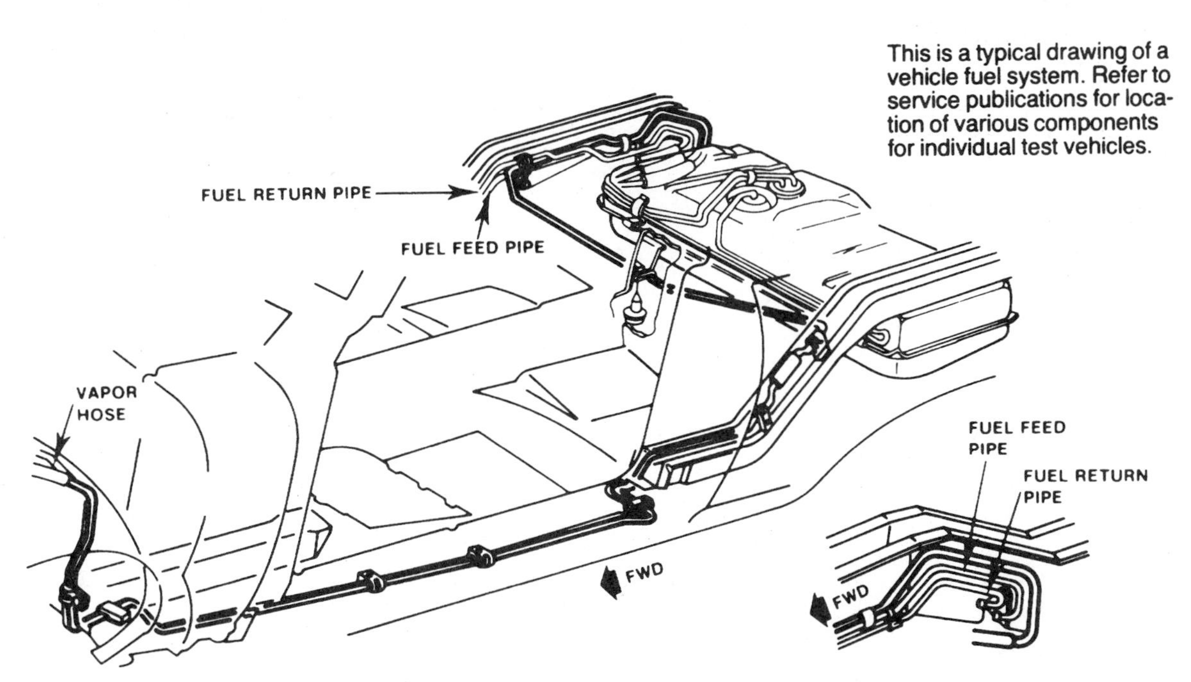

Figure 10-16 Use an exhaust gas analyzer to check for pressure leaks (courtesy Environmental Systems Products, Inc.).

There are disadvantages to these manuals. If the shop works on a variety of products, buying factory manuals can get very expensive. They cover one automotive year, and in most cases, one model. These manuals are written before the vehicle is put into production; therefore, you can expect to find some errors in these manuals as a result of last-minute updates and corrections.

2. *Aftermarket manuals* are written by independent publishers who compile the Original Equipment Manufacturer's (OEM) specifications, testing/repair procedures, and flat-rate standards into independent service and repair manuals. In some cases, the independent publishers do their own research and development of repair procedures. Specifications usually come directly from OEM factory manuals. Multiple years and models are covered in one manual.
3. *Data retrieval systems* are the newest information systems available to the aftermarket. They are based on Compact Disc (CD-ROM) technology, which means they require a computer to access the information. These systems provide quick access to a tremendous amount of information. The information and diagrams can be printed out, which gives the technician a hard copy to take to the vehicle. Another advantage of these systems is that the information is updated quarterly, providing the latest information and changes.
4. *Trade magazines* are a good source for training information. They are published monthly, and usually include articles on specific driveability topics. In addition, they usually include a section that highlights some of the OEM manufacturer's most recent technical bulletins. Trade magazines are available to the independent shop/technician at a very reasonable price.
5. *Technical Service Bulletins (TSB)* and *hotline phone* and *computer services,* for a fee, provide access to the latest changes direct from the OEM manufacturers. These bulletins are designed to provide the technician with changes to factory manuals, service changes, new part numbers, and quick fixes for common problems to a specific make, model, engine, etc. These bulletins can address factory recalls for specific vehicles. Without this type of service, technicians can spend needless hours trying to find a fix for a specific problem.
6. *Training books* and *video tapes* are available from all domestic manufacturers and numerous independent publishers. These books and tapes provide the theory and operation of specific components and systems. This can be a relatively inexpensive way to update technicians. Knowing the theory of how a system operates can reduce diagnostic time.

With the complexity of modern-day vehicles and the tremendous amount of information required to repair them, a technician cannot be expected to remember it all. Knowing what reference to use and where to find the information quickly will decrease diagnostic and repair time.

CHAPTER 11

Diagnosing Computer-Related Problems

A SYSTEMATIC DIAGNOSTIC APPROACH

Troubleshooting a driveability complaint requires a few basic steps to help with finding the problem in a reasonable amount of time. When approaching a complaint regarding a computer-controlled system, it is important to realize that the computer system is the least likely problem. Words of caution: *do not forget the basics.* First, pretend that the computer is not there before automatically assuming that it is the cause of the problem.

The word **VECTOR** should be part of the technician's systematic approach to diagnosing driveability complaints:

- **V** = **V**erify what the customer's concern is. The average customer may not be technically oriented. They will describe things as they understand them. An engine miss is often described as *jerking* or *chugging.* Have the customer describe the problem, and pay close attention to their *body language* (gestures); they may even make faces or try to imitate vehicle noises. Identify the driving conditions under which the problem occurs. Communication between the customer, service writer, and/or the technician is essential to accurately diagnosing driveability complaints. Figure 11-1 is sample check sheet that can be used to guide the technician through this step.
- **E** = **E**xamine the vehicle for obvious problems. Perform a visual inspection and check if the *check engine* light (Malfunction Indicator Lamp or MIL) is on with the engine running. Check under the hood for loose connections, disconnected wires, or vacuum hoses that have fallen off or are cracked. Listen for noises (i.e., vacuum leaks). Look for signs that might indicate that someone has recently worked on the vehicle. Attempting to diagnose the computer system before taking care of the basics will consume time and cause much aggravation.
- **C** = **C**onfirm that the problem(s) exist. Road test the vehicle to confirm the problem, trying to duplicate the conditions the customer described. Once the problem is confirmed, the next step is to begin testing. If the problem cannot be duplicated, this is called an intermittent problem. For additional help, refer to driveability symptoms charts, technical bulletins, or a technical assistance hotline (see Chapter 2).
- **T** = **T**est the circuit or component(s) in question to pinpoint the problem. Before testing the system in question, remember three things:
  1. Have a good understanding of how the system is designed to operate.
  2. Refer to the proper reference manuals to locate diagnostic procedures and specifications.
  3. Use the proper test equipment.

Vehicle: CAMRY	Model: LE	Engine: 3S-FE	Year: 91
VIN: S V Z Z E X X X X X X X X X X X X		Dealer No.: 9 9 9 9 9	
Customer Last Name: SMITH		Mileage: 3,250	

Frequency of Problem:	☑ Continuous ☐ Intermittent (_____times/per _____day/month)	☐ Once Only ☐ Other	Brand of Fuel Used: LOWEST PRICE	☐ Premium ☑ Regular

CONDITIONS AT TIME OF OCCURRENCE:

Weather	Outdoor Temperature	Place	Engine Temperature	Engine Operation
☐ Sunny ☐ Cloudy ☐ Rain ☐ Snow ☑ No Effect	☐ Hot (Approx. _______°F) ☐ Warm ☐ Cool ☐ Cold (Approx. _______°F)	☐ Freeway ☐ Highway ☐ Suburbs ☐ Inner City ☐ Hill ☐ Up ☐ Down ☐ Rough Road ☐ Other	☐ Cold ☐ Warming up ☑ After warming up ☑ Normal ☐ Other	☐ Starting ☐ Just after starting ☐ Idling ☐ Racing without load ☑ Driving ☑ Constant ☐ Acceleration ☐ Deceleration ☐ Other

PROBLEM SYMPTOMS:

☐ Engine Does Not Start	☑ Difficult to Start	☑ Poor Idling	☑ Poor Driveability	☐ Engine Stall	☐ Other
☐ Engine does not crank ☐ No initial combustion ☐ No complete combustion	☐ Engine cranks slowly ☐ Engine cranks too quickly ☑ Other MAINLY WHEN WARM	☐ Incorrect first idle ☐ Idling rpm is abnormal ☐ High ☐ Low (_____rpm) ☑ Rough idling ☐ When A/C ON ☐ When elect. load ON ☐ Other	☑ Hesitation ☐ Backfire ☑ Surging ☐ Knocking ☐ Lack of power ☐ Other SURGE AT LIGHT THROTTLE	☐ After starting ☐ After accelerator pedal depressed ☐ After accelerator pedal released ☐ During A/C operation ☐ When N to D shift ☐ When coming to a stop ☐ Other	☐ Poor mileage ☐ Sulphur smell

CONDITION OF "CHECK" ENGINE WARNING LIGHT	DIAGNOSTIC CODE INSPECTION	
	Normal Mode	Test Mode (Optional)
☐ Remains on ☑ Sometimes lights up ☐ Does not light up LIGHT STEADY THROTTLE SUBURBAN DRIVING	☐ Normal ☑ Malfunction code Codes 25	☐ Normal ☐ Malfunction code Codes_______

Figure 11-1 Asking the right questions can lead the technician to the problem.

- **O** = **O**rganize the information collected. Analyze the data by comparing it to specifications, prior experiences, etc., and decide what course of action is required to fix the problem. Identify the cause of the problem; do not just fix the result of the problem.
  Example: The computer is bad. What caused the computer to go bad? Did a diode across an output solenoid short, did a solenoid short, was there another type of voltage spike, is there a bad power ground? Remember that the computer is an expensive fuse.
- **R** = **R**epair, **R**eplace, or **R**ebuild the cause of the problem and then **R**oad test. Once the problem has been identified and the repair is done, retest the vehicle to confirm that the problem is fixed.
  Example #1: If the problem set off a trouble code, be sure to clear the computer's memory. Test drive the vehicle to confirm that the trouble code does not reset. The test drive allows the computer to adjust its adaptive learning to the completed repairs.
  Example #2: The technician injects propane into the intake and the O_2 climbs very slowly to 0.950 volts. The technician creates a large vacuum leak and the O_2 sensor only drops to 0.400 volts. The technician proceeds by installing a new sensor because the old sensor is carbon-fouled. The question is, was the O_2 sensor the cause or the result of another problem? If this is not the cause of the problem, the customer will be back because the problem has not been correctly repaired.

WHAT THE COMPUTER DOESN'T KNOW

A little bit of knowledge goes a long way. Technicians today attend many workshops, classes, trade shows, and spend many hours reading about the *high-tech* world of the modern automobile. As a result, many technicians are too quick to focus attention on computer diagnostics.

The complaint may be a surge, hesitation or stumble, stalling, poor gas mileage, or any number of related symptoms. Before taking on the computer, *go for the basics* and do a visual inspection first. Remember that the computer responds to what it sees through input sensors and switches. The input sensors and switches react to mechanical, hydraulic, and physical changes to the engine. The expression *garbage in, garbage out* enforces the need to check the basics before taking on the computer.

Before getting buried in the computer and its various components, go for the basics, and include a thorough visual inspection. *What the computer doesn't know* is what should be checked first. The computer doesn't know:

1. *Engine Compression* — Low or uneven compression affects engine vacuum, air fuel mixture, and fuel pressure on a fuel-injected engine. The MAP or MAF sensor must respond to this compression problem, and the O_2 sensor must read the byproduct of this poor compression, which results in low vacuum and poor engine performance.
2. *Engine Oil Consumption* — Worn engines that burn oil can foul spark plugs, cause carbon buildup in the combustion chambers, contaminate oxygen sensors, or ruin a catalytic converter.
3. *Restricted Exhaust System* — Excessive backpressure from the exhaust system will decrease engine vacuum and performance, cause a no-start condition, or even cause a backpressure EGR to open at the wrong time.
4. *Induction System Air Leaks* — Air leaks after the vane air flow meter, or mass air flow sensor, will deliver erratic signals to the computer, resulting in poor engine performance. On vehicles equipped with a MAP sensor, a vacuum leak will usually cause a higher-than-normal idle speed. This is because the MAP measures the additional air and the computer compensates by increasing fuel and spark advance. If the system is equipped with an air flow meter, the engine will idle rough or possibly stall. This is because the air entering is not measured by the air flow sensor, so the computer cannot compensate.
5. *Secondary Ignition* — The computer does not monitor secondary ignition and does not know if the coil is defective, the rotor is arcing or burned, spark plugs are worn-out or improperly gapped, etc.
6. *Initial (Base) Timing* — The computer is programmed to add spark advance to the initial (base) timing. If base timing is too advanced,

pinging will be heard. An engine equipped with a knock sensor will suffer a loss of power because spark advance timing is not being maximized. If base timing is retarded, engine power will suffer, engine overheating may occur, etc.

7. *Cam Timing*— Misalignment of the cam/crank timing marks can affect manifold vacuum, cylinder filling, and distributor timing. This can be especially misleading when working on distributorless ignition engines with camshaft sensors.
8. *Fuel Pressure* — Fuel pump control (relay) circuits are sometimes monitored, but the computer does not know if fuel pressure is high or low. It cannot tell if an injector is restricted or clogged. While the pressure may be adequate, the quality and quantity of fuel are unknown.
9. *Minimum Idle Speed Setting* — A dirty throttle body restricts air flow around the closed throttle plates. This can result in erratic idle speed and/or stalling. Sometimes technicians remove the anti-tamper plug and turn up the idle speed screw to correct this problem. This is the incorrect way to fix the problem, and can lead a technician away from the original problem.
10. *Actuator Solenoids* — The computer can monitor these circuits for electrical switching, but does not know if the purge canister is loaded with fuel, if the air pump switched the air downstream to the converter, if the heat riser is stuck closed by trapped vacuum, or if the EGR is plugged or leaking. Any of these devices could have a split vacuum hose or an improperly-routed hose. A stuck solenoid pintle cannot be detected by the computer.
11. *Electrical Connections* — Loose or dirty pin connectors, dirty or broken ground eyelets, or missing or altered grounds can lead to unnecessary replacement of good components (Figure 11-2).
12. *Charging System* — The charging system voltage can go through wide swings depending on the electrical load, engine idle speed, and temperature of the battery. Any of these can adversely affect the computer.

This list does not cover every possible problem; it is intended to help the technician when covering the basics. Remember that the computer reacts to data supplied to it through a sensor network, and the computer is only as good as the data it receives.

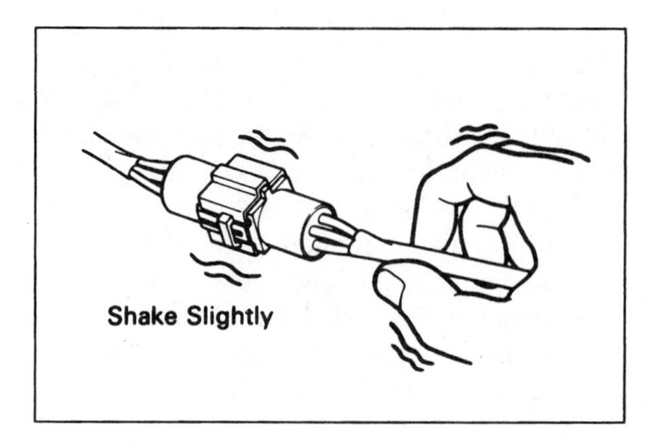

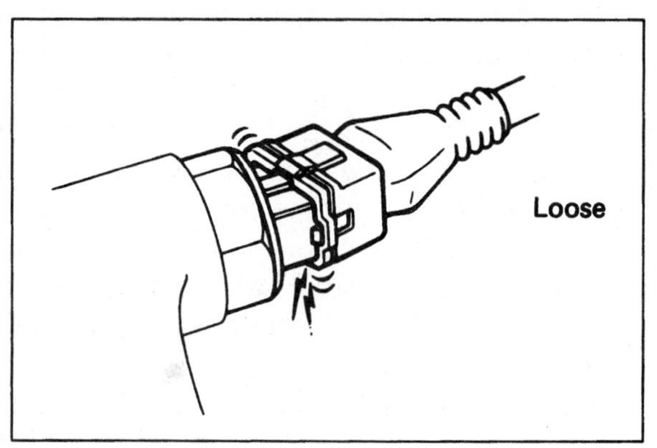

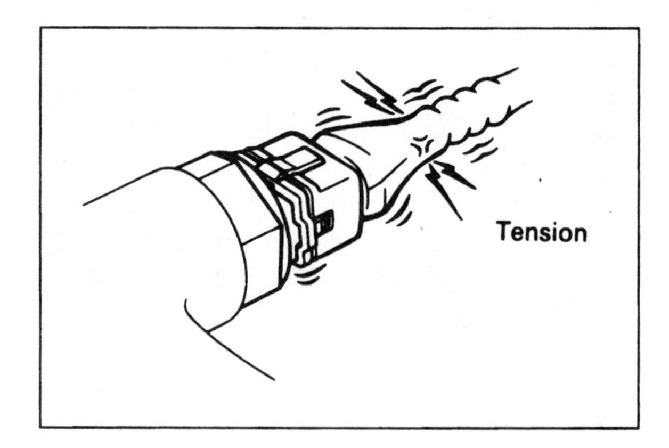

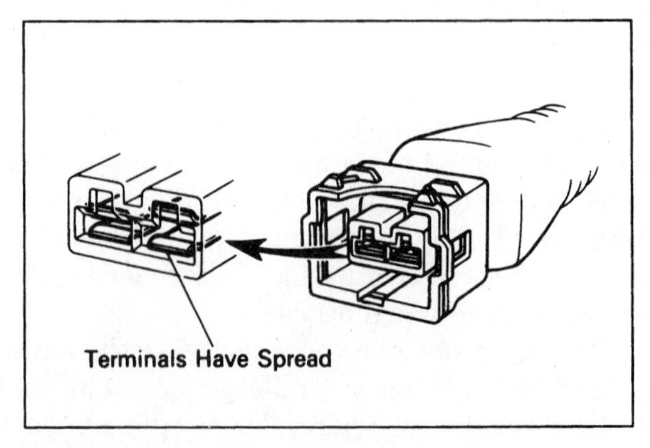

Figure 11-2 Types of checks to perform on electrical connections.

INTERMITTENT (NON-COMPUTER) COMPLAINTS

Most manufacturers include charts for intermittent driveability complaints and provide suggestions on how to handle them. They are **not** to be confused with trouble trees. Some common intermittent complaints and possible causes of non-computer type problems are:

1. *Surges (Chuggles)* — This is an engine power variation under steady throttle or cruise. It feels like the engine speeds up and slows down with no change to the accelerator pedal position. Possible causes might include:
   a. Improper EGR operation.
   b. Carburetor float incorrectly set.
   c. Fuel pressure too high.
   d. Secondary ignition problems.
   e. Poor quality or contaminated gasoline.
2. *Lack of Power, Sluggish, or Spongy Performance* — This is when the engine delivers less power than expected. It requires excessive throttle during acceleration, hill climbing, or trailer pulling. There is little or no increase in vehicle speed when the accelerator pedal is pushed part-way to the floor. Possible causes might include:
   a. Improperly-adjusted throttle linkage.
   b. Dirty air filter.
   c. Improper ignition timing.
   d. Exhaust system restriction.
   e. Internal engine mechanical problems.
3. *Detonation or Spark Knock* — This is a mild to severe pinging, usually worse under acceleration. The engine makes a sharp metallic knock that changes with throttle opening. Possible causes might include:
   a. Improper EGR valve operation.
   b. Advanced ignition timing.
   c. TAC door stuck closed to heated air.
   d. Excessive carbon buildup in the combustion chamber.
   e. Engine overheating.
4. *Hesitation/Sag or Stumble* — This is a momentary lack of response between the time the throttle is depressed and when the engine responds. Usually, it is more noticeable when first trying to accelerate the vehicle from a stop. If it is severe enough, the engine may stall. Possible causes might include:
   a. Low carburetor float level.
   b. Carburetor accelerator pump defective or misadjusted.
   c. EGR valve opening too soon.
   d. Retarded ignition timing.
   e. Low fuel pressure.
5. *Rough, Unstable, or Incorrect Idle Speed* — Occurs when the engine runs unevenly or vibrates excessively at idle. If the idle RPM varies, it is called *hunting*. Either condition may be severe enough to cause stalling. Possible causes might include:
   a. Vacuum leaks.
   b. Improperly-adjusted idle speed screw or motor.
   c. Improperly-adjusted air/fuel mixture or duty-cycle.
   d. Ignition system misfire.
   e. Internal engine mechanical problems.
   f. PCV valve leaking.
6. *Engine Misses (Cuts Out)* — The symptoms are a pulsation or jerking that happens as engine speed increases. It is usually more pronounced as engine load increases. A spitting or popping sound at idle or low speed can be felt or heard at the tailpipe. Possible causes might include:
   a. Primary ignition problems: pickup coil/module.
   b. Secondary ignition problems.
   c. Internal engine mechanical problems (e.g., burned valve).
   d. Exhaust system restriction.
   e. Excessively lean air/fuel mixture.

This information is important when attempting to diagnose an intermittent fault code or driveability complaint where no codes are stored in memory.

WHAT THE COMPUTER KNOWS

The computer knows if sensors are open or shorted, if sensor readings are within specified voltage limits, and if output circuit coils and relays are switching on and off. The computer monitors electrical signals only.

The computer compares this information to data stored in its memory. It then uses this information to perform calculations and to issue commands to control various outputs.

COMPUTER SYSTEM DIAGNOSIS

Understanding computer systems also means learning how to repair them. While problems on mechanical systems are often more obvious because they can be examined visually, computer systems require a different means of diagnosis. Computer diagnosis relies heavily on test equipment and a good understanding of the system to find otherwise invisible malfunctions.

As explained in Chapters 5 and 6, all computer systems are a makeup of input devices, output devices, and computer running programs from preprogrammed memories (Figure 11-3). The similarity between computer systems allows the technician to approach problem diagnosis in the same way for any system. Usually, it may not be the computer, and it may not even be the sensors, switches, or actuators. It is quite likely to be in the wiring, which means a short, an open circuit, or a poor connection.

The computer may express these problems as diagnostic codes (Figure 11-4). Half the battle is won by looking up the code in the repair manual. The code is an input to technicians to help locate where the problem is and direct them to the correct diagnostic chart, which guides the technician through the diagnosis.

Since the computer system is made up of many parts that interact with one another, use a logical approach to diagnosis. If the problem is computer-related, the check engine light is the most obvious indication, but not every code activates the check engine light. In addition, not every vehicle is equipped with a check engine light.

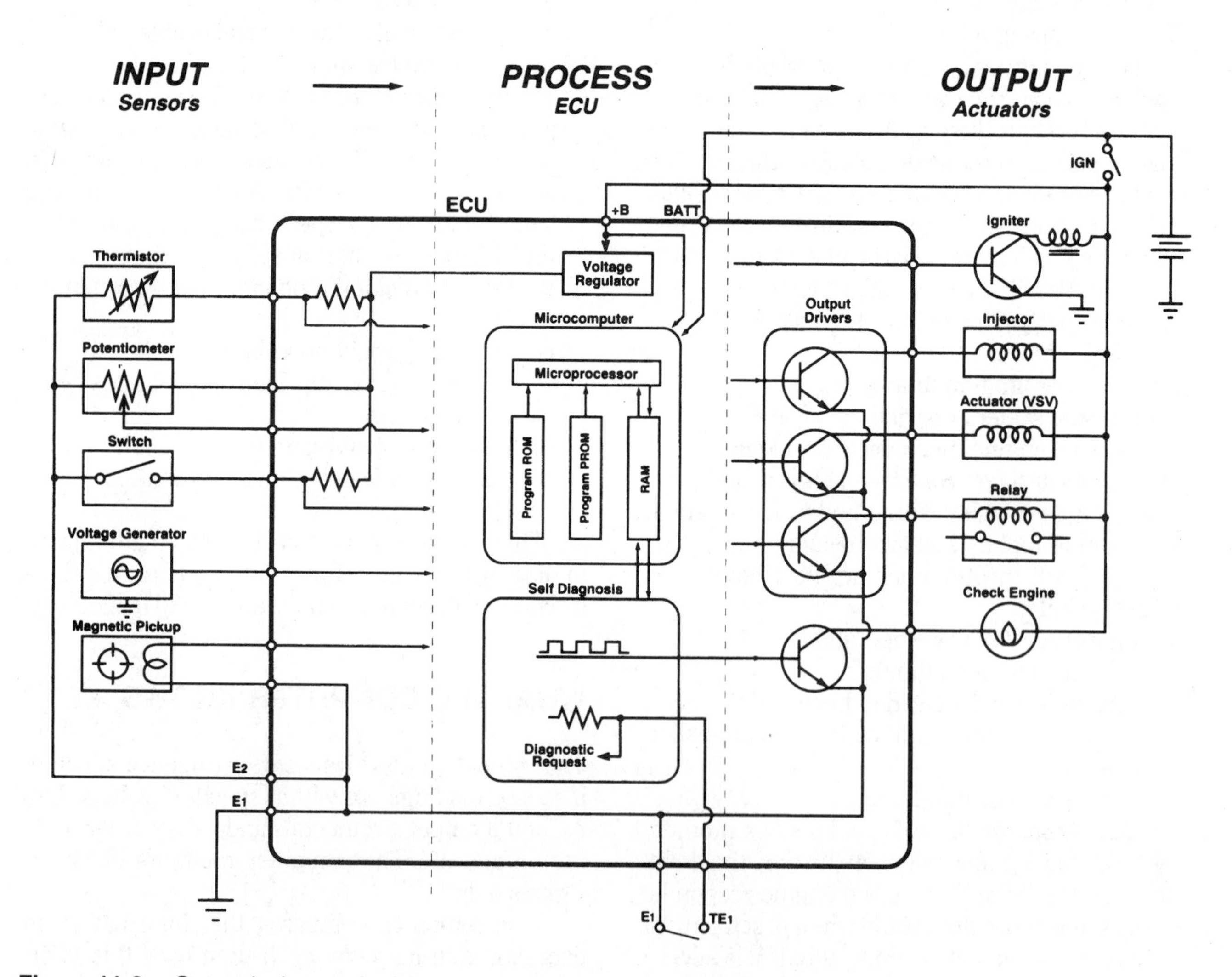

Figure 11-3 Output is the result of the computer processing information from inputs.

Service Code Numbers

No.	Indicator flashing pattern	Diagnosed circuit	Condition	Point	Memorized	Page
02	ON OFF	NE2 signal crank angle sensor 2	No NE2 signal	• Crank angle sensor 2 connector • Wiring from crank angle sensor 2 to ECU • Crank angle sensor 2	Yes	F2–79
03	ON OFF	G signal crank angle sensor 1	No. G signal input while a specified number of Ne1 or Ne2 signal pulses are input	• Distributor connector • Wiring from main relay to distributor • Wiring from distributor to ECU • Crank angle sensor 1	Yes	F2–80
04	ON OFF	NE1 signal crank angle sensor	No NE1 signal		Yes	F2–81
05	ON OFF	Knock sensor	Open or short circuit	• Knock sensor connector • Wiring from knock sensor to ECU • Knock sensor	Yes	F2–82
08	ON OFF	Airflow meter	Short circuit	• Airflow meter connector • Wiring from airflow meter to ECU • Airflow meter	Yes	F2–83
09	ON OFF	Water thermosensor	Open or short circuit	• Water thermosensor connector • Wiring from water thermosensor to ECU • Water thermosensor resistance	Yes	F2–84
10	ON OFF	Intake air thermosensor		• Airflow meter connector • Wiring from airflow meter to ECU • Intake air thermosensor resistance	Yes	F2–85
12	ON OFF	Throttle sensor		• Throttle sensor connector • Wiring from throttle sensor to ECU • Throttle sensor	Yes	F2–86
14	ON OFF	Atmospheric pressure sensor		• ECU	Yes	F2–87
15	ON OFF	Oxygen sensor (Left side)	Sensor output continues less than 0.55V 100 sec. after engine exceeds 1,500 rpm	• Oxygen sensor connector • Wiring from oxygen sensor to ECU • Oxygen sensor	Yes	F2–87
16	ON OFF	EGR position sensor	Open or short circuit	• EGR position sensor connector • Wiring from EGR position sensor to ECU • EGR position sensor	Yes	F2–88
17	ON OFF	Feed Back system (Left side)	Sensor output not changed 50 sec. after engine exceeds 1,500 rpm	• Fuel pressure • Injection fuel leakage • Ignition system • Air leakage • ECU	Yes	F2–89

Figure 11-4 Diagnostic codes give the technician a *clue* about the problem. Notice that the code conditions are related to electrical problems (courtesy Mazda Motor of America).

On many vehicles, engineers have designed the computers with additional artificial intelligence so the computer can monitor its operating systems more closely. If the computer sees a problem, it will turn on the check engine light and substitute a default value using data from other sensors and from memory. This results in an activated check engine light, if equipped, and complaints of poor gas mileage. Usually though, the customer experiences very little change in the way the vehicle drives.

Fault Codes

The computer uses specific *Arming Conditions* to set codes. These conditions will vary for each sensor or actuator, but four conditions may be used for all circuits:

1. *Time Interval* — Circuit voltage is monitored for a specific time looking for specific voltages.
2. *Temperatures* — Circuit voltage is monitored during some or all temperature ranges, depending on the sensor.
3. *Engine Conditions* — Circuits are monitored during different engine conditions. Not every sensor is monitored all of the time.
4. *Fault Code Parameter Cycle Counter* — A fault code parameter cycle counter is activated once an out-of-range voltage is identified. The computer then counts the number of input cycles from the sensor in question. Once a specific number is reached without interruption, a fault code is set and the voltage from this circuit is no longer used.

When a code is present at the time the system is tested, it is called a *hard fault.* A code is only an indication of which component or circuit is at fault.

Arming conditions and fault parameter cycle counts will vary for different sensors and between manufacturers. The following paragraphs describe examples of specific arming conditions.

Chrysler MAP Sensor Code 14

1. Engine must be between 400 and 1500 RPM.
2. Sensor voltage must be less than 2 volts or greater than 4.67 volts.
3. Out-of-range voltage must be present for 1.76 seconds (20 fault code parameter cycles at 88 milliseconds per cycle).

To set a fault code for a Chrysler MAP sensor, the MAP sensor is monitored between 400 and 1500 RPM. The voltage must be lower than 0.2 volts or higher than 4.67 volts. The fault code parameter cycle requires 20 uninterrupted cycles before the code will be set.

If the out-of-value sensor voltage returns to normal, then the fault code cycle counter will count in the opposite direction until it reaches zero. Once it reaches a zero value, the computer will again respond to the reading from the sensor and normal driveability will be restored. This is called *recoverable limp-in.* The code will remain in memory and now becomes an intermittent problem. Usually the check engine light will be turned off, but some Chrysler products keep the light on until the ignition switch is turned off.

Figure 11-5 shows a snapshot of a Chrysler MAP sensor circuit failure. The customer is complaining of intermittent surging at higher engine RPM. There are no codes stored in memory. Frames -7, -6, 3, 4, 5, and 13-15 show the MAP voltage higher than 4.67 volts. The MAP voltage indicates a heavy engine load, and the computer responds by increasing the pulse-width and decreasing the spark advance because of the load. This is the cause of the intermittent surging because the mixture goes rich-lean-rich. Also, the spark advance is erratic.

Why didn't this set a code? The arming condition was not met because engine speed stayed above 1500 RPM. If the engine speed had dropped below 1500 RPM, it still would not have set a code because the fault code parameter cycle was interrupted. It appears the problem is a bad MAP or intermittent electrical connection.

Toyota's Two-Trip Detection Logic

Toyota's Computer Control Systems (TCCS) use a *two-trip detection logic* to set trouble codes. This is done to prevent the setting of *false* trouble codes. Figure 11-6 shows a list of two-trip detection trouble codes.

Trouble codes that use two-trip detection logic require a malfunction to occur twice to set a trouble code (Figure 11-7, page 185). A trouble code is stored in memory the first time the problem is detected. During the next *key-on* cycle, if the same problem is detected, a code is set and the check engine light is activated.

88VJ2501 E-Z EVENT PLUS
CHRYSLER 1990 2.5 MINIVAN VIN
EVENT 1 TAG -- FRAME -30
FAULT CODES: NO

	-30	-29	-28	-27	-26	-25	-24	-23	-22	-21	-20	-19	-18	-17	-16	-15
RPM	2535	2598	2640	2667	2699	2697	2742	2795	2797	2844	2840	2836	2850	2860	2875	2876
MAP VOLTS	1.40	1.40	1.40	1.38	1.40	1.38	1.44	1.52	1.44	1.50	1.46	1.44	1.44	1.40	1.36	1.44
THROT POS SENSOR	1.26	1.26	1.26	1.26	1.30	1.26	1.28	1.28	1.28	1.28	1.28	1.28	1.28	1.28	1.28	1.28
COOLANT TEMP	142	142	142	142	142	142	142	142	142	142	142	142	142	144	144	144
BARO PRESSURE VOL	4.74	4.74	4.74	4.74	4.74	4.74	4.74	4.74	4.74	4.74	4.74	4.74	4.74	4.74	4.74	4.74
O2 VOLTS	0.32	0.80	0.76	0.14	0.16	0.26	0.80	0.14	0.20	0.48	0.08	0.74	0.56	0.06	0.02	0.10
INJ PULSE WIDTH	01.4	01.3	01.3	01.3	01.4	01.3	01.2	01.4	01.3	01.4	01.3	01.3	01.3	01.3	01.3	01.3
ELECTRONIC ADV/RE	025	024	024	024	024	024	024	024	024	024	024	024	024	024	024	024

	-15	-14	-13	-12	-11	-10	-9	-8	-7	-6	-5	-4	-3	-2	-1	0
RPM	2876	2888	2885	2900	2886	2867	2844	2899	2925	2463	2061	2299	2551	2762	2754	2754
MAP VOLTS	1.44	1.36	1.40	1.40	1.40	1.48	1.46	1.52	4.98	5.02	1.86	2.00	1.68	0.94	1.64	1.64
THROT POS SENSOR	1.28	1.28	1.28	1.28	1.28	1.28	1.28	1.30	1.30	1.30	1.30	1.32	1.30	1.32	1.32	1.32
COOLANT TEMP	144	144	144	144	144	144	144	144	144	144	144	144	144	144	144	144
BARO PRESSURE VOL	4.74	4.74	4.74	4.74	4.74	4.74	4.74	4.74	4.80	5.02	5.02	5.02	5.02	5.02	5.02	5.02
O2 VOLTS	0.10	0.82	0.80	0.10	0.02	0.06	0.28	0.82	0.86	0.66	0.60	0.94	0.92	0.94	0.14	0.14
INJ PULSE WIDTH	01.3	01.2	01.2	01.3	01.3	02.2	01.4	01.4	09.9	06.4	02.1	02.0	01.7	00.8	01.7	01.7
ELECTRONIC ADV/RE	024	024	023	024	024	019	024	025	009	008	027	030	028	019	028	028

	0	1	2	3	4	5	6	7	8	9	10	11	12	13	14	15
RPM	2754	2978	2978	3018	3018	3018	2410	2410	2418	2655	2655	2655	2981	2876	2427	2019
MAP VOLTS	1.64	1.56	1.56	4.94	4.94	4.94	1.78	1.78	1.78	1.76	1.76	1.76	1.62	5.02	5.02	5.02
THROT POS SENSOR	1.32	1.32	1.32	1.30	1.30	1.30	1.32	1.32	1.32	1.34	1.34	1.34	1.34	1.32	1.34	1.34
COOLANT TEMP	144	144	144	144	144	144	144	144	144	144	144	144	144	144	144	144
BARO PRESSURE VOL	5.02	5.02	5.02	5.02	5.02	5.02	5.02	5.02	5.02	5.02	5.02	5.02	5.02	5.02	5.02	5.02
O2 VOLTS	0.14	0.84	0.84	0.68	0.68	0.68	0.64	0.64	0.64	0.96	0.96	0.96	0.90	0.64	0.56	0.58
INJ PULSE WIDTH	01.7	01.3	01.3	08.8	08.8	08.8	02.0	02.0	02.0	01.6	01.6	01.6	01.2	06.5	05.4	05.2
ELECTRONIC ADV/RE	028	027	027	010	010	010	028	028	028	028	028	028	028	009	008	007

	15	16	17	18	19	20	21	22	23	24	25	26	27	28	29	30
RPM	2019	2265	2617	2872	3090	3121	2694	2565	2884	3115	3299	3344	3309	3300	3206	3189
MAP VOLTS	5.02	2.30	1.92	1.76	1.78	4.58	5.02	2.08	1.92	1.86	1.72	1.66	1.50	1.54	1.66	1.70
THROT POS SENSOR	1.34	1.36	1.36	1.36	1.36	1.36	1.36	1.38	1.38	1.38	1.38	1.36	1.34	1.34	1.36	1.36
COOLANT TEMP	144	144	144	144	144	144	144	144	144	144	144	144	144	144	144	144
BARO PRESSURE VOL	5.02	5.02	5.02	5.02	5.02	5.02	5.02	5.02	5.02	5.02	5.02	5.02	5.02	5.02	5.02	5.02
O2 VOLTS	0.58	0.94	0.96	0.94	0.92	0.24	0.64	0.98	0.90	0.86	0.88	0.32	0.02	0.00	0.00	0.00
INJ PULSE WIDTH	05.2	01.9	01.5	01.4	01.5	09.8	05.2	01.6	01.4	01.4	01.2	01.1	01.0	01.1	01.1	01.2
ELECTRONIC ADV/RE	007	028	031	029	029	012	009	030	029	029	029	028	026	027	028	028

Figure 11-5 Driveability complaints do not always result in a code being set.

DTC #	Trouble Code Definition
Code 21	O2 Sensor signal
Code 25	Air/fuel lean malfunction
Code 26	Air/fuel rich malfunction
Code 27	Sub-O2 sensor signal
Code 71	EGR malfunction
Code 78	Fuel pump control malfunction (Lexus only)
Code P0116	Engine Coolant Temperature circuit range problem
P0135/141/155	O2S heater malfunction
P0171	Fuel system too lean
P0172	Fuel system tool rich
P0300	Random misfire detected
P0301-306	Misfire detected, specific cylinder
P0335	Crankshaft position sensor circuit malfunction (Cranks)
P0336	Crankshaft position sensor circuit range/performance problem
P0340	Camshaft position sensor circuit malfunction (Cranks)
P0402	EGR excessive flow detected
P0505	Idle control system malfunction
P0510	Closed Throttle Position switch malfunction
P1780	Park/Neutral Position switch malfunction

Figure 11-6 Toyota two-trip detection codes.

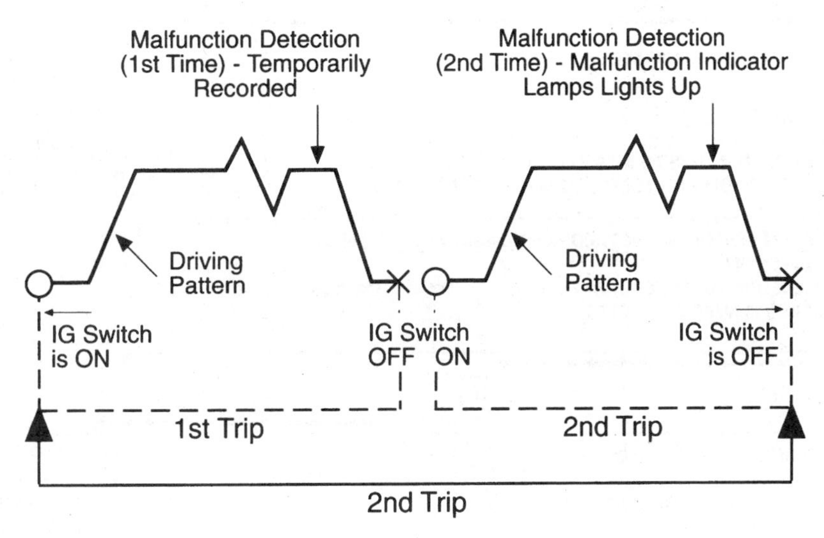

Figure 11-7 Toyota two-trip detection process.

Diagnostic Trouble Trees

Diagnostic trouble trees are used when a hard fault code exists. If a diagnostic trouble tree is used for an intermittent code, a good component might be unintentionally replaced.

Once it is learned that a hard fault exists, and the use of a trouble tree is necessary, remember these basic rules:

1. Start at the beginning of the diagnostic trouble tree and read everything (Figure 11-8). Some manufacturers include some preliminary steps and precautions before performing the procedures listed in the diagnostic trouble tree.
2. Most diagnostic trouble trees are directly related to electrical problems only.
3. *DO NOT skip any steps.* A diagnostic trouble tree should be followed until the problem is found. Skipping or misunderstanding steps will lead to misdiagnosis.
4. A diagnostic trouble tree is like following directions when driving a car. For example, the driver comes to a stop at an intersection where only a left or right turn is possible. The directions require a left turn based on where the driver intends to go. Voltage and resistance readings found while using a diagnostic trouble tree require the technician to do the same thing. Not following directions could get the driver or technician lost and cost a considerable amount of time (Figure 11-9, page 187).
5. When the last step indicates a bad computer, make sure to check power and sensor ground circuits before replacing it.

Intermittent Computer-Related Problems

If the check engine light is not on while the engine is running, but there is a trouble code stored in memory, two possible problems exist. The first problem could be an intermittent failure that is no longer occurring.

The second problem could be a hard fault, but the system is designed not to turn on the check engine light for this code. For example, a Chrysler vehicle speed sensor code does not turn on the light. This type of failure can cause the customer to complain about poor economy and driveability.

Follow a few simple steps when diagnosing intermittent problems:

1. Record all trouble codes stored in memory.
2. Clear codes. Codes can be cleared by removing the battery negative cable for thirty seconds, removing the memory fuse, using a scan tool, etc.

   CAUTION: Removing the battery cable to clear codes will also clear other items in memory, such as:

   a. Clock and radio station settings.
   b. Seat position settings.

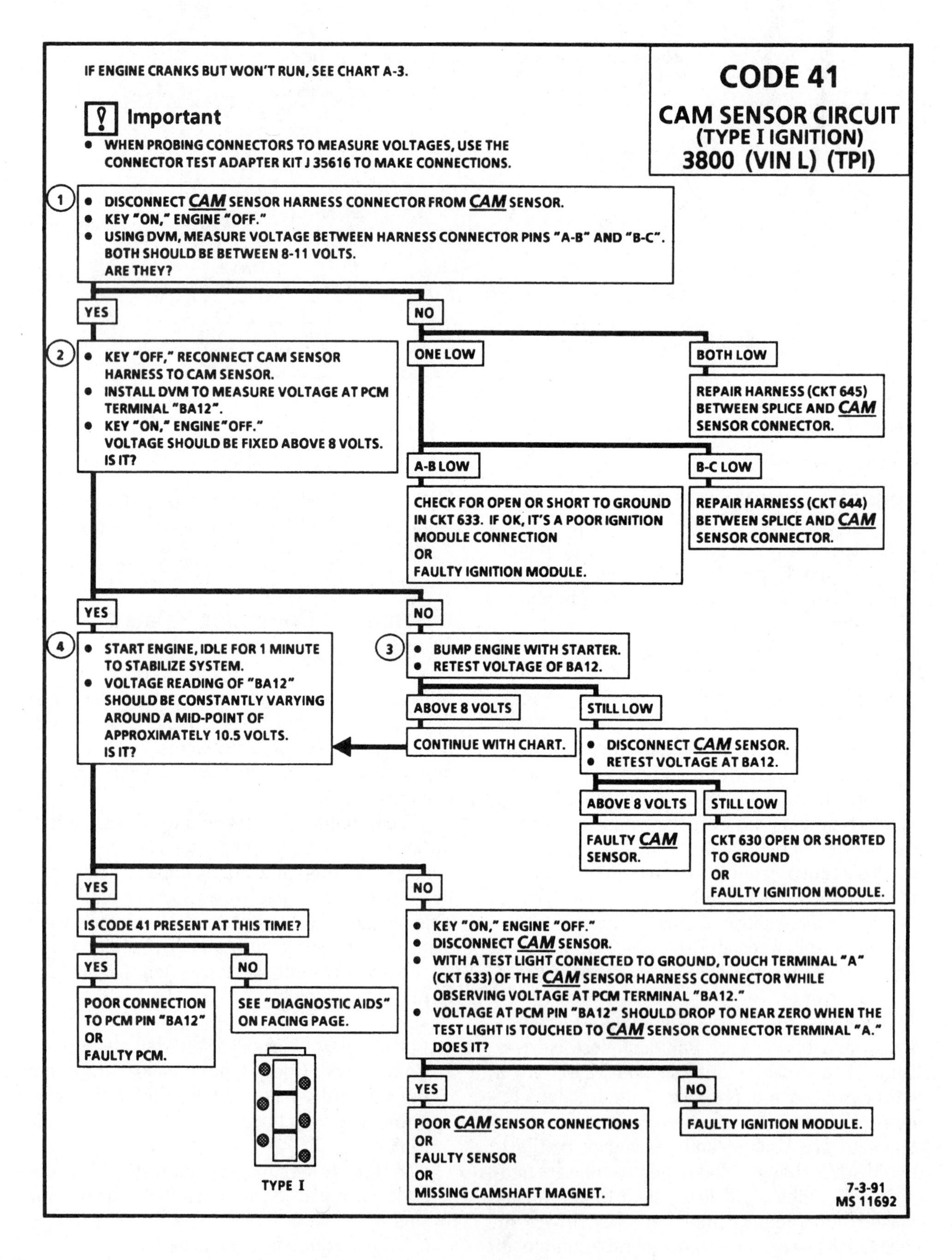

Figure 11-8 Do not skip the information at the top of the diagnostic trouble tree above box #1 (courtesy GM).

<table>
<tr><th>CODE No.</th><th colspan="3">08 (AIRFLOW METER)</th></tr>
<tr><th>STEP</th><th>INSPECTION</th><th colspan="2">ACTION</th></tr>
<tr><td rowspan="2">1</td><td rowspan="2">Does airflow meter circuit have a poor connection?</td><td>Yes</td><td>Repair or replace connector</td></tr>
<tr><td>No</td><td>Go to next step</td></tr>
<tr><td rowspan="2">2</td><td rowspan="2">Is Code No. 10 also present?</td><td>Yes</td><td>Look for open circuit in wiring from airflow meter wire-terminal (B/LG) to ground</td></tr>
<tr><td>No</td><td>Go to next step</td></tr>
<tr><td rowspan="2">3</td><td rowspan="2">Is resistance of airflow meter OK?
<table><tr><th>Airflow meter</th><th>Fully closed (Ω)</th><th>Fully open (Ω)</th></tr><tr><td>B (LG/R)—A (R)</td><td>20—600</td><td>20—1,000</td></tr><tr><td>B (LG/R)—D (B/BR)</td><td colspan="2">200—400</td></tr></table></td><td>Yes</td><td>Go to next step</td></tr>
<tr><td>No</td><td>Replace airflow meter</td></tr>
<tr><td rowspan="2">4</td><td rowspan="2">Does wire harness between airflow meter and ECU have continuity?
<table><tr><th>Airflow meter</th><th>ECU</th></tr><tr><td>B (LG/R)</td><td>2I (LG/R)</td></tr><tr><td>A (R)</td><td>2B (R)</td></tr></table></td><td>Yes</td><td>Go to next step</td></tr>
<tr><td>No</td><td>Repair or replace wire harness</td></tr>
<tr><td rowspan="2">5</td><td rowspan="2">Are ECU terminal 3D, 2I, and 2B voltages OK? ☞ page F2–144</td><td>Yes</td><td>Replace ECU ☞ page F2–143</td></tr>
<tr><td>No</td><td>Look for a short circuit in wiring from airflow meter to ECU</td></tr>
</table>

2PU0F2-067

Figure 11-9 The results of an inspection give the technician directions to the next step (courtesy Mazda Motor of America).

c. "Learned" air/fuel values (Block Learn, Fuel Trim, etc.), idle speed settings, transmission shift patterns, and so on.

The loss of settings and values through battery cable removal is an irritant to customers. To maintain good public relations, explain this to the customer ahead of time.

3. Perform a visual inspection of the component, connections, grounds, hoses, temperature, etc. of the circuit in question.
4. Do not use trouble code charts for intermittent problems, or good components will needlessly be replaced. Code charts are to be used for **hard faults only.**
5. If the vehicle is equipped with a data stream (information entering and leaving the computer) that can be read with a scan tool, connect the scan tool and set it up to record. Test drive the vehicle to recreate the complaint. Keep in mind that code parameters and conditions that cause codes to be set can differ between manufacturers and different engine calibrations. Once the problem is recreated, the technician presses the trigger or record button to store the data present at the time. The technician returns to the shop and reviews the recorded data.
6. If a scan tool is not available, or if a scan tool cannot be used on a particular vehicle, test drive the vehicle to recreate the complaint. If the technician is fortunate enough to recreate the problem, a check engine light should be on if the system is equipped with one. Return to the shop and perform a diagnostic circuit check to read the code or codes stored in memory.
7. If no codes are stored or set, return to the basics. The technician should think about the engine RPM, load, temperature, vehicle speed, etc., when the problem was experienced. Other factors to

consider: was the air conditioning on, the power steering under load, the brake pedal depressed, or the charging system under load?

The technician is diagnosing the problem through a process of elimination. By eliminating the systems that function correctly, the technician can isolate the problem to a defective system, circuit, or device that needs to be adjusted, repaired, or replaced.

Computer Wiring

To function properly, the computer needs the proper operating voltage and good connections. The operating voltage range for most computers is approximately 10 to 16 volts. If the voltage should exceed 16 volts, or drop below 10 volts, then the computer might go into a *limp-in* mode. This will prevent normal computer operation and cause poor driveability.

Most computer systems have a minimum of two and sometimes three power feeds (Figure 11-10). The first power feed will always be hot. This allows the computer to retain the information stored in volatile memory, such as stored codes and learned values. The second power feed is supplied to the computer once the ignition switch is turned on.

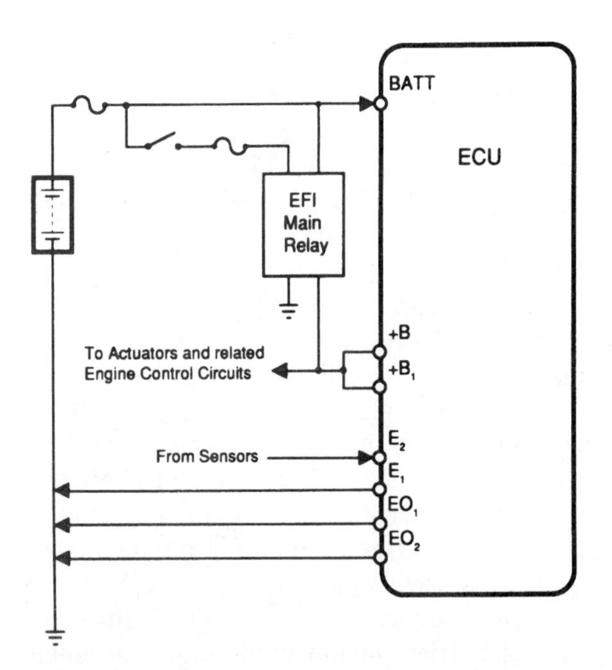

Figure 11-10 Multiple power feeds and grounds are used by most computers.

This will feed the computer with the voltage necessary to run all of its systems. A third power feed is sometimes used and works when the ignition switch is on.

All power feed wires are protected by a fusible link or fuse. Fusible links are usually located near the battery or starter relay. Fuses present another challenge—finding them. They can be located in a fuse block, remote relay center, or under the hood in a cover that looks like a relay. General Motors likes to use remote relay centers (Figure 11-11), or install fuses inside a cover that looks like a relay. When working on some Japanese vehicles, a high-amperage main fuse, fusible link, and fuse may be used to protect the same circuit (Figure 11-12 on page 190).

When checking power feeds to the computer, do not forget to check the supply of voltage from the battery and the charging system. The charging system voltage can go through wide swings in voltage, depending on the electrical load, engine idle speed, and temperature of the battery. Any of these can adversely affect the computer.

Grounds are many times forgotten when looking for electrical problems. They provide the circuit with a path back to the battery. Without them, an electrical circuit cannot operate. The number of grounds can vary from as few as two to as many as six.

Grounds can be of two types, power and sensor. Figure 11-10 shows sensor grounds as E_1 and E_2. Power grounds are used by the computer for outputs such as motors and solenoids and are shown in Figure 11-10 as EO_1 and EO_2. Sensor grounds return to the computer and then back to the engine, transmission bell housing, or fender well (Figure 11-13, page 190).

When testing ground circuits, always read the voltage drop across the ground side of the circuit. This can be accomplished by connecting the positive lead of the voltmeter to the ground side of the sensor or output. This will be the lead that returns to the computer. Connect the negative lead of the voltmeter to the battery negative terminal. A sensor's ground voltage should not exceed 0.1 volts. Power grounds should not exceed 0.3 volts. If the voltage is too high, begin by moving the negative lead from the battery to the engine. If the voltage reading is now normal, the problem is between the battery and the engine block. If the reading is still high, then the problem is between the engine, computer, and sensor.

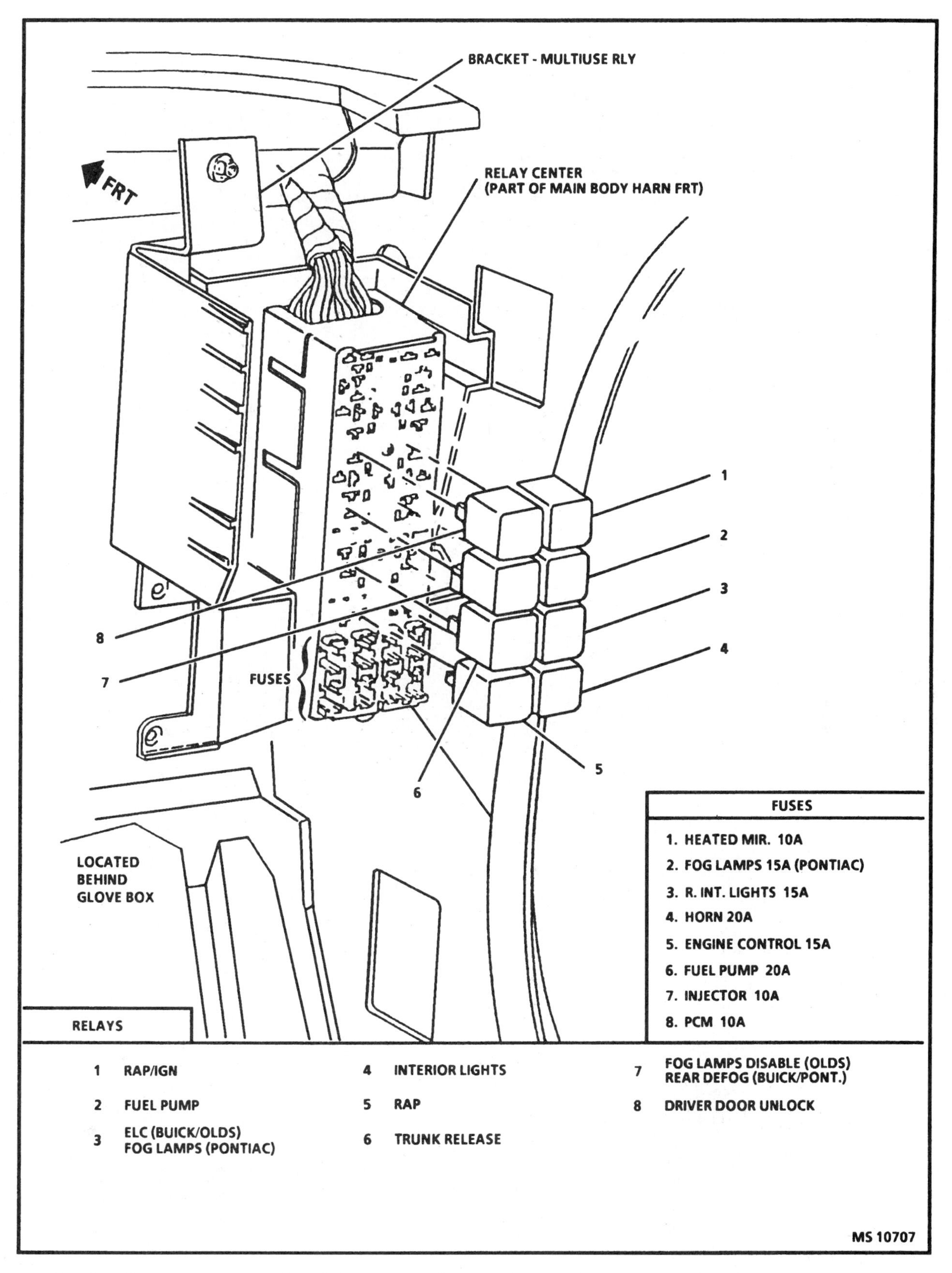

Figure 11-11 Some computer-related fuses and relays are located in a remote relay center behind the glove box (courtesy GM).

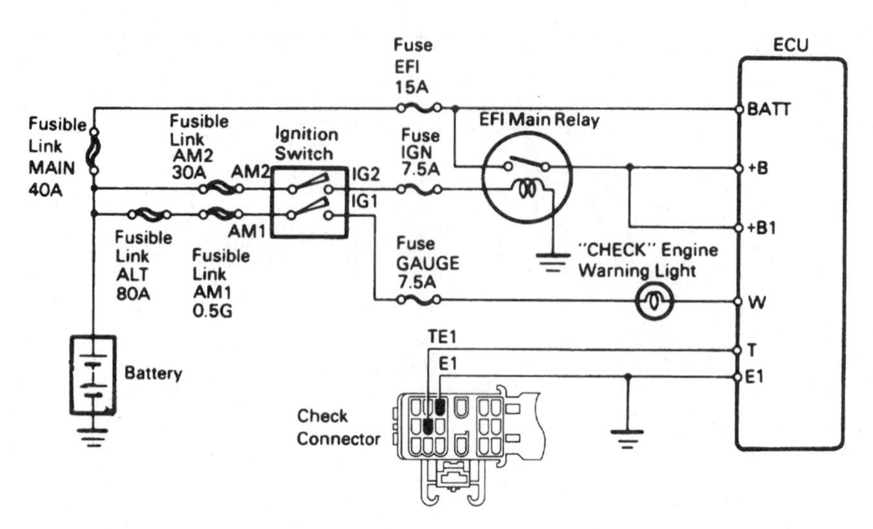

Figure 11-12 A fusible link and fuse can be used in combination to protect a circuit.

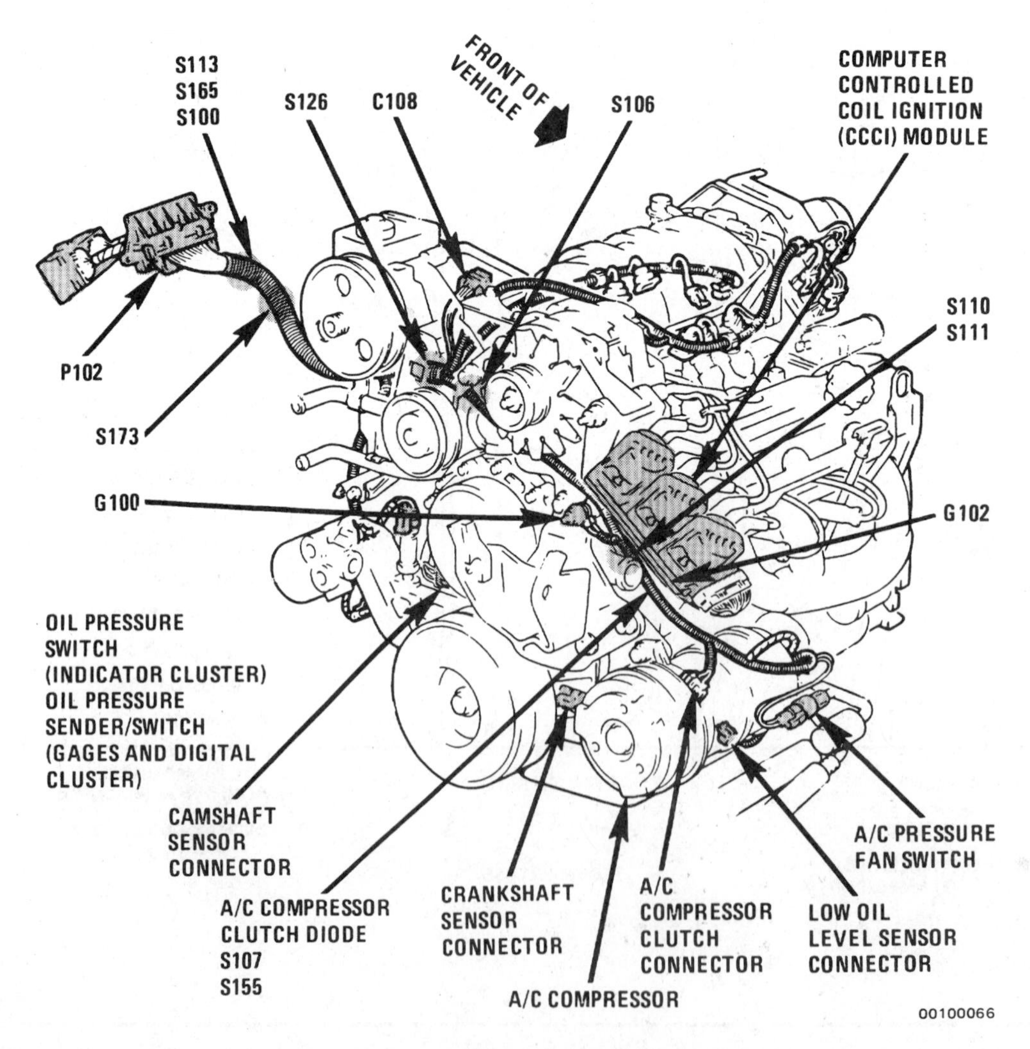

Figure 11-13 A component locator can help the technician locate the hard-to-find grounds connections (courtesy GM).

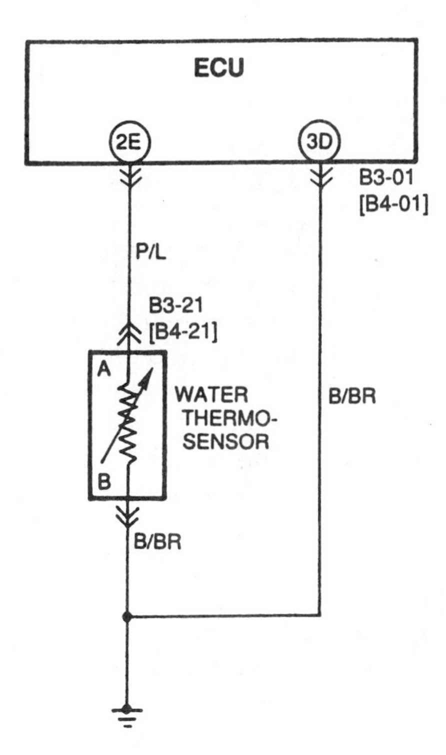

Figure 11-14 A high-resistance ground connection will affect sensor voltage measured by the computer (courtesy Mazda Motor of America).

If ground-side resistance is higher than normal, this will decrease the available voltage in the circuit. The computer will see a lower-than-normal return signal voltage from the sensor. Figure 11-14 shows the circuit for a Mazda water thermosensor (coolant). If resistance at the ground symbol is higher than normal, the voltage drop across the water thermosensor will decrease. The decrease in voltage drop will cause the computer to think the engine is warmer than it really is. This would probably result in a poor cold driveability. If a power ground, such as for a fuel injector(s), has high ground resistance, it will operate slower than normal and open for a shorter time, or not at all, creating a lean air/fuel mixture. Since voltage drop can vary depending on the source voltage, it is a good practice to isolate the connection and check resistance.

When working on an engine or transmission where ground connections must be removed, always remember to:

1. Reinstall grounds at the same place where they were originally connected.
2. Do not alter the length of a ground wire.
3. Route ground wires around the engine compartment exactly the way they were routed from the factory. This can prevent unwanted electrical interference (RFI – Radio Frequency Interference) which might affect the computer.
4. Check ground connector eyelets for loose connections or broken wire strands.

Other things to remember about computer wiring are:

1. Computer connectors, terminal location, and colors are not standard, so do not assume anything. **Always** use a wiring diagram.
2. A bare wire wrapped around the wires from the distributor to the computer is a shielding wire. The shielded wire is connected to ground. It absorbs any RFI that might disrupt the reference signal (RPM) to the computer. When looking at shielded wires in a schematic, they look like a dashed (- - - -) lasso.
3. Because current will take the path of least resistance, pay close attention to splices in parallel circuits. An example of this would be a shorted MAP sensor; this could affect the TPS because they are usually on the same circuit.

The computer is only as good as the electrical signals it receives and sends out. An unreliable power supply can wreak havoc with the computer and cause some unique driveability complaints. Therefore, the integrity of the power feed and ground connections must always be considered.

Scan Tool Diagnosis

Throughout the 1980s, GM and Chrysler were the two manufacturers that consistently made provisions for testing with a scan tool. Starting in the late 1980s, other manufacturers began adding a serial data line to many of their systems. Also, automotive manufacturers began to make serial data access protocols available to the manufacturers of aftermarket scan tools.

Because of these changes and the introduction of On-Board Diagnostics Second Generation (OBD-II), the scan tool will play a more important role in diagnosing driveability complaints. Chapter 7 covered the features and basics of using a scan tool. Scan tool data is an invaluable tool, providing the technician understands the information covered in Chapters 5 through 8. If the following material is unclear, refer to those chapters for review.

There are four steps to troubleshooting with a scan tool. By following each of these steps in order, the customer's complaint can be resolved in the shortest time:

1. Scan diagnostic circuit inspection.
2. Serial data quick inspection (Figure 11-15).
3. Troubleshooting by symptom (Figures 11-16 through 11-20).
4. Repair confirmation (Figure 11-21).

The *scan diagnostic circuit inspection* confirms three important things:

1. By confirming that the check engine light functions normally and the scan tool is communicating with the computer, this establishes that the on-board diagnostic system is functional.
2. In the event that communications cannot be established, the problem must be identified before proceeding with the diagnosis.
3. If there are fault codes detected during the diagnostic circuit check, refer to the appropriate diagnostic trouble tree or driveability symptoms chart.

The *serial data quick inspection* is used because no fault codes are present to guide the technician to a faulty circuit. Because a scan tool allows quick access to a lot of data, a systematic scan of the data is the next logical step.

A systematic scan of the data establishes an order of importance for sensors, switches, and actuators. The first step is to look at key-on, engine-off (KOEO) data:

1. *Coolant Temperature and Intake Air Temperature* — If the vehicle has sat long enough for the engine to cool down, then the sensors should display similar readings. Do they correspond to the surrounding ambient air temperature?
2. *Baro Sensor* — If the system is equipped with a separate barometric sensor, the information is used to establish barometric air pressure to calculate spark timing and air/fuel. It is initially used at start-up.
3. *MAP Sensor* — If the vehicle is equipped with a separate Baro sensor, are the MAP and Baro readings similar? If the vehicle is equipped with only a MAP sensor (no Baro sensor), this is a Barometric/Manifold Absolute Pressure (BMAP) sensor. In the KOEO position, the computer uses this sensor to obtain the barometric pressure reading needed to calculate spark timing and air/fuel.
4. *Throttle Position Sensor* — If the vehicle is fuel-injected, TPS voltage should be approximately 1 volt or under. If TPS voltage exceeds approximately 3 volts in this mode, a clear flood mode may be set, resulting in a no-start condition.
5. *Park/Neutral Switch* — It should indicate that the transmission selector is in Park/Neutral. If out of adjustment, this could affect idle quality and tailpipe emissions.
6. *Fuel Injector Pulse-Width* — A number may or may not be present. If a number is present, it indicates that the computer is anticipating the next start-up. If no pulse-width is indicated, the vehicle may be in a clear flood mode, or a pulse-width may not be available until the key is turned to the crank position.
7. *Idle Air Control (IAC)* — Verify that a number is present because this indicates that the computer has positioned the IAC for start-up. (Each manufacturer uses a different step count to prepare the engine for the next restart.)

Figure 11-15 shows the normal key-on, engine-off readings for a 1987 Dodge Caravan. How does it match up against the inputs to check with the key-on, engine-off? (Not all of the listed information is available from every computer.)

Figure 11-16 shows the same vehicle in a no-start condition during cranking. (Code 31, Purge Solenoid, was not related to this problem.) The data shown is listed in order of priority. Move down the list and check each input for an abnormal reading. The first input on the list that is out of specification is the TPS. The voltage is above 3.0 volts; this puts the computer into a clear flood mode. This can be verified by moving down the list to the Injector Pulse-Width. The pulse-width is 00.0 ms because a clear flood mode prevents fuel to the cylinders to allow the spark plugs time to dry out.

Once an engine is running, prioritize the inputs by the effect they have on engine operation (in descending order):

```
KOEOGOOD                  E-Z EVENT PLUS
CHRYSLER 1987 3.0 MFI VIN 3
EVENT 1    TAG  --    FRAME -30
FAULT CODES:  31


                     0    1    2    3    4
RPM                0000 0000 0000 0000 0000
COOLANT TEMP       058  058  058  058  058
CHARGE TEMP        059  059  059  059  059
MAP VOLTS          4.64 4.64 4.64 4.64 4.64
BARO PRESSURE VOL4.84 4.84 4.84 4.84 4.84
THROT POS SENSOR 1.10 1.10 1.10 1.10 1.10
INJ PULSE WIDTH  21.2 21.2 21.2 21.2 21.2
```

Figure 11-15 Always begin scan tool diagnostics in the key-on, engine-off mode. Compare the temperature and voltages of common sensors.

```
NOSTART                   E-Z EVENT PLUS
CHRYSLER 1987 3.0 MFI VIN 3
EVENT 2    TAG  --    FRAME -30
FAULT CODES:  31


                   -30  -29  -28  -27  -26
RPM                0000 0000 0000 0000 0000
MAP VOLTS          4.64 4.64 4.64 4.64 4.64
THROT POS SENSOR 3.86 3.86 3.86 3.86 3.86
COOLANT TEMP       058  058  058  058  058
BARO PRESSURE VOL4.84 4.84 4.84 4.84 4.84
CHARGE TEMP        059  059  059  059  059
INJ PULSE WIDTH  00.0 00.0 00.0 00.0 00.0
```

Figure 11-16 Prioritize data and scan the data from top to bottom looking for out-of-range sensors, switches, and outputs.

1. *RPM* — Check this during a no-start. If the vehicle is fuel injected and this signal is not generated, the engine cannot start. If the signal disappears while the engine is running, the engine will stall. If the vehicle is carbureted it will continue to run, but the check engine light will be on.
2. *Load Sensor* — MAP, Vacuum, or Mass Air Flow sensors tell the computer how much air is in the engine. This sensor has a major affect on the air/fuel ratio and ignition timing.
3. *Throttle Position Sensor* — It can be a sensor or switch that indicates a closed throttle, steady throttle, acceleration, or deceleration. Some systems display throttle position as an angle or a mode. If the throttle is closed, does the throttle angle indicate 0°? If throttle position is displayed as a mode, it shows C/T for a closed throttle. This indicates the computer has recognized that the throttle is closed.
4. *Coolant Temperature Sensor* — Affects a variety of engine operations; for that reason it should always be monitored.
5. *Barometric Sensor* — Compensating for increases in altitude (less pressure = less air) is important because less air requires less fuel, and leaner air/fuel ratios require additional spark advance.
6. *Manifold Air Temperature Sensor* — It allows the computer to fine tune the air/fuel ratio based on the density of the air in the intake air stream.
7. *Oxygen Sensor* — Indicates the condition of exhaust oxygen content. It can be read in volts or cross counts. Once the vehicle is operating in closed loop, does the sensor rapidly cycle between 0.2 to 0.8 volts? If the voltage is cycling between 0.0 and 0.6 volts, then the computer is correcting for a lean exhaust. If the voltage is cycling between 0.4 and 1.0 volts, then the computer is correcting for a rich exhaust.
8. *Battery Sense Input* — Most computers operate effectively when battery/charging voltage is from 10 to 16 volts. If voltage is not in this range, the computer might revert to a limp-in mode.
9. *Switch Circuits* — They can be in one of two states, on/off or high/low. Most switches indicate that a load is being placed on the engine.
10. *Vehicle Speed Sensor* — Use this input when the vehicle is being driven on a test drive. It can be displayed in miles per hour (mph) or kilometers per hour (kph).

Remember, output commands are the result of the computer processing input data from sensors and switches. When looking at the outputs, it is not important to prioritize them:

1. *Idle Air Control (IAC)* — The number should increase when the throttle is opened and decrease as the throttle is closed. Additional engine loads should increase the IAC number: i.e., when the air conditioning is turned on; the vehicle is shifted into gear; or when a heavy electrical demand is placed on the charging system. IAC

counts can be used to pinpoint air/fuel-related problems that affect idle speed.

2. *Fuel Injector Pulse-Width* — The pulse-width should increase or decrease with fuel demand.
3. *Fuel Control* (Refer to Chapter 8) — Different terms are used for this, such as: Mixture Control Dwell (MC), Block Learn and Block Integrator, Short-Term Fuel Trim and Long-Term Fuel Trim, and Adaptive Memory and Additive Fuel Factor. Remember, on one side of the mid-point, the computer is adding fuel, and on the other side, it is subtracting fuel.
4. *Spark Advance* — It should change with coolant temperature, throttle opening, engine load, and spark knock.
5. *Emissions Controls* (Refer to Chapter 6) — These systems are controlled by the computer, based on input from the various sensors. They indicate what the computer is attempting to do, not necessarily what is actually occurring.

Figure 11-17 shows a snapshot (approximately five seconds of time) captured while test driving the vehicle. The customer's complaint was that the vehicle would jerk or miss intermittently. Two codes are present, but only Code 37 (Torque Converter Clutch) is related to this problem.

The engine RPM for this vehicle is erratic. The RPM signal is an input, but it can also reflect a computer or output problem. The engine load, based on throttle position and MAP, is relatively steady. Vehicle Speed Sensor (VSS) readings show the vehicle speed changing erratically from 18 to 200 mph. A VSS signal is an input, but the VSS data shown on the scan tool is the computer calculated output of the vehicle's speed.

C88MPIVSS E-Z EVENT PLUS
CHRYSLER 1988 3.0 MFI VIN 3
EVENT 1 TAG -- FRAME -15
FAULT CODES: 12 37

	-15	-14	-13	-12	-11	-10	-9	-8	-7	-6	-5	-4	-3	-2	-1	0
RPM	1624	1691	1712	1732	1736	1776	1802	1798	1805	1809	1798	1833	1322	1871	1819	1819
MAP VOLTS	3.06	3.04	3.04	3.00	2.98	2.98	2.98	2.96	2.96	2.92	2.90	2.88	3.16	2.88	2.98	2.98
THROT POS SENSOR	1.56	1.56	1.56	1.58	1.58	1.58	1.58	1.58	1.58	1.58	1.58	1.58	1.58	1.58	1.60	1.60
COOLANT TEMP	178	178	178	178	178	178	178	178	178	178	178	178	178	178	178	178
BARO PRESSURE VOL	4.70	4.70	4.70	4.70	4.70	4.70	4.70	4.70	4.70	4.70	4.70	4.70	4.70	4.70	4.70	4.70
CHARGE TEMP	098	098	098	098	098	098	098	098	098	098	099	099	099	099	099	099
O2 VOLTS	0.06	0.06	0.06	0.06	0.12	0.16	0.16	0.16	0.18	0.20	0.22	0.24	0.18	0.14	0.10	0.10
BATTERY VOLTS	14.3	14.1	14.1	14.2	14.3	14.3	14.2	14.1	14.4	14.2	14.3	14.2	14.4	14.3	14.2	14.2
INJ PULSE WIDTH	06.2	06.2	06.3	06.3	06.4	06.4	06.4	06.4	06.4	06.5	06.5	06.5	06.9	05.9	06.4	06.4
ELECTRONIC ADV/RE	030	031	000	000	000	000	001	001	001	001	001	001	029	002	001	001
VEHICLE SPEED SEN	019	018	025	044	029	026	025	045	100	088	101	115	089	085	107	107

	0	1	2	3	4	5	6	7	8	9	10	11	12	13	14	15
RPM	1819	1240	1240	1240	1652	1652	1652	1783	1783	1783	1580	1580	1580	1778	1793	1800
MAP VOLTS	2.98	3.42	3.42	3.42	3.38	3.38	3.38	3.08	3.08	3.08	3.12	3.12	3.12	3.30	3.16	3.16
THROT POS SENSOR	1.60	1.60	1.60	1.60	1.60	1.60	1.60	1.60	1.60	1.60	1.60	1.60	1.60	1.62	1.64	1.64
COOLANT TEMP	178	178	178	178	178	178	178	178	178	178	178	178	178	178	178	178
BARO PRESSURE VOL	4.70	4.70	4.70	4.70	4.70	4.70	4.70	4.70	4.70	4.70	4.70	4.70	4.70	4.70	4.70	4.70
CHARGE TEMP	099	099	099	099	099	099	099	099	099	099	099	099	099	099	099	099
O2 VOLTS	0.10	0.08	0.08	0.08	0.06	0.06	0.06	0.04	0.04	0.04	0.04	0.04	0.04	0.02	0.04	0.02
BATTERY VOLTS	14.2	14.1	14.1	14.1	14.1	14.1	14.1	14.2	14.2	14.2	14.1	14.1	14.1	14.3	14.3	14.1
INJ PULSE WIDTH	06.4	08.5	08.5	08.5	08.4	08.4	08.4	06.2	06.2	06.2	06.9	06.9	06.9	07.4	06.5	06.6
ELECTRONIC ADV/RE	001	028	028	028	029	029	029	000	000	000	030	030	030	031	031	000
VEHICLE SPEED SEN	107	170	170	170	082	082	082	091	091	091	200	200	200	066	074	084

Figure 11-17 One advantage of a scan tool is the *snapshot* feature that can be used during a test drive. The computer problem shown here could not be found while working with the vehicle in Park or Neutral.

The first thing to test is the Vehicle Speed Sensor input signal to the computer; here it proved to be good. Because the VSS input is good, and the output data does not correspond to the input, the computer must be defective. After all grounds were checked, the computer was replaced. Code 37 (Torque Converter Clutch) and the erratic electronic spark advance are a result of the bad computer.

Figure 11-18 shows a snapshot of a high-speed surge problem with no code present. A surge is usually related to a rich air/fuel condition. This vehicle was equipped with a Pressure Differential (vacuum) sensor (represented by the MAP volts), and the high voltage indicates a light engine load.

When the vehicle was tested at idle, everything checked out okay. Once the engine RPM was increased, the mixture control (MC) solenoid dwell increased to 57°, showing the computer was trying to subtract fuel. The oxygen sensor confirms the rich condition. A large vacuum leak was created and the 0_2 voltage dropped to 0.050 volts, confirming the sensor was operating correctly.

When checking MC dwell, do not disconnect anything. If the MC dwell indicates a problem, disconnect components one at a time. The problem has been found when the reading returns to normal. Here, the air filter was removed and no change was noticed. Once the purge canister line was disconnected, the dwell dropped to 25° and began to drift between 25° and 30°. The computer was again in air/fuel control. A defective purge valve proved to be the problem. The problem did not occur at idle because the purge was controlled by ported vacuum.

Figure 11-19 shows a snapshot taken from a vehicle with a higher-than-normal idle speed. Normal idle for this vehicle was approximately 550-650 RPM. The vehicle was equipped with a MAP sensor and Idle Air Control (IAC) motor.

The oxygen sensor readings show wide voltage swings between 0.09 and 0.91 volts. The Block Integrator and Block Learn show that the computer was adding fuel to compensate for a lean exhaust condition. The IAC count dropped from 32 down to zero. This shows the computer is attempting to lower the

84VZ2805 E-Z EVENT PLUS
G.M. 1984 FULL VIN Z
EVENT 5 TAG -- FRAME -30
DIAGNOSTIC STATE 8 ALCL
TROUBLE CODES: NO

	-30	-29	-28	-27	-26	-25	-24	-23	-22	-21	-20	-19	-18	-17	-16	-15
RPM	2975	2975	2925	2950	2975	3200	3500	3525	3600	3625	3525	3400	3400	3400	3450	3475
MAP VOLTS	3.96	3.96	3.96	3.94	3.94	3.84	3.86	3.92	3.96	3.96	4.00	4.02	3.98	3.96	3.94	3.96
THROT POS SENSOR	0.82	0.82	0.82	0.82	0.82	0.92	1.00	1.04	1.04	1.04	1.02	0.96	0.96	0.98	1.00	1.00
COOLANT TEMP	190	190	192	192	192	192	192	194	194	194	194	194	195	195	195	195
O2 VOLTS	0.78	0.18	0.80	0.74	0.80	0.66	0.76	0.69	0.79	0.69	0.72	0.68	0.77	0.67	0.29	0.72
O2 CROSS COUNTS	012	002	002	004	002	002	000	000	000	000	000	001	005	002	004	000
M/C SOLENOID DWEL	45 D	45 D	49 D	48 D	51 D	52 D	56 D	57 D	57 D	57 D	57 D	57 D	57 D	57 D	56 D	57 D

	-15	-14	-13	-12	-11	-10	-9	-8	-7	-6	-5	-4	-3	-2	-1	0
RPM	3475	3500	3500	3475	3525	3500	3525	3525	3550	3475	3525	3575	3550	3525	3575	3575
MAP VOLTS	3.96	3.98	3.98	3.98	3.98	3.98	3.98	3.98	3.98	3.98	3.98	3.98	3.98	3.98	3.98	4.00
THROT POS SENSOR	1.00	1.00	1.00	1.00	1.00	1.00	1.00	1.00	1.00	1.00	1.00	1.00	1.00	1.00	1.00	1.00
COOLANT TEMP	195	197	197	197	197	197	199	199	199	199	199	201	201	201	201	203
O2 VOLTS	0.72	0.72	0.76	0.80	0.81	0.76	0.75	0.73	0.76	0.66	0.81	0.66	0.76	0.69	0.72	0.68
O2 CROSS COUNTS	000	002	000	000	000	002	000	000	000	000	000	002	000	000	002	000
M/C SOLENOID DWEL	57 D	57 D	57 D	57 D	57 D	57 D	57 D	57 D	57 D	57 D	57 D	57 D	57 D	57 D	57 D	57 D

Figure 11-18 A 57° MC dwell indicates a rich exhaust condition. Drive the system full lean to verify proper feedback circuit operation.

92VK5705 E-Z EVENT PLUS
GM 1992 TRUCK TBI VIN K
EVENT 5 TAG -- FRAME -30
DIAGNOSTIC STATE 9 ROAD
TROUBLE CODES: NO

	-30	-29	-28	-27	-26	-25	-24	-23	-22	-21	-20	-19	-18	-17	-16
RPM	0675	0675	0650	0650	0675	0650	0650	0675	0650	0650	0650	0675	0675	0650	0650
MAP VOLTS	0.97	0.93	0.97	0.97	0.95	1.00	0.97	0.95	0.95	0.95	0.97	0.95	0.95	0.95	0.97
THROT POS SENSOR	0.54	0.54	0.54	0.54	0.54	0.54	0.54	0.54	0.54	0.54	0.54	0.54	0.54	0.54	0.54
COOLANT TEMP	177	177	179	179	179	179	179	179	179	179	179	179	179	179	181
O2 VOLTS	0.73	0.66	0.12	0.11	0.17	0.10	0.18	0.86	0.88	0.87	0.87	0.89	0.88	0.81	0.79
LOOP STATUS	CLOS	CLOS	CLOS	CLOS	CLOS	CLOS	CLOS	CLOS	CLOS	CLOS	CLOS	CLOS	CLOS	CLOS	CLOS
INTEGRATOR	128	127	127	127	127	128	129	130	130	130	130	130	129	128	127
BLOCK LEARN MULT	125	125	125	125	125	125	125	125	125	125	125	125	125	125	125
IDLE AIR CONTROL	032	032	032	032	032	032	032	032	032	032	032	032	032	032	031
AIR SWITCH	OFF	OFF	OFF	OFF	OFF	OFF	OFF	OFF	OFF	OFF	OFF	OFF	OFF	OFF	OFF
AIR DIVERT	OFF	OFF	OFF	OFF	OFF	OFF	OFF	OFF	OFF	OFF	OFF	OFF	OFF	OFF	OFF

	-15	-14	-13	-12	-11	-10	-9	-8	-7	-6	-5	-4	-3	-2	-1
RPM	0650	0875	1000	0950	0950	0925	0925	0900	0900	0875	0875	0875	0875	0875	0875
MAP VOLTS	1.01	1.31	0.83	0.89	0.89	0.89	0.89	0.89	0.87	0.87	0.85	0.85	0.85	0.85	0.83
THROT POS SENSOR	0.54	0.54	0.54	0.54	0.54	0.54	0.54	0.54	0.54	0.54	0.54	0.54	0.54	0.54	0.54
COOLANT TEMP	181	181	181	181	181	181	181	181	181	181	181	181	181	181	181
O2 VOLTS	0.10	0.93	0.11	0.10	0.09	0.10	0.10	0.11	0.12	0.12	0.89	0.91	0.91	0.91	0.91
LOOP STATUS	CLOS	CLOS	CLOS	CLOS	CLOS	CLOS	CLOS	CLOS	CLOS	CLOS	CLOS	CLOS	CLOS	CLOS	CLOS
INTEGRATOR	127	127	127	127	128	129	130	131	133	134	135	136	136	136	136
BLOCK LEARN MULT	125	125	125	125	125	125	125	126	127	128	129	129	130	131	132
IDLE AIR CONTROL	031	031	028	024	021	018	014	011	007	004	001	000	000	000	000
AIR SWITCH	OFF	OFF	OFF	OFF	OFF	OFF	OFF	OFF	OFF	OFF	OFF	OFF	OFF	OFF	OFF
AIR DIVERT	OFF	OFF	OFF	OFF	OFF	OFF	OFF	OFF	OFF	OFF	OFF	OFF	OFF	OFF	OFF

Figure 11-19 The problem shown here can be seen in frames -14 to -1. Monitoring Idle Air Control data can help the technician identify an air/fuel problem.

idle speed by reducing the amount of air bypassing the closed throttle plates. For this to occur, additional air has to be entering the engine after the closed throttle plates.

A visual inspection found the problem. By disconnecting and placing a thumb over the PCV, the engine idle speed momentarily dipped and then returned to normal. The valve was leaking internally, allowing extra air from the crankcase to enter the engine.

The engine did not idle rough because the MAP sensor responded to the extra air by adding fuel and advancing the spark slightly. This allowed the engine to run normal, but with a higher-than-normal speed. Because the idle speed was higher than normal, the computer decreased the IAC counts in an attempt to reach the desired 550-650 RPM idle speed.

Figure 11-20 shows a snapshot taken because of an intermittent check engine light. The test drive failed to turn on the check engine light. A code stored in memory revealed a Code 13 (Lazy Oxygen Sensor).

The next step was to look at O_2 sensor voltage, O_2 cross counts, and MC dwell at idle and cruise. The O_2 sensor voltage remained at 0.450 volts, cross counts were zero, and the dwell remained fixed at 30°. Attempts were made to drive the O_2 sensor lean and rich, but none were successful. The oxygen sensor was removed and did not appear to be contaminated, which meant that it was the most likely cause of the problem. The light did not turn on during the

87VH0203 E-Z EVENT PLUS
GM 1987 FULL VIN G
EVENT 3 TAG -- FRAME -30
DIAGNOSTIC STATE 8 ALCL
TROUBLE CODES: 13

	0	1	2	3	4
RPM	2550	2550	2550	2525	2550
COOLANT TEMP	156	156	156	158	158
M/C SOLENOID DWEL	29 D	29 D	29 D	29 D	29 D
02 CROSS COUNTS	000	000	000	000	000
02 VOLTS	0.45	0.44	0.45	0.45	0.45

Figure 11-20 The feedback circuit should always be monitored at higher RPM to ensure that the engine and oxygen sensors have reached operating temperature.

87VHO204 E-Z EVENT PLUS
GM 1987 FULL VIN G
EVENT 4 TAG -- FRAME -30
DIAGNOSTIC STATE 8 ALCL
TROUBLE CODES: NO

	0	1	2	3	4	5	6	7	8	9	10
RPM	2650	2650	2700	2700	2725	2700	2725	2700	2725	2700	2700
COOLANT TEMP	168	167	168	168	168	168	168	168	168	168	168
M/C SOLENOID DWEL	36 D	37 D	36 D	36 D	37 D	37 D	37 D	37 D	36 D	37 D	36 D
02 CROSS COUNTS	003	003	003	002	004	001	006	003	004	005	001
02 VOLTS	0.61	0.67	0.39	0.39	0.57	0.48	0.52	0.49	0.93	0.43	0.95

Figure 11-21 Use the scan tool to verify that the repair has been successful.

test drive because the fault code arming conditions were not met.

Figure 11-21 shows how the scan tool can be used to verify that the repair has been successful. The vehicle used in Figure 11-21 was tested following the replacement of the O_2 sensor. The snapshot shows that the O_2 sensor voltage, cross counts, and MC dwell were normal. The computer was again in control of the air/fuel ratio.

The problems discussed here do not address every type of problem, but are intended to help the technician develop a logical process to scan tool diagnostics. The scan tool is just part of the process because the technician has to interpret the data and choose the necessary repair. The more familiar the technician becomes with looking at scan tool data, the quicker the problem will be identified and repaired.

CHAPTER 12

On-Board Diagnostics Two (OBD-II)

ON-BOARD DIAGNOSTICS (OBD)

The California Air Resources Board (CARB) first developed On-Board Diagnostics (OBD) regulations in 1985. These regulations were known as OBD-I, and took effect beginning with the 1988 model year. The regulations require the vehicle's on-board computer to monitor the vehicle's oxygen sensor, the EGR valve, and the evaporative purge solenoid for proper operation. In addition to monitoring specific systems, all vehicles certified for sale in California were required to be equipped with a Malfunction Indicator Light (MIL), usually known as a *check engine* light, on the dash to notify the driver that a computer-related failure had occurred. Other names were used instead of *check engine*:

1. Power loss.
2. Service engine now.
3. Service engine soon.

Unfortunately, OBD-I does not cover several critical emissions-related components (e.g., catalyst, evaporative system vapor leaks). Also, it did not require sufficient sensitivity to detect significant malfunctions of monitored components. The California Air Resources Board found that by the time an emissions system component failed and caused the Malfunction Indicator Light (MIL) to illuminate, the vehicle may have been emitting excess emissions for some time. In an attempt to identify partially-functioning systems, the CARB developed enhanced system monitoring requirements called On-Board Diagnostics II (OBD-II).

OBD-II Legislation

On November 15, 1990, the Clean Air Act was amended (CAAA), directing the Environmental Protection Agency (EPA) to develop new regulations pertaining to OBD systems. The CAAA requires that all 1994 and later model year light-duty vehicles (LDVs), and light-duty trucks (LDTs) contain an OBD system. All LDVs sold in the United States must meet OBD-II requirements by the 1996 model year. By 1997, all LDTs must meet these requirements. Phase-in of OBD-II began with some 1994 model year vehicles.

California's OBD-II regulations differ slightly from the EPA's regulations. CARB has taken more of a design-oriented approach, and the EPA places more emphasis on emissions performance standards. CARB's OBD-II requirements include catalyst monitoring, evaporative emission control system leak detection, monitoring of the operational characteristics of the oxygen sensor, and detection of engine misfire.

The EPA elected to adopt California OBD-II standards for federal emissions certification, effective with the 1996 model year. The EPA expects that manufacturers can probably develop one system that complies with both the EPA and California OBD-II regulations. However, the EPA will accept compliance with the California OBD-II requirements as an alternative means of meeting certain portions of the federal requirements through the 1998 model year. Beginning with the 1998 model year, a new federal OBD standard will be adopted, thus eliminating the differences between California and federal emissions certification (Figure 12-1).

Under these regulations, a vehicle's OBD system must detect a malfunction or the deterioration of an emissions-related component or system. OBD-II requires automobile manufacturers to install on-board diagnostic systems that can:

1. Identify component degradation or a malfunction of major emissions-related systems that could prevent a vehicle from complying with federal emissions standards.
2. Alert the vehicle operator of the need to maintain and/or repair emissions-related components and/or systems.
3. Store *standardized* Diagnostic Trouble Codes (DTC) and provide access to the vehicle's on-board information with a generic scan tool.
4. Make available to all interested parties all necessary emissions maintenance and repair information.

For LDVs and LDTs, regulations require that the OBD system monitor the performance and detect malfunctions and deterioration of the catalytic converter and oxygen sensor, and also detect engine misfire. When the OBD system makes such a detection, a trouble code must be stored in the Powertrain Control Module (PCM). The following conditions will illuminate the Malfunction Indicator Light (MIL):

1. Catalytic converter deterioration that would result in an increase greater than 0.4 g/mi HC above the 0.6 g/mi HC FTP standards.
2. Engine misfire that would result in an increase greater than 0.4 g/mi HC, 3.4 g/mi CO, or 1.0 g/mi NO_x above FTP standards.
3. Oxygen sensor deterioration that would result in an increase greater than 0.2 g/mi HC, 1.7 g/mi CO, or 0.5 g/mi NO_x above FTP standards.

The system must also detect a misfire and store a code showing which cylinder is misfiring or that multiple cylinders are misfiring. Besides detecting oxygen sensor deterioration, the system must

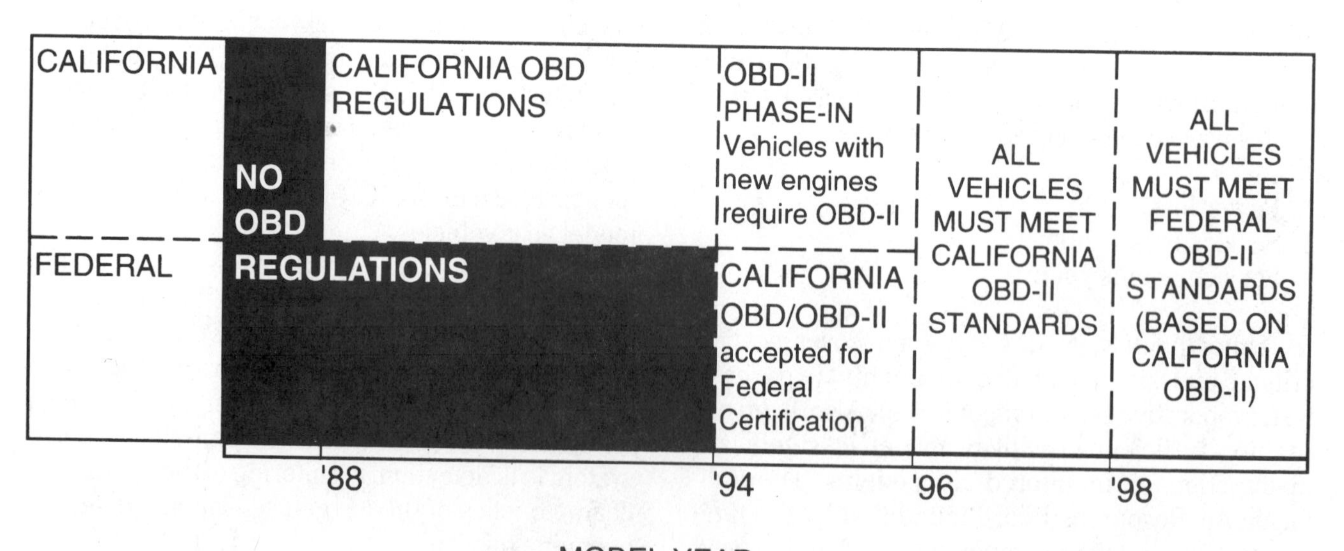

Figure 12-1 OBD regulations.

detect any malfunction of the sensor that renders it incapable of satisfactorily performing its functions as part of the OBD system.

DIAGNOSTIC EXECUTIVE OR TASK MANAGER

There are many components responsible for controlling vehicle emissions. The PCM must coordinate their operation and keep them functioning in the most efficient manner possible. The PCM determines if the systems are operating properly. A new piece of software, designed specifically for this job, is called the *Diagnostic Executive* or *Task Manager.* In this text, it will be called the *Diagnostic Executive.*

The Diagnostic Executive is responsible for the following tasks (see Figure 12-2):

1. Trip monitor and test sequence.
2. Passive, active, and intrusive diagnostic tests.
3. Comprehensive system monitoring.
4. Diagnostic Trouble Codes (DTC) and Malfunction Indicator Light (MIL).
5. Freeze frame data storage and erasure.
6. Failure records (GM only).

Trip Monitor and Test Sequence

The Diagnostic Executive reviews the PCM inputs during each key-on, engine run, key-off cycle. It then determines if the conditions for a diagnostic test have been met before it allows the test to be

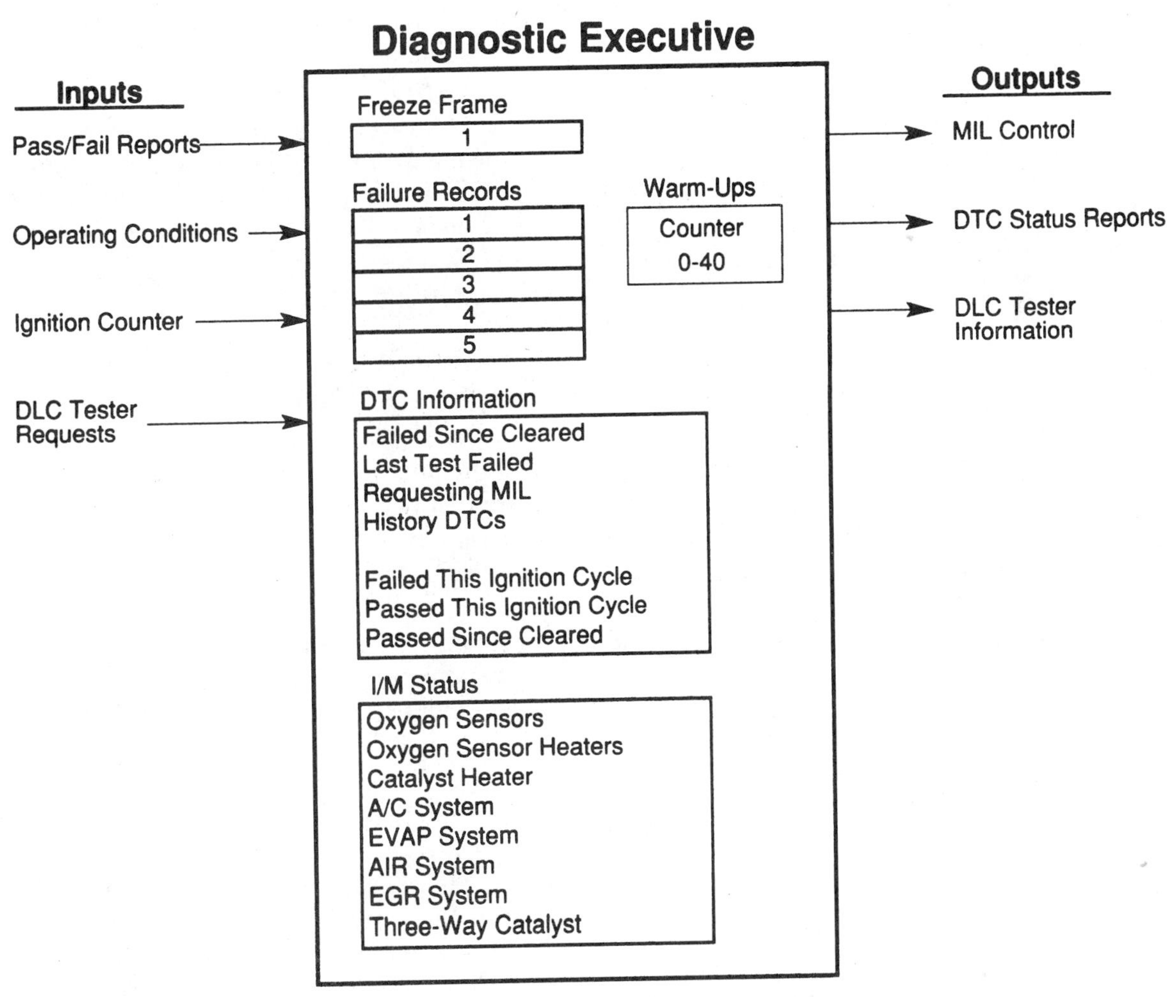

Figure 12-2 The Diagnostic Executive stores data and controls diagnostic test operation (courtesy GM).

performed. Requirements for each test will vary. They typically include information such as elapsed time since start-up, engine coolant temperature, RPM, calculated load, throttle position, and vehicle speed.

A diagnostic test is a series of steps which has a beginning and end, the result of which is a pass or fail reported to the Diagnostic Executive. When a diagnostic test reports a pass result, the Diagnostic Executive records that the:

1. Diagnostic test has been completed since the last ignition cycle.
2. Diagnostic test passed during the current ignition cycle.
3. Fault identified by the diagnostic test is not currently active.

When a diagnostic test reports a fail result, the Diagnostic Executive records:

1. That the diagnostic test has been completed since the last ignition cycle.
2. That the fault identified by the diagnostic test is currently active.
3. That the fault has been active during this ignition cycle.
4. The operating conditions at the time of the failure.

Each key-on, engine run, key-off cycle that meets specific criteria is called a *trip*. A trip is a difficult concept to define because the requirements for a trip will vary depending on the test being run (Figure 12-3). These conditions can include unrelated items such as an individual's driving style, length of trip, and ambient temperature. The minimum requirement for a trip is that it includes one key-on/off cycle and some drive time before a test is performed.

The term *enable criteria* describes the conditions necessary for a given diagnostic test to run. A trip is official when the enable criteria are met for a given diagnostic routine. Enable criteria can be compared to the fault code arming conditions covered in Chapter 11. Each diagnostic routine has a specific list of conditions required for the diagnostic to run. Since the enable criteria varies from one component or system to another, the definition of trip will also vary.

The concept of a trip is similar to Toyota's two-trip code detection covered in Chapter 11. A vehicle must fail a test more than once before the MIL is illuminated and a DTC is stored in memory. A two-trip monitor allows the system to double check itself, preventing a false MIL. If the conditions are not met on two consecutive key-on cycles, the information is stored in the Diagnostic Executive and remains there until the appropriate conditions exist. If a two-trip fault fails during the first trip, a malfunction on the second consecutive trip illuminates the MIL. The MIL does not illuminate and no DTC is stored if a malfunction does not occur during a second trip.

There are times when a test is held up *pending* the results of a related problem indicated by the MIL. Testing the system or component at this time guarantees that it will fail erroneously, so the Diagnostic Executive does not bother running the test.

At times there are other tests running or existing faults that *conflict* with a specific test operation. Then, the Diagnostic Executive chooses not to run the test and no trip is completed. A test is *suspended* until another monitor with a higher priority is run and passed.

Figure 12-4 (page 204) shows a list of *enable criteria* for a Chrysler Catalyst Monitoring Test. As shown in the figure, the test will be *pending* if any of the DTCs listed are present. The system is said to be in *conflict* if any of the listed conditions are present. The test is *suspended* until the O_2 sensor has passed its monitor test.

Passive, Active, and Intrusive Diagnostic Tests

To pinpoint a problem before illuminating the MIL, the PCM performs a variety of tests in an attempt to verify the operation of a specific system or component:

1. A passive test is a diagnostic test that monitors a system or component and no action is taken.
2. An active test takes some sort of action when performing diagnostic functions, and does not affect performance. Here, the PCM performs an action for which it expects to see a specific reaction. Active tests are performed when a passive test fails.
3. Intrusive testing is done when the PCM does not see the expected results after performing the passive and active test. The intrusive test will affect the vehicle's performance and emissions levels.

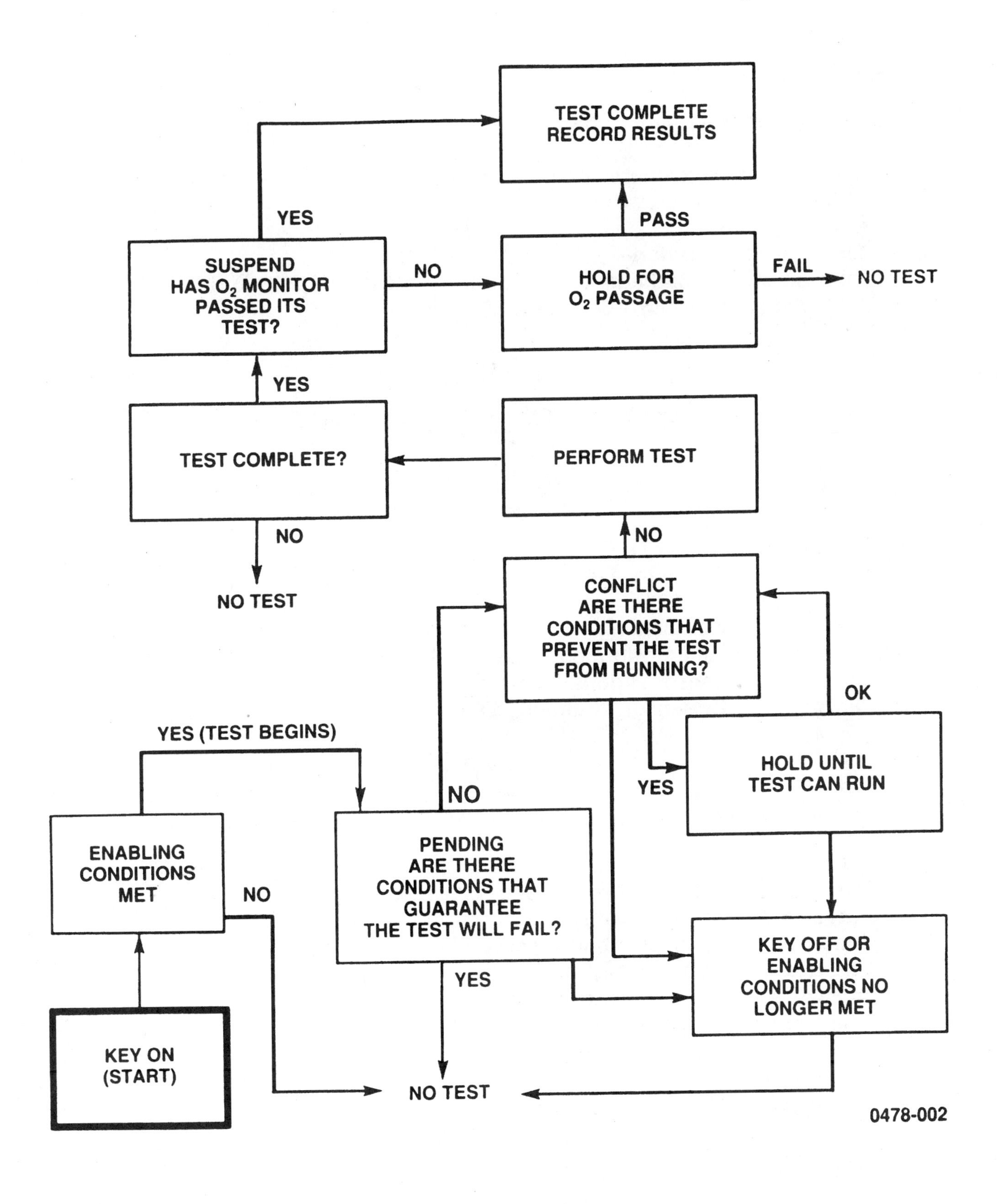

Figure 12-3 Sequencing of diagnostic tests during a typical trip (courtesy Chrysler Corp.).

Enabling Conditions

The following conditions must be met before the catalyst monitor will run:

- Engine coolant temperature greater than 170° F
- Vehicle speed greater than 20 mph for 2 minutes
- Open throttle
- Closed loop operation
- Rpm between 1,248 and 1,952 (auto), or between 1,248 and 2,400 (manual)
- MAP voltage between 1.50 and 2.60

Pending

The catalyst monitor does not run if the MIL is illuminated due to one of the following:

- Misfire DTC
- O_2 monitor DTC
- Upstream O_2 heater DTC
- Downstream O_2 heater DTC
- Fuel system rich DTC
- Fuel system lean DTC
- Vehicle is in the limp-in mode due to MAP, TPS, or engine temperature DTC
- Upstream O_2 sensor rationality DTC
- Downstream O_2 sensor rationality DTC

Conflict

The monitor does not run if any of the following are present:

- EGR monitor is in progress
- Fuel system rich intrusive test is in progress
- Purge monitor is in progress
- Time since start is less than 60 seconds
- One trip misfire maturing code
- One trip O_2 monitor maturing code
- One trip upstream O_2 heater maturing code
- One trip downstream O_2 heater maturing code
- One trip fuel system rich maturing code
- One trip fuel system lean maturing code

Suspend

Results of the monitor are not recorded until the O_2 monitor passes.

Figure 12-4 Catalyst monitoring criteria (courtesy Chrysler Corp.).

Comprehensive System Monitoring

OBD-II requires the monitoring of all vehicle powertrain emissions-related systems and components providing input to, or receiving output from, the PCM. Comprehensive systems monitoring is a series of diagnostic tests of each emission control system that evaluates whether a component or system is operating properly.

In only a few instances was the addition of components necessary to perform the required monitoring. The major changes were software changes to the PCM and refinement of existing systems. Comprehensive systems monitoring includes:

1. Rigorous oxygen sensor evaluation.
2. Closely-monitored air/fuel ratios.
3. Evaluation of catalyst conversion efficiency.
4. Misfire monitoring.
5. Adequate Exhaust Gas Recirculation (EGR) system flow rate.
6. Proper secondary air injection performance.
7. System integrity and function of the evaporative control system.

Usually, a malfunction is determined to have occurred if the deterioration or failed component would cause emissions to exceed 1.5 times the applicable emissions standard. The following paragraphs cover the aspects of OBD-II monitoring requirements pertaining to maintaining low emissions levels.

Oxygen Sensor Monitoring — OBD-II systems use an upstream and downstream oxygen sensor for each catalyst, as shown in Figure 12-5. V-Type engine arrangements with a single exhaust include one oxygen sensor for each engine bank, and one downstream of the catalyst. True dual-exhaust systems include upstream and downstream oxygen sensors for each side of the exhaust. All oxygen sensors are heated (HO_2S). This allows for quicker warm-up and prevents the sensor from cooling down during prolonged idle conditions. HO_2S monitoring includes:

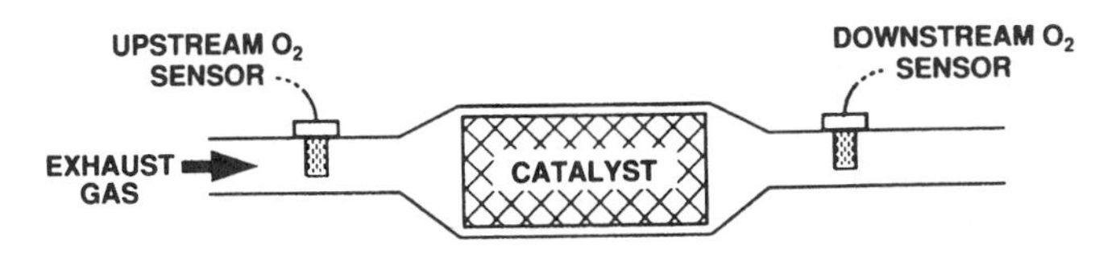

Figure 12-5 Typical oxygen sensor locations (courtesy Chrysler Corp.).

1. The HO_2S must detect changes quickly when exposed to different levels of oxygen; this is called the *response time test.* Figure 12-6 shows the lean-to-rich transition times of an HO_2S. The lean-to-rich response time is typically less than the rich-to-lean response time. Comparing the average transition time of the upstream HO_2S to a calibrated failure threshold will determine if the sensor is operating properly.

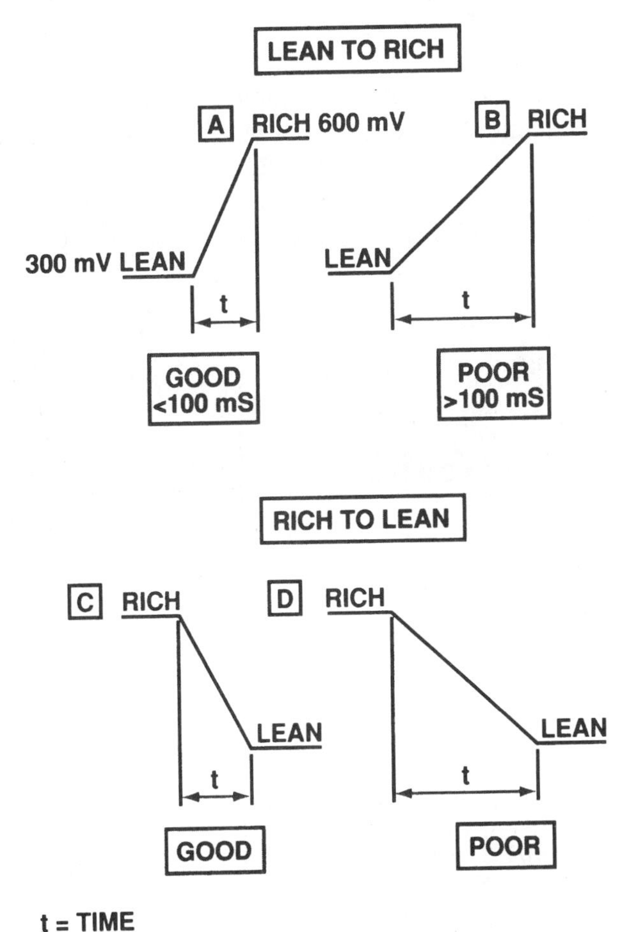

Figure 12-6 Oxygen sensor response test (courtesy GM).

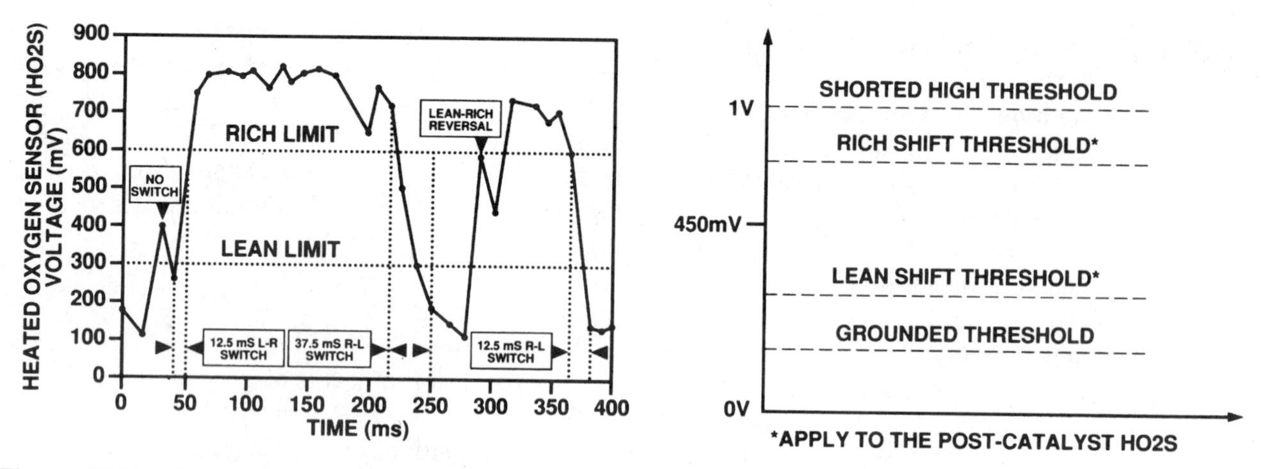

Figure 12-7 Oxygen sensor threshold voltages (courtesy GM).

2. A properly-operating HO_2S must be capable of generating an output voltage within its operating range of 0 to 1 volt. HO_2S voltage must exceed a minimum and a maximum voltage threshold as shown in Figure 12-7.

   The HO_2S is checked for an open circuit, shorted circuit, or switching frequency. HO_2S switching frequency is the number of times the voltage crosses over the midpoint of 450 mv in a specific time. The PCM compares the switching frequency to the required minimum stored in memory. Any number below the minimum indicates that the HO_2S is showing signs of deterioration and must be serviced. (See Figure 12-8.)
3. Internal oxygen sensor heater performance is monitored during a cold start and warm-up. The PCM looks at the activity of the HO_2S during this mode and compares the activity to a calibrated

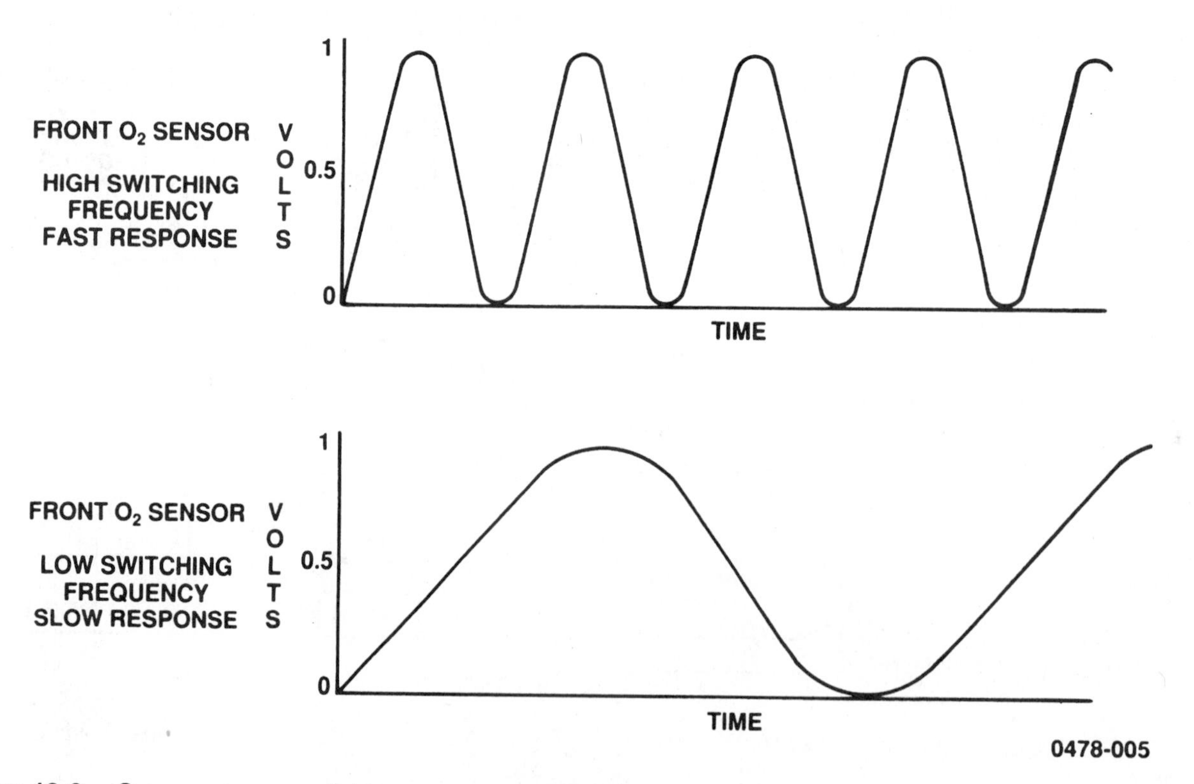

Figure 12-8 Oxygen sensor switching response test (courtesy Chrysler Corp.).

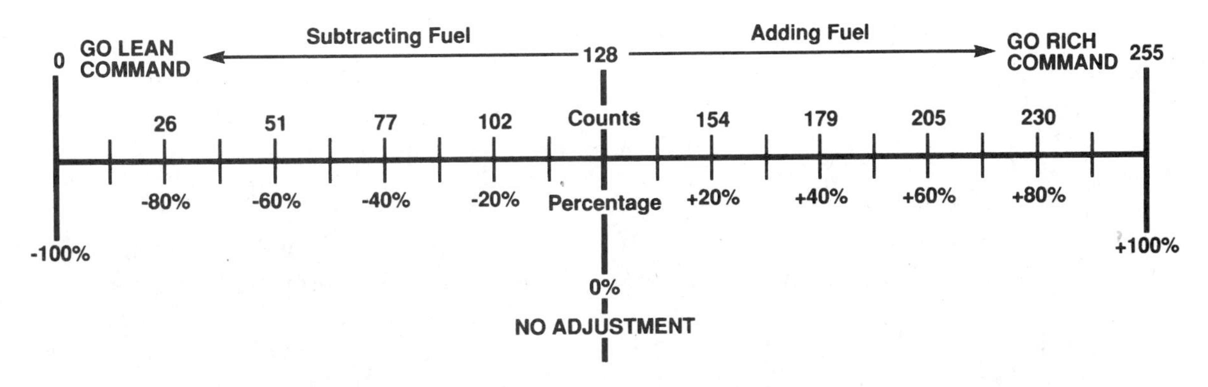

Figure 12-9 Fuel trim conversions between OBD-I and OBD-II (courtesy GM).

rate programmed into memory. Heater deterioration will increase the amount of time it takes for the HO_2S to generate the proper operating voltages and switching frequency required of a properly-functioning sensor.

Fuel System Monitoring — The ability of the PCM to correct the air/fuel ratio is closely monitored. The fuel system monitor can detect when the system is no longer able to maintain a stoichiometric air/fuel ratio. This monitor should detect the problem before emissions increase to 1.5 times the emissions standards.

Under OBD-I, a variety of names are used to indicate the amount of correction to the desired air/fuel ratio of 14:7 to 1. Short-Term Fuel Trim and Long-Term Fuel Trim are now the standard names for fuel system monitoring. Previously, the amount of correction was measured in counts, percentages, or dwell. The standard unit of measurement under OBD-II is percentage (%). Figure 12-9 shows the relationship between the fuel counts previously used and the percentages used in OBD-II. The most important thing to remember is that one direction adds fuel and the other subtracts fuel. Chapter 8 covered the operation and testing of air/fuel control under OBD-I.

Catalyst System — For catalyst systems, OBD-II monitors the front catalyst in the system. Because the front catalyst is critical to achieving low emissions, OBD-II regulations require the front catalyst to be monitored. When the front catalyst is a small volume unit, the next catalyst in the system will be monitored as well.

Because OBD-II only requires front catalyst monitoring, it is possible that physical damage to the unmonitored rear catalyst could remain undetected by the OBD-II system. On the other hand, damage to the rear catalyst caused by excessive temperature, such as misfire, as opposed to physical damage, could not occur without first affecting the front catalyst, which is monitored.

The PCM monitors front catalyst efficiency by comparing the switching activity of the main (upstream) and the sub (downstream) oxygen sensors (Figure 12-10). Since a good catalyst stores oxygen, the downstream HO_2S voltage should remain flat. As the catalyst deteriorates and loses its ability to store oxygen, the switch rate of the downstream oxygen sensor begins to mirror the upstream sensor. At this point, exhaust emissions are projected to exceed the legal limit.

Misfire Monitoring — Misfire is the result of either poor combustion or no combustion in the cylinder due to the absence of spark, poor fuel metering, or compression. The Misfire Detection Monitor detects fuel, ignition, or mechanically-induced misfires. Early detection and correction of engine

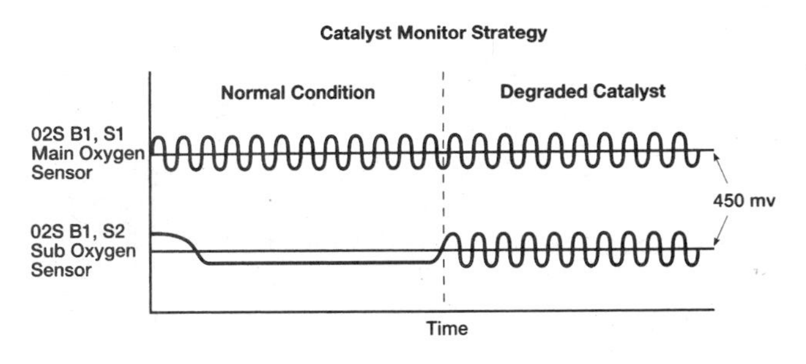

Figure 12-10 Comparing the switching rate of oxygen sensors.

misfires are important to reduce emissions and prevent catalyst damage.

Using a high-frequency crankshaft position sensor signal, the PCM closely monitors crankshaft speed variations during each individual cylinder's power stroke. When an engine is firing cleanly on all cylinders, the crankshaft speeds up with each power stroke. When a misfire occurs, the crankshaft speed increase is affected for that cylinder (Figure 12-11). The threshold for determining what amount of RPM change indicates misfire will vary with engine speed and load. This is because as engine speed increases or load decreases, the general effect of a single cylinder misfire diminishes due to the momentum of the crankshaft.

The misfire monitor has an adaptive strategy that can allow for component wear, sensor fatigue, and machining tolerances. The PCM does this by monitoring the variation between cylinders during normal operation. This information is then used to calculate the threshold of RPM variation that indicates a misfire.

After the *enable criteria* have been met, the Diagnostic Executive monitors the engine for misfire that could cause damage to the three-way catalyst (TWC). If the monitor detects misfire in more than 15% of cylinder firing opportunities during any 200-revolution segments (Figure 12-12), the MIL immediately begins flashing. The system defaults to an open-loop mode to prevent the adaptive fuel controls from adding additional fuel to the cylinder. When the misfire detection drops below 15%, the MIL stops flashing but the light remains illuminated.

The Diagnostic Executive also monitors the engine for misfire that could cause the exhaust emissions to exceed the FTP standards by 1.5 times. If the misfire occurs in 2% of the cylinder firing opportunities during 1000-revolution segments, a temporary DTC is set. The last 200 revolutions of the 1000-revolution segments are stored in freeze frame data. This type of misfire requires two-trip detection, preventing the MIL from illuminating until the failure occurs a second time. To assure the accuracy of this test, the Diagnostic Executive maintains a

Crankshaft Speed Misfire Detection

The ECM monitors crankshaft speed and position through Ne and G inputs. Because crankshaft speed normally increases during firing events, the ECM can monitor this change from one firing event to the next to monitor for the presence and degree of misfire.

When partial misfire occurs, the rate of crankshaft speed increase is reduced. If a total misfire occurs, there will be no crank speed increase at all.

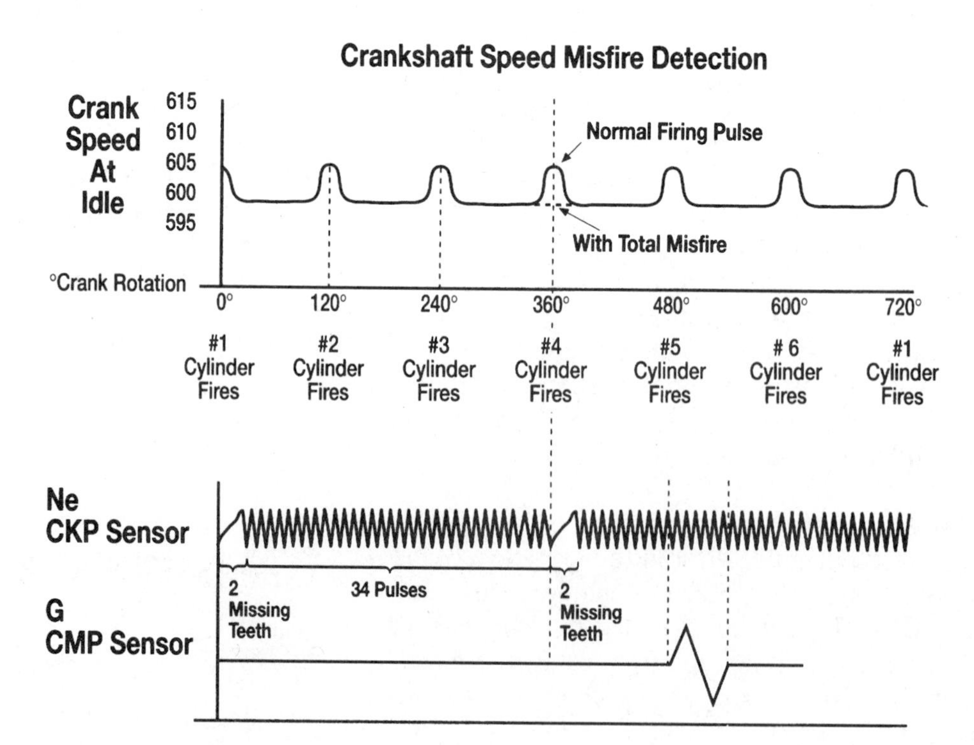

Figure 12-11 Detecting misfire through crankshaft speed.

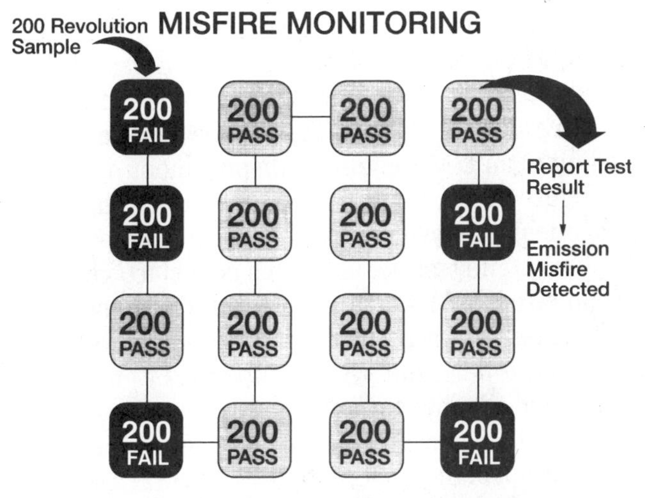

Figure 12-12 Fuel trim conversions between OBD-I and OBD-II (courtesy GM).

record of the previous 3200 engine revolutions. Figure 12-12 shows a 3200-revolution monitor separated into sixteen 200-revolution blocks.

EGR System Monitoring — Enhanced monitoring of EGR characteristics include the ability to detect flow rates that are above or below the design flow rate for a given engine operating condition. Three different methods are used to monitor EGR flow rate:

1. One method of accomplishing this is to monitor the change in temperature on the intake side of the EGR passage. The temperature rises as exhaust gases enter the intake manifold. The amount of temperature change allows the PCM to calculate the amount of EGR flow.
2. Another method is to measure the degree of rich correction as the EGR valve is closed during a steady-state cruise condition. Because the EGR creates a rich air/fuel condition, the PCM compensates by leaning out the air/fuel. When the EGR valve is closed, the air/fuel ratio moves to the lean side, forcing the PCM to make a rich correction. The amount of correction allows the computer to calculate the amount of EGR flow.
3. The EGR diagnostic system will force the EGR valve open during closed throttle deceleration and/or force the EGR valve closed during a steady-state cruise. Opening the valve on deceleration will result in an increase in manifold pressure (more air). Closing the valve during the cruise test results in a decrease in manifold vacuum (less air). The EGR flow is measured over several cycles, allowing the PCM to average the readings and prevent false codes.

Secondary Air-Injection Monitoring — For a secondary air-injection system, OBD-II requires the actual air pump flow rate to be monitored. Two methods are used to verify the operation of the secondary air-injection system:

1. It is monitored in closed loop by monitoring the Short-Term Fuel Trim (STFT) corrections. Air injection is switched upstream in front of the first oxygen sensor. Introducing this false air should drive the O_2 sensor full lean. An increase in the STFT value indicates that the secondary air-injection system is functioning.
2. The Short-Term Fuel Trim (STFT) correction is temporarily disabled and for 18 seconds approximately 10% more fuel is added to the pulsewidth. When the downstream O_2 sensor switches rich, the secondary air solenoid is energized. If the secondary air injection is functioning properly, energizing the solenoid drives the O_2 sensor reading to the lean direction. This must occur in less than 19 seconds.

Evaporative System Monitoring — The evaporative system monitor required by OBD-II detects leaks or reductions in purge system flow rate. The regulations specify that a vacuum or pressure test is performed on-board to determine if any leaks are present. Leaks from holes as small as 0.04 inches in diameter should be detected. Proper purge volume is determined by the time needed to achieve a specified vacuum/pressure within a predetermined time limit.

Because OBD-II evaporative requirements provide low cost, frequent monitoring, it should be more effective at identifying evaporative emissions control failures. Currently, the IM240 is the only test designed to detect evaporative system leaks and purge flow volume. OBD-II systems are not presently designed to check for evaporative system leaks and purge flow volume. However, OBD-II can monitor purge system operation. Two methods can be used to check the purge system operation:

1. A normally closed diagnostic switch (Figure 12-13) can be installed between the evaporative canister and the purge valve. When the purge valve is open (purge on) and sufficient vacuum is present in the manifold, the switch should indicate a vacuum. When the purge valve is closed (no purge), the switch should not indicate a vacuum.
2. By monitoring the oxygen sensor and injection pulse-width while the canister is purging, the PCM can detect the reduction of exhaust oxygen content and decrease the pulse-width to correct for a momentary rich condition.

OBD-II Drive Cycle — The OBD-II diagnostic system continually monitors for misfire and fuel system faults. It also performs functional tests on the catalyst, EGR system, and oxygen sensors once during every *driving cycle* or *trip*.

The OBD-II Drive Cycle is used to run on-board diagnostic routines to set *flags* for the Inspection and Maintenance (I/M) Readiness Test. Flags indicate that certain required diagnostic routines have been successfully completed, and the vehicle system or component has been diagnosed. I/M Readiness Tests are not concerned with whether the emissions system passed or failed the test, only that the on-board diagnosis is completed.

Certain driving conditions must be encountered before these systems can be confirmed as operating. Under most circumstances the flags will be set through normal driving conditions. There will be times when the vehicle may be driven, but some flags are not set. Figure 12-14 shows a scan tool reading of an I/M Readiness display menu.

Figure 12-15 shows a Ford Drive Cycle and the components or systems tested during each portion of the trip. The test begins with a cold start. During this time, the PCM monitors misfire and adaptive fuel shifts. After the warm-up is complete and the vehicle is allowed to idle, the system continues to monitor misfire and adaptive fuel, and now begins to monitor the heated O_2 and other comprehensive monitors. Near the end of the cycle, with the vehicle at a steady cruise, the catalyst monitor is now completed. The technician can repeat any portion of the cycle if the drive cycle is interrupted before completion.

While all manufacturers are required to have a test, the actual drive cycle is not standard. The driving modes cover idle, acceleration, steady throttle, and deceleration. A General Motors Drive Cycle is shown in Figure 12-16 (page 212). Slight time differences exist between the Ford Drive Cycle (Figure 12-15) and the GM Drive Cycle. Also, while the same components and systems are monitored, they are monitored in a different sequence

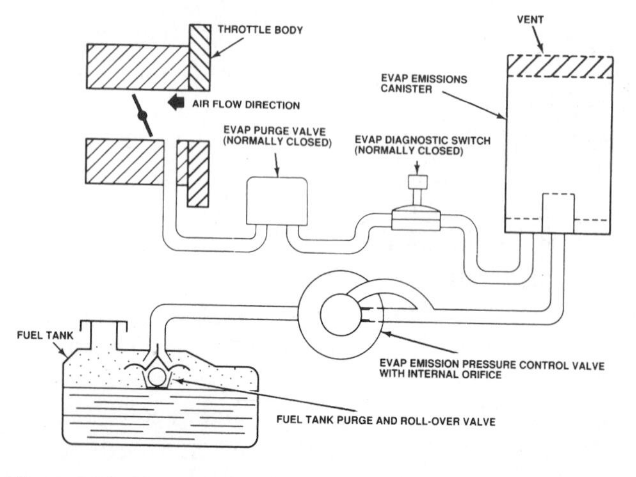

Figure 12-13 Measuring purge vacuum using a diagnostic switch (courtesy GM).

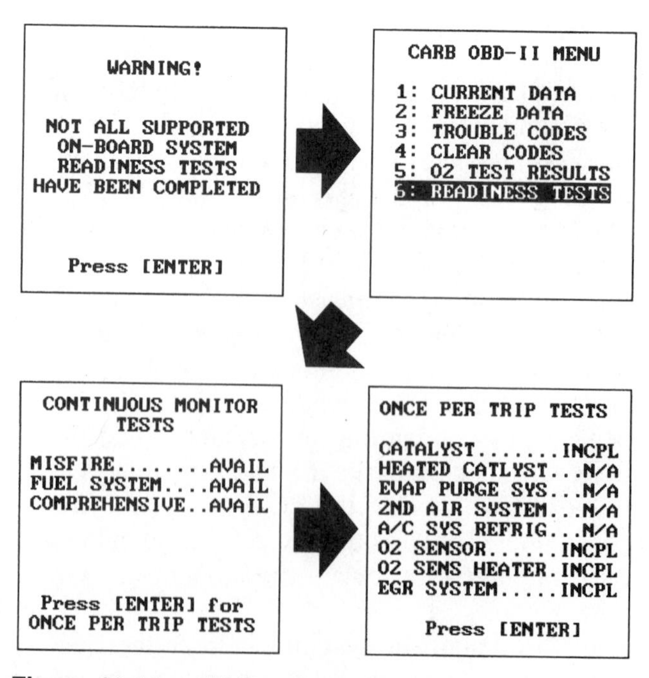

Figure 12-14 I/M Readiness Test flags.

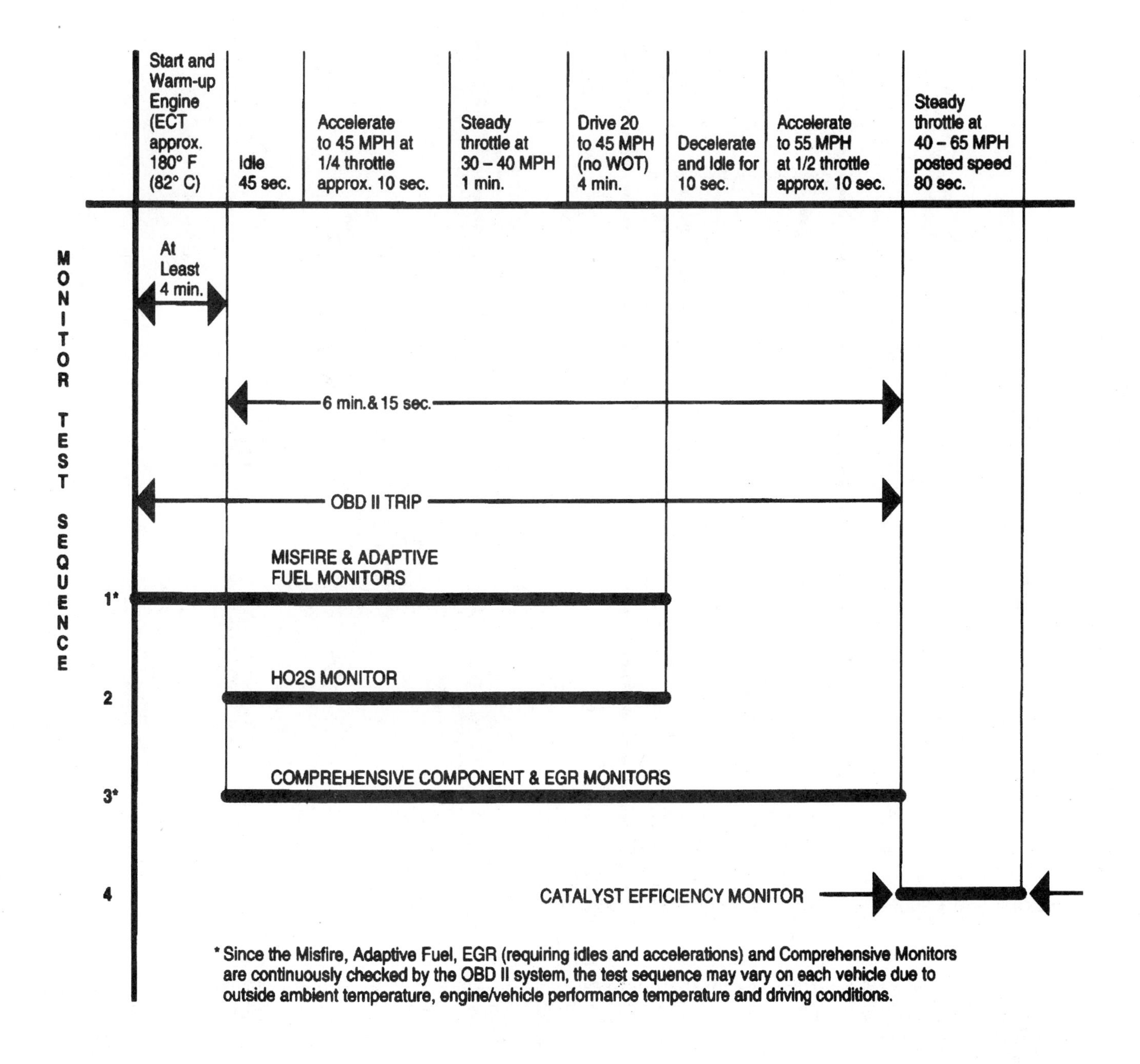

Figure 12-15 Ford I/M Readiness Drive Cycle (courtesy Ford).

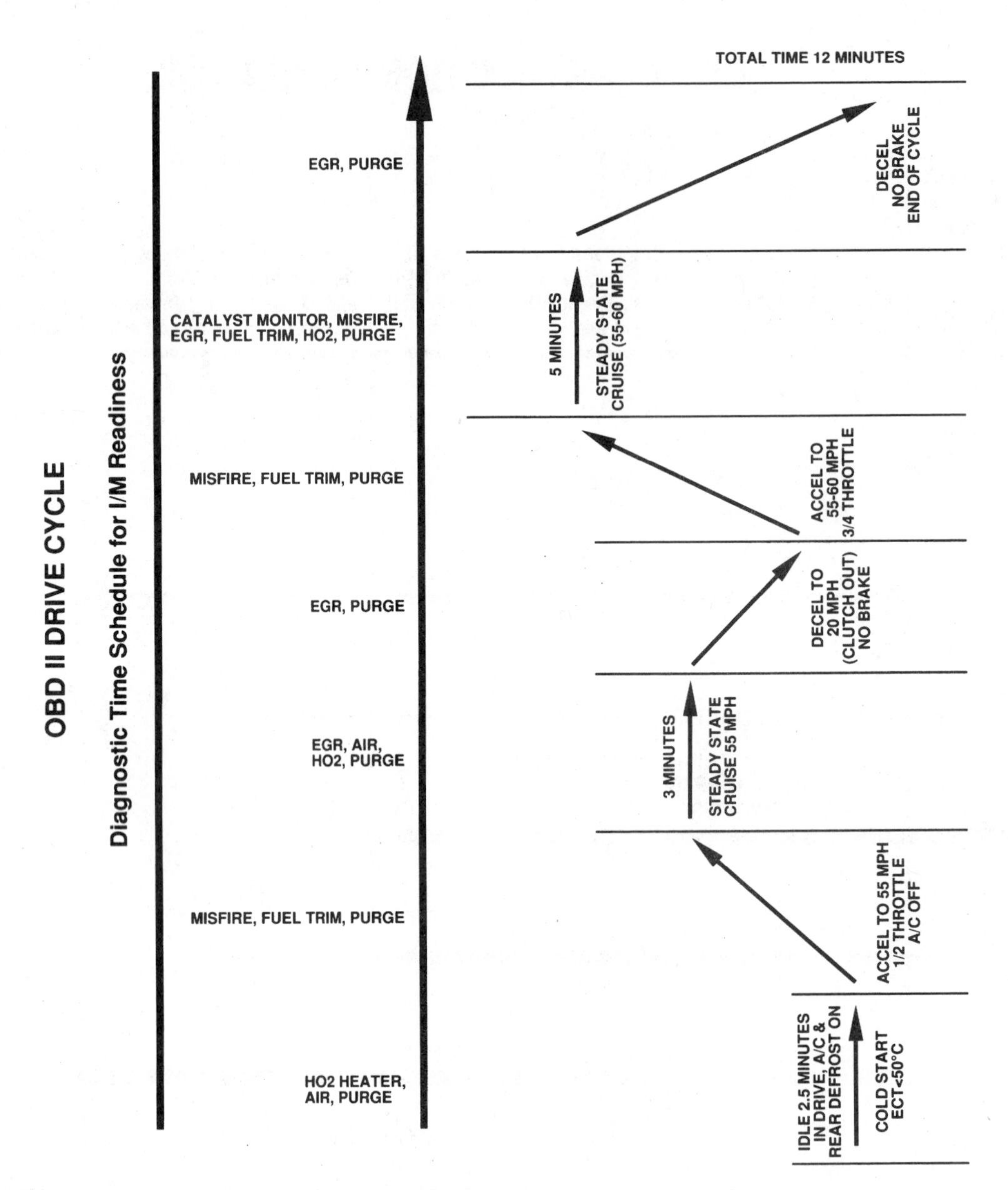

Figure 12-16 General Motors I/M Readiness Drive Cycle (courtesy GM).

and during different driving modes. At the completion of either drive cycle, I/M flags should be set.

Diagnostic Trouble Codes (DTC) and the Malfunction Indicator Light (MIL)

OBD-I regulations required manufacturers to install a *check engine* light to notify the driver that an emissions-related trouble code was present. The trouble code provided the technician with a *clue* about what the problem might be. This is as far as the standardization went. OBD-I did not require:

1. A standard diagnostic test connector in a standard location. Different-shape connectors located in a variety of places required different tools and procedures for reading and erasing fault codes.
2. Manufacturers to use the same code identification system for the same problem. The same number code had different meanings between the manufacturers. Some problems were represented by single-digit codes, two-digit codes, and three-digit codes.
3. That the diagnostic connectors provide serial data information that could be read using a scan tool.

4. Standard names and terminology for similar or even identical components or systems.

Under OBD-II regulations, the automotive industry and the Society of Automotive Engineers (SAE) have established regulations that standardize most OBD-II procedures. These regulations address the limitations of OBD-I that will help the technician develop a logical approach to diagnostics because they are standard to all manufacturers.

SAE Standard J2012 establishes a standard diagnostic connector in a general location. The provision for communicating with the PCM is the Data Link Connector (DLC). It will be located under the dash. It does not need to be in the normal line of sight, but it must be visible as the technician looks under the dash and must be accessible without the use of tools.

The connector has a standard shape, number of pins, and assignment for each pin (Figure 12-17). The connector is D-shaped and has two rows of eight pins, numbered one to eight and nine to sixteen. Seven of the sixteen pins have a common assignment and location. Vehicle manufacturers can use the remaining nine pins at their discretion.

SAE Standard J1978 establishes the standards for scan tool manufacturers that outlines the technical requirements for a generic scan tool. Among other things, these standards must be met or exceeded before a scan tool manufacturer can claim that their scan tool is OBD-II compatible. SAE Standard J1850 recommends the use of a common serial data protocol network interface. Serial data is transmitted through diagnostic connector pins #2 and #10 using standards established by the International Standards Organization (ISO). Regulation ISO 9141 establishes an alternate communications network interface to the data bus link at pins #7 and #15.

SAE Standard J2012 establishes a system of Diagnostic Trouble Codes (DTC) and fault identification messages. It allows manufacturers the flexibility to establish their own unique diagnostic codes and procedures while standardizing common codes by circuit and type of failure.

The recommended procedure for reading DTCs stored in memory is by using a diagnostic scan tool plugged into the DLC. Figure 12-18 shows a partial list of generic codes used by all manufacturers. Each manufacturer has the option of providing additional codes that are specific to their systems.

All DTCs are represented by a five-digit alphanumeric code. The first letter indicates the function of the monitored component or system that has failed:

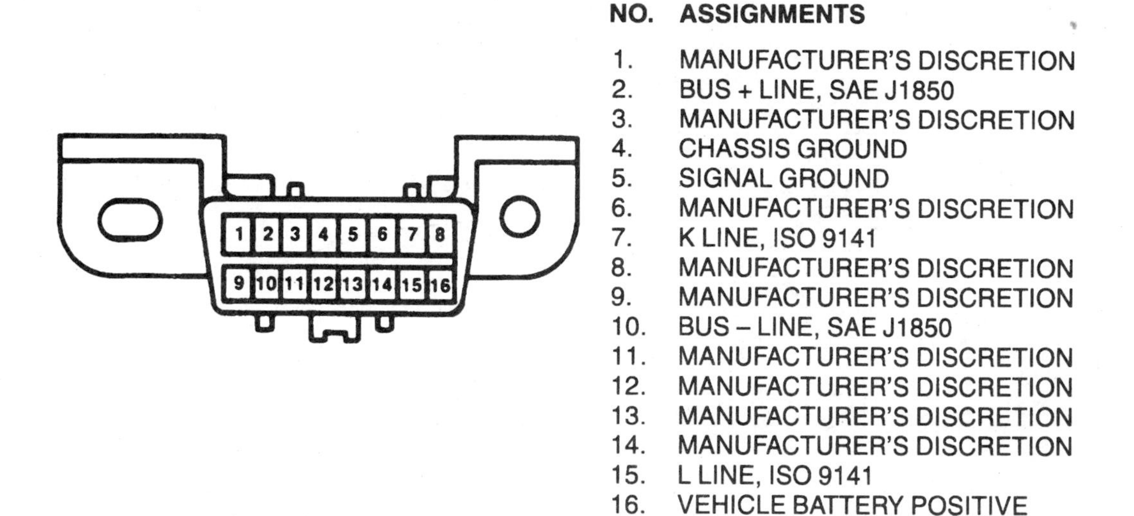

PIN NO.	ASSIGNMENTS
1.	MANUFACTURER'S DISCRETION
2.	BUS + LINE, SAE J1850
3.	MANUFACTURER'S DISCRETION
4.	CHASSIS GROUND
5.	SIGNAL GROUND
6.	MANUFACTURER'S DISCRETION
7.	K LINE, ISO 9141
8.	MANUFACTURER'S DISCRETION
9.	MANUFACTURER'S DISCRETION
10.	BUS – LINE, SAE J1850
11.	MANUFACTURER'S DISCRETION
12.	MANUFACTURER'S DISCRETION
13.	MANUFACTURER'S DISCRETION
14.	MANUFACTURER'S DISCRETION
15.	L LINE, ISO 9141
16.	VEHICLE BATTERY POSITIVE

Figure 12-17 OBD-II diagnostic connector and pin assignments (courtesy Snap-On Tools Corp.).

DTC	Definitions
P0102	Mass Air Flow (MAF) sensor circuit low input
P0103	Mass Air Flow (MAF) sensor circuit high input
P0112	Intake Air Temperature (IAT) sensor circuit low input
P0113	Intake Air Temperature (IAT) sensor circuit high input
P0117	Engine Coolant Temperature (ECT) sensor circuit low input
P0118	Engine Coolant Temperature (ECT) sensor circuit high input
P0122	Throttle Position (TP) sensor circuit low input
P0123	Throttle Position (TP) sensor circuit high input
P0125	Insufficient Coolant Temperature to enter closed loop fuel control
P0132	Upstream Heated Oxygen sensor (HO2S 11) circuit high voltage (Bank # 1)
P0135	Heated Oxygen sensor heater (HTR 11) circuit malfunction
P0138	Downstream Heated Oxygen sensor (HO2S 12) circuit high voltage (Bank # 1)
P0140	Heated Oxygen sensor (HO2S 12) circuit no activity detected (Bank # 1)
P0141	Heated Oxygen sensor heater (HTR 12) circuit malfunction
P0152	Upstream Heated Oxygen sensor (HO2S 21) circuit high voltage (Bank #2)
P0155	Heated Oxygen sensor heater (HTR 21) circuit malfunction
P0158	Downstream Heated Oxygen sensor (HO2S 22) circuit high voltage (Bank #2)
P0160	Heated Oxygen sensor (HO2S 22) circuit no activity detected (Bank #2)
P0161	Heated Oxygen sensor heater (HTR 22) circuit malfunction
P0171	System (adaptive fuel) too lean (Bank # 1)
P0172	System (adaptive fuel) too rich (Bank # 1)
P0174	System (adaptive fuel) too lean (Bank #2)
P0175	System (adaptive fuel) too rich (Bank #2)
P0300	Random Misfire detected
P0301	Cylinder # 1 Misfire detected
P0302	Cylinder #2 Misfire detected
P0303	Cylinder #3 Misfire detected
P0304	Cylinder #4 Misfire detected
P0305	Cylinder #5 Misfire detected
P0306	Cylinder #6 Misfire detected
P0307	Cylinder #7 Misfire detected
P0308	Cylinder #8 Misfire detected
P0320	Ignition Engine Speed (Profile Ignition Pickup (PIP)) input circuit malfunction
P0340	Camshaft Position (CMP) sensor circuit malfunction (CID)
P0402	Exhaust Gas Recirculation (EGR) flow excess detected (valve open at idle)
P0420	Catalyst system efficiency below threshold (Bank # 1)
P0430	Catalyst system efficiency below threshold (Bank #2)
P0443	Evaporative emission control system Canister Purge Control Valve (CANP) circuit malfunction
P0500	Vehicle Speed Sensor (VSS) malfunction
P0505	Idle Air Control (IAC) system malfunction
P0605	Powertrain Control Module (PCM) - Read Only Memory (ROM) test error
P0703	Brake On / Off switch (BOO) input malfunction
P0707	Manual Lever Position (MLP) sensor circuit low input
P0708	Manual Lever Position (MLP) sensor circuit high input

Figure 12-18 Diagnostic trouble code definitions (courtesy Ford).

P = Powertrain
B = Body
C = Chassis
U = Indicates a network or data link code

The second digit is represented by a number that indicates who is responsible for the code:

0 = Society of Automotive Engineers (SAE)
1 = Manufacturer-specific

The third digit indicates the specific system in question. Numbers one through seven indicate a powertrain-related problem. The number eight is reserved for non-powertrain related problems:

0 = Total system
1 = Air/fuel metering control
2 = Air/fuel metering control for injector circuit malfunctions only
3 = Ignition system or misfire
4 = Emissions control
5 = Vehicle speed control and idle control system
6 = ECM and computer input/output circuits
7 = Transmission
8 = Non-powertrain related

The fourth and fifth digits represent the component, system, or area experiencing the problem.

Using DTC, P0306 shown in Figure 12-17 is described as:

P = Powertrain-related
0 = SAE-defined DTC
3 = Ignition system or misfire-related problem
06 = Cylinder #6 misfire detected

An Inspection and Maintenance Readiness DTC, P1000, indicates that all emissions-related control system monitors have not been completed since the PCM memory was last cleared. When a vehicle is in for normal service, the P1000 DTC does not need to be cleared from memory unless it must be done to pass an I/M test.

If the code is present during an I/M test, it will be necessary to clear the code to continue with the inspection. To clear this code the technician must successfully complete the OBD-II drive cycle covered earlier in the chapter. DTC P1000 is automatically set when:

1. A new vehicle from the factory has not completed the OBD-II drive cycle.
2. An OBD-II monitor occurs before the completion of the OBD-II drive cycle.
3. PCM DTCs have been cleared during the course of a normal repair procedure. This can be accomplished with a generic scan tool or by disconnecting the PCM power for 30 seconds on some systems.

The Diagnostic Executive must be able to acknowledge when all emissions-related comprehensive monitoring diagnostic tests have reported a pass or fail condition since the last ignition cycle (Figure 12-19). Each diagnostic monitor is separated into four categories:

1. *Type A* — This is an emissions-related failure. When the Diagnostic Executive identifies a failed comprehensive monitor test during a first trip, it commands the MIL to illuminate.
2. *Type B* — This is an emissions related failure. When the Diagnostic Executive identifies a failed comprehensive monitor test during two consecutive drive cycles (two-trip detection), it commands the MIL to illuminate.
3. *Type C* — This is a non-emissions related failure. The MIL is commanded to illuminate when the problem occurs during the first-trip. This type of code is reserved for use in the future.
4. *Type D* — This is a non-emissions related failure, and the MIL is not illuminated at any time.

When the Diagnostic Executive commands the MIL to turn on, a history DTC is also recorded for the diagnostic test. The requirement for erasing a history Type A or Type B DTC involves the system passing 40 consecutive warm-up cycles during which a Type A or Type B diagnostic test has not resulted in another failure.

Misfire or fuel system malfunctions are special cases or Type B diagnostics. Each time a misfire or fuel system malfunction is detected, engine load, engine RPM, and coolant temperature conditions are recorded. Just before the key is turned off, the last recorded data is stored.

If the malfunction occurs within two consecutive trips, the Diagnostics Executive treats the failure as a normal Type B diagnostic, and does not use the stored conditions. However, if a misfire occurs on two non-consecutive trips under similar driving conditions, the stored conditions are compared to the current conditions. The MIL is commanded on

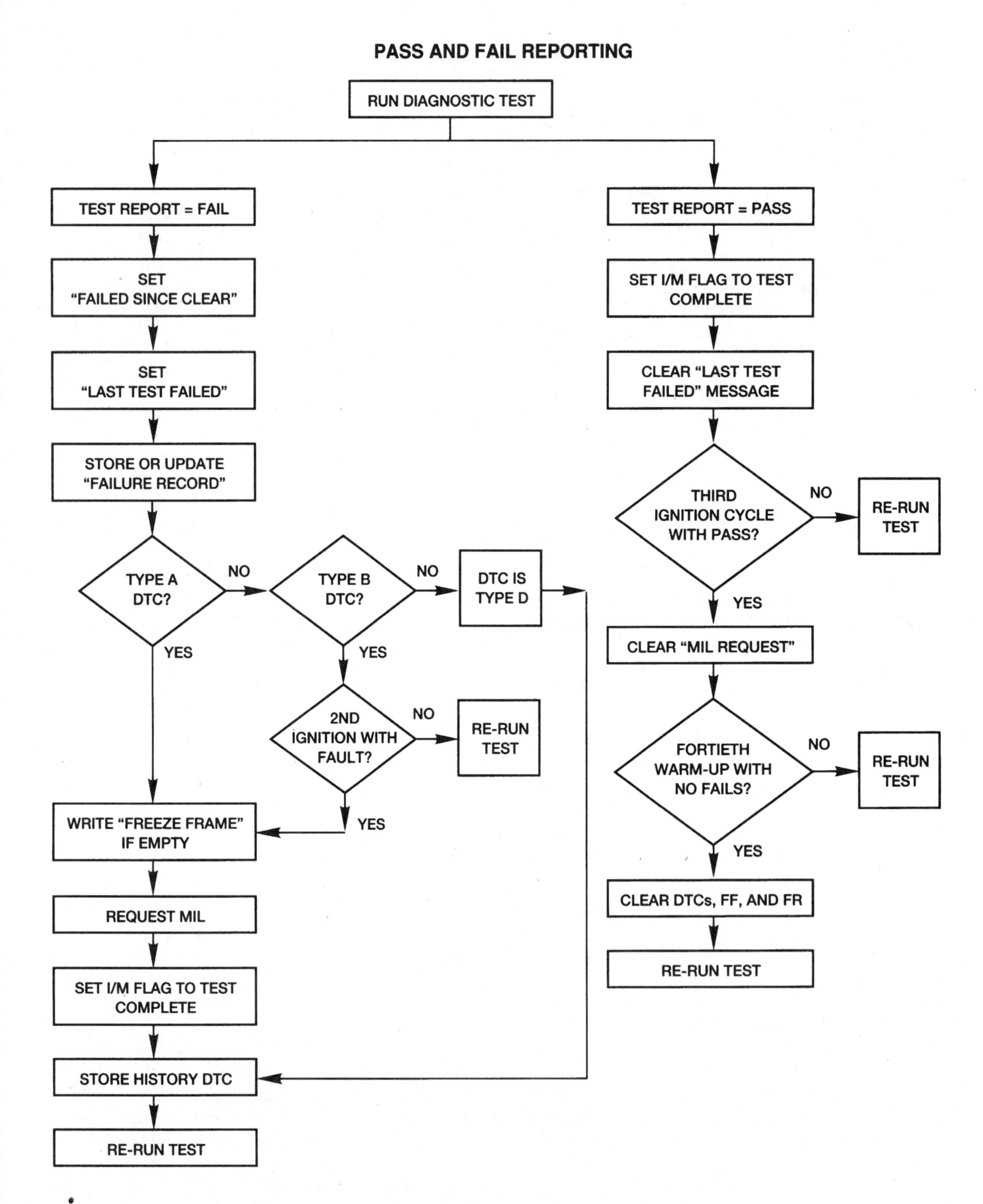

Figure 12-19 Diagnostic test sequencing (courtesy GM).

when the engine condition is within 10%, engine RPM is within 375, and the coolant temperature is in the same range as the last recorded failure. After the second trip, if similar conditions are not duplicated, the saved conditions are updated to those most recently recorded.

If the MIL is on, the Diagnostic Executive will command the MIL off after three consecutive trips report a *test passed* for the diagnostic test that originally commanded the MIL on. The DTC will remain in memory until the diagnostic test completes 40 consecutive warm-up cycles without failure.

Freeze Frame Data

Before OBD-II, a technician could program a scan tool to take a snapshot of the data when a problem was detected. The snapshot could be programmed to trigger manually or automatically when a code was set.

The *freeze frame* feature of OBD-II allows the Diagnostic Executive to capture data the moment an emissions-related DTC is stored and the MIL is commanded on. No matter how often the diagnostic test is repeated and failed, the data is not updated (refreshed).

When multiple codes are present, the first code to occur is stored in the freeze frame. The exceptions are for a fuel trim or misfire DTC. They have priority and can write new freeze frame data over stored freeze frame data. Information in the freeze frame (Figure 12-20) can include, but is not limited to:

1. Diagnostic Trouble Codes (DTC).
2. Engine RPM.
3. Engine load.
4. Fuel trim (short-term and long-term).
5. Engine Coolant Temperature (ECT).
6. Calculated load (MAP, MAF).
7. Operation mode (closed loop).
8. Vehicle speed.
9. Number of trips.

Freeze frame data is retained in memory and can be retrieved by the technician using a generic scan tool and OBD-II cartridge. The freeze frame data will not be erased unless the history DTC is cleared for the malfunction recorded in freeze frame.

Failure Modes – GM Only

This mode is similar to the snapshot feature of a GM Tech 1 or MPSI scan tool. These scan tools allow the snapshot to trigger when any fault code is set. *Failure Records Mode* provides the same features, and it will store critical operating conditions at the instant any diagnostic test fails.

Failure records are updated every time a diagnostic test is repeated and failed. The PCM can store up to five individual failure records that can be retrieved by the technician using a scan tool. Failure records will clear any time the associated DTC is cleared from memory.

SAE Standard J1930 — SAE Standard J1930 establishes standard names, acronyms, and terminology for all electric and electronic systems related to engine and emissions systems. Beginning with the 1993 model year, under OBD-II regulations, all service information was required to use the new technical jargon. Figure 12-21 shows a list of old terminology and the newer SAE J1930 standards.

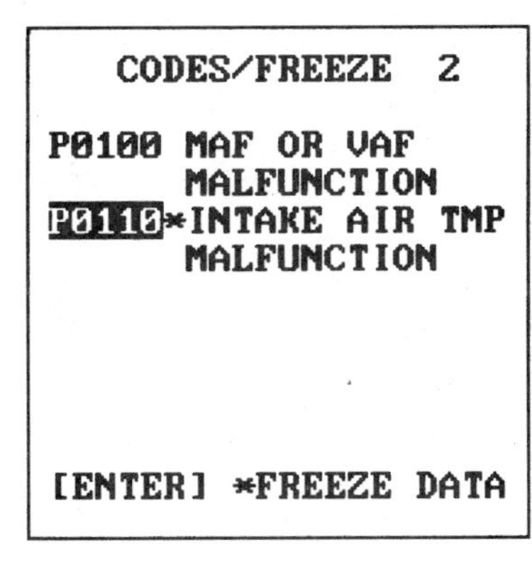

TROUBLE CODE 110
FUEL SYS #1 .... OPEN
FUEL SYS #2 .... OPEN
CALC LOAD 0.0%
COOLANT TEMP .... 203°F
SHORT FT #1 .... 0.0%
LONG FT #1 .... 0.0%
SHORT FT #2 .... 0.0%
LONG FT #2 .... 0.0%
ENGINE SPD 0r/min
VEHICLE SPD 0MPH
TRIPS 1
STARTER SIG .... OFF
CTP SW .... ON
A/C SIG .... OFF
PNP SW .... ON

Figure 12-20 Typical freeze frame data.

SAE J1930 Revised JUN93

Existing Usage	Acceptable Usage	Acceptable Acronized Usage
3GR (Third Gear)	Third Gear	3GR
4GR (Fourth Gear)	Fourth Gear	4GR
A/C (Air Conditioning)	Air Conditioning	A/C
A/C Cycling Switch	Air Conditioning Cycling Switch	A/C Cycling Switch
A/T (Automatic Transaxle)	Automatic Transaxle	A/T
A/T (Automatic Transmission)	Automatic Transmission	A/T
AC (Air Conditioning)	Air Conditioning	A/C
ACC (Air Conditioning Clutch)	Air Conditioning Clutch	A/C Clutch
Accelerator	Accelerator Pedal	AP
ACCS (Air Conditioning Cyclic Switch)	Air Conditioning Cycling Switch	A/C Cycling Switch
ACH (Air Cleaner Housing)	Air Cleaner Housing1	ACL Housing[1]
ACL (Air Cleaner)	Air Cleaner[1]	ACL[1]
ACL (Air Cleaner) Element	Air Cleaner Element[1]	ACL Element[1]
ACL (Air Cleaner) Housing	Air Cleaner Housing[1]	ACL Housing[1]
ACL (Air Cleaner) Housing Cover	Air Cleaner Housing Cover[1]	ACL Housing Cover[1]
ACS (Air Conditioning System)	Air Conditioning System	A/C System
ACT (Air Charge Temperature)	Intake Air Temperature[1]	IAT[1]
Adaptive Fuel Strategy	Fuel Trim[1]	FT[1]
AFC (Air Flow Control)	Mass Air Flow[1]	MAF[1]
AFC (Air Flow Control)	Volume Air Flow[1]	VAF[1]
AFS (Air Flow Sensor)	Mass Air Flow Sensor[1]	MAF Sensor[1]
AFS (Air Flow Sensor)	Volume Air Flow Sensor[1]	VAF Sensor[1]
After Cooler	Charge Air Cooler[1]	CAC[1]
AI (Air Injection)	Secondary Air Injection[1]	AIR[1]
AIP (Air Injection Pump)	Secondary Air Injection Pump[1]	AIR Pump[1]
AIR (Air Injection Reactor)	Pulsed Secondary Air Injection[1]	PAIR[1]
AIR (Air Injection Reactor)	Secondary Air Injection[1]	AIR[1]
AIRB (Secondary Air Injection Bypass)	Secondary Air Injection Bypass[1]	AIR Bypass[1]
AIRD (Secondary Air Injection Diverter)	Secondary Air Injection Diverter[1]	AIR Diverter[1]
Air Cleaner	Air Cleaner[1]	ACL[1]
Air Cleaner Element	Air Cleaner Element[1]	ACL Element[1]
Air Cleaner Housing	Air Cleaner Housing[1]	ACL Housing[1]
Air Cleaner Housing Cover	Air Cleaner Housing Cover[1]	ACL Housing Cover[1]
Air Conditioning	Air Conditioning	A/C
Air Conditioning Sensor	Air Conditioning Sensor	A/C Sensor
Air Control Valve	Secondary Air Injection Control Valve[1]	AIR Control Valve[1]
Air Flow Meter	Mass Air Flow Sensor[1]	MAF Sensor[1]
Air Flow Meter	Volume Air Flow Sensor[1]	VAF Sensor[1]
Air Intake System	Intake Air System[1]	IA System[1]
Air Flow Sensor	Mass Air Flow Sensor[1]	MAF Sensor[1]
Air Management 1	Secondary Air Injection Bypass[1]	AIR Bypass[1]
Air Management 2	Secondary Air Injection Diverter[1]	AIR Diverter[1]
Air Temperature Sensor	Intake Air Temperature Sensor[1]	IAT Sensor[1]
Air Valve	Idle Air Control Valve[1]	IAC Valve[1]
AIV (Air Injection Valve)	Pulsed Secondary Air Injection[1]	PAIR[1]
ALCL (Assembly Line Communication Link)	Data Link Connector[1]	DLC[1]
Alcohol Concentration Sensor	Flexible Fuel Sensor[1]	FF Sensor[1]
ALDL (Assembly Line Diagnostic Link)	Data Link Connector[1]	DLC[1]
ALT (Alternator)	Generator	GEN
Alternator	Generator	GEN
AM1 (Air Management 1)	Secondary Air Injection Bypass[1]	AIR Bypass[1]

Recommended Terms and Recommended Acronyms See Figure 2
[1] Emission-Related Term

Figure 12-21A SAE J1930 names and terminology (courtesy SAE International).

SAE J1930 Revised JUN93

Existing Usage	Acceptable Usage	Acceptable Acronized Usage
AM2 (Air Management 2)	Secondary Air Injection Diverter[1]	AIR Diverter[1]
APS (Absolute Pressure Sensor)	Barometric Pressure Sensor[1]	BARO Sensor[1]
ATS (Air Temperature Sensor)	Intake Air Temperature Sensor[1]	IAT Sensor[1]
Automatic Transaxle	Automatic Transaxle[1]	A/T[1]
Automatic Transmission	Automatic Transmission[1]	A/T[1]
B+ (Battery Positive Voltage)	Battery Positive Voltage	B+
Backpressure Transducer	Exhaust Gas Recirculation Backpressure Transducer[1]	EGR Backpressure Transducer[1]
BARO (Barometric Pressure)	Barometric Pressure[1]	BARO[1]
Barometric Pressure Sensor	Barometric Pressure Sensor[1]	BARO Sensor[1]
Battery Positive Voltage	Battery Positive Voltage	B+
Block Learn Matrix	Long Term Fuel Trim[1]	Long Term FT[1]
BLM (Block Learn Memory)	Long Term Fuel Trim[1]	Long Term FT[1]
BLM (Block Learn Multiplier)	Long Term Fuel Trim[1]	Long Term FT[1]
BLM (Block Learn Matrix)	Long Term Fuel Trim[1]	Long Term FT[1]
Block Learn Memory	Long Term Fuel Trim[1]	Long Term FT[1]
Block Learn Multiplier	Long Term Fuel Trim[1]	Long Term FT[1]
BP (Barometric Pressure) Sensor	Barometric Pressure Sensor[1]	BARO Sensor[1]
C3I (Computer Controlled Coil Ignition)	Electronic Ignition[1]	EI[1]
CAC (Charge Air Cooler)	Charge Air Cooler[1]	CAC[1]
Camshaft Position	Camshaft Position[1]	CMP[1]
Camshaft Position Sensor	Camshaft Position Sensor[1]	CMP Sensor[1]
Camshaft Sensor	Camshaft Position Sensor[1]	CMP Sensor[1]
Canister	Canister[1]	Canister[1]
Canister	Evaporative Emission Canister[1]	EVAP Canister[1]
Canister Purge Valve	Evaporative Emission Canister Purge Valve[1]	EVAP Canister Purge Valve[1]
Canister Purge Vacuum Switching Valve	Evaporative Emission Canister Purge Valve[1]	EVAP Canister Purge Valve[1]
Canister Purge VSV (Vacuum Switching Valve)	Evaporative Emission Canister Purge Valve[1]	EVAP Canister Purge Valve[1]
CANP (Canister Purge)[1]	Evaporative Emission Canister Purge[1]	EVAP Canister Purge[1]
CARB (Carburetor)	Carburetor[1]	CARB[1]
Carburetor	Carburetor[1]	CARB[1]
CCC (Converter Clutch Control)	Torque Converter Clutch[1]	TCC[1]
CCO (Converter Clutch Override)	Torque Converter Clutch[1]	TCC[1]
CDI (Capacitive Discharge Ignition)	Distributor Ignition[1]	DI[1]
CDROM (Compact Disc Read Only Memory)	Compact Disc Read Only Memory[1]	CDROM[1]
CES (Clutch Engage Switch)	Clutch Pedal Position Switch	CPP Switch
Central Multiport Fuel Injection	Central Multiport Fuel Injection[1]	Central MFI[1]
CFI (Continuous Fuel Injection)	Continuous Fuel Injection[1]	CFI[1]
CFI (Central Fuel Injection)	Throttle Body Fuel Injection[1]	TBI[1]
Charcoal Canister	Evaporative Emission Canister	EVAP Canister[1]
Charge Air Cooler	Charge Air Cooler[1]	CAC[1]
Check Engine	Service Reminder Indicator[1]	SRI[1]
Check Engine	Malfunction Indicator Lamp[1]	MIL[1]
CID (Cylinder Identification) Sensor	Camshaft Position Sensor	CMP Sensor[1]
CIS (Continuous Injection System)	Continuous Fuel Injection[1]	CFI[1]
CIS-E (Continuous Injection System-Electronic)	Continuous Fuel Injection[1]	CFI[1]
CKP (Crankshaft Position)	Crankshaft Position[1]	CKP[1]
CKP (Crankshaft Position) Sensor	Crankshaft Position Sensor[1]	CKP Sensor[1]
CL (Closed Loop)	Closed Loop[1]	CL[1]
Closed Bowl Distributor	Distributor Ignition[1]	DI[1]
Closed Throttle Position	Closed Throttle Position[1]	CTP[1]

Recommended Terms and Recommended Acronyms See Figure 2
[1] Emission-Related Term

Figure 12-21B SAE J1930 names and terminology (courtesy SAE International).

SAE J1930 Revised JUN93

Existing Usage	Acceptable Usage	Acceptable Acronized Usage
Closed Throttle Switch	Closed Throttle Position Switch[1]	CTP Switch[1]
CLS (Closed Loop System)	Closed Loop[1]	CL[1]
Clutch Engage Switch	Clutch Pedal Position Switch[1]	CPP Switch[1]
Clutch Pedal Position Switch	Clutch Pedal Position Switch[1]	CPP Switch[1]
Clutch Start Switch	Clutch Pedal Position Switch[1]	CPP Switch[1]
Clutch Switch	Clutch Pedal Position Switch[1]	CPP Switch[1]
CMFI (Central Multiport Fuel Injection)	Central Multiport Fuel Injection[1]	Central MFI[1]
CMP (Camshaft Position)	Camshaft Position[1]	CMP[1]
CMP (Camshaft Position) Sensor	Camshaft Position Sensor[1]	CMP Sensor[1]
COC (Continuous Oxidation Catalyst)	Oxidation Catalytic Converter[1]	OC[1]
Condenser	Distributor Ignition Capacitor[1]	DI Capacitor[1]
Continuous Fuel Injection	Continuous Fuel Injection[1]	CFI[1]
Continuous Injection System	Continuous Fuel Injection System[1]	CFI System[1]
Continuous Injection System-E	Electronic Continuous Fuel Injection System[1]	Electronic CFI System[1]
Continuous Trap Oxidizer	Continuous Trap Oxidizer[1]	CTOX[1]
Coolant Temperature Sensor	Engine Coolant Temperature Sensor[1]	ECT Sensor[1]
CP (Crankshaft Position)	Crankshaft Position[1]	CKP[1]
CPP (Clutch Pedal Position)	Clutch Pedal Position[1]	CPP[1]
CPP (Clutch Pedal Position) Switch	Clutch Pedal Position Switch	CPP Switch[1]
CPS (Camshaft Position Sensor)	Camshaft Position Sensor[1]	CMP Sensor[1]
CPS (Crankshaft Position Sensor)	Crankshaft Position Sensor[1]	CKP Sensor[1]
Crank Angle Sensor	Crankshaft Position Sensor[1]	CKP Sensor[1]
Crankshaft Position	Crankshaft Position[1]	CKP[1]
Crankshaft Position Sensor	Crankshaft Position Sensor[1]	CKP Sensor[1]
Crankshaft Speed	Engine Speed[1]	RPM[1]
Crankshaft Speed Sensor	Engine Speed Sensor[1]	RPM Sensor[1]
CTO (Continuous Trap Oxidizer)	Continuous Trap Oxidizer[1]	CTOX[1]
CTOX (Continuous Trap Oxidizer)	Continuous Trap Oxidizer[1]	CTOX[1]
CTP (Closed Throttle Position)	Closed Throttle Position[1]	CTP[1]
CTS (Coolant Temperature Sensor)	Engine Coolant Temperature Sensor[1]	ECT Sensor[1]
CTS (Coolant Temperature Switch)	Engine Coolant Temperature Switch[1]	ECT Switch[1]
Cylinder ID (Identification) Sensor	Camshaft Position Sensor[1]	CMP Sensor[1]
D-Jetronic	Multiport Fuel Injection[1]	MFI[1]
Data Link Connector	Data Link Connector[1]	DLC[1]
Detonation Sensor	Knock Sensor[1]	KS[1]
DFI (Direct Fuel Injection)	Direct Fuel Injection[1]	DFI[1]
DFI (Digital Fuel Injection)	Multiport Fuel Injection[1]	MFI[1]
DI (Direct Injection)	Direct Fuel Injection[1]	DFI[1]
DI (Distributor Ignition)	Distributor Ignition[1]	DI[1]
DI (Distributor Ignition) Capacitor	Distributor Ignition Capacitor[1]	DI Capacitor[1]
Diagnostic Test Mode	Diagnostic Test Mode[1]	DTM[1]
Diagnostic Trouble Code	Diagnostic Trouble Code[1]	DTC[1]
DID (Direct Injection - Diesel)	Direct Fuel Injection[1]	DFI[1]
Differential Pressure Feedback EGR (Exhaust Gas Recirculation) System	Differential Pressure Feedback Exhaust Gas Recirculation System[1]	Differential Pressure Feedback EGR System[1]
Digital EGR (Exhaust Gas Recirculation)	Exhaust Gas Recirculation[1]	EGR[1]
Direct Fuel Injection	Direct Fuel Injection[1]	DFI[1]
Direct Ignition System	Electronic Ignition System[1]	EI System[1]
DIS (Distributorless Ignition System)	Electronic Ignition System[1]	EI System[1]
DIS (Distributorless Ignition System) Module	Ignition Control Module[1]	ICM[1]

Recommended Terms and Recommended Acronyms See Figure 2

[1] Emission-Related Term

Figure 12-21C SAE J1930 names and terminology (courtesy SAE International).

SAE J1930 Revised JUN93

Existing Usage	Acceptable Usage	Acceptable Acronized Usage
Distance Sensor	Vehicle Speed Sensor[1]	VSS[1]
Distributor Ignition	Distributor Ignition[1]	DI[1]
Distributorless Ignition	Electronic Ignition[1]	EI[1]
DLC (Data Link Connector)	Data Link Connector[1]	DLC[1]
DLI (Distributorless Ignition)	Electronic Ignition[1]	EI[1]
DS (Detonation Sensor)	Knock Sensor[1]	KS[1]
DTC (Diagnostic Trouble Code)	Diagnostic Trouble Code[1]	DTC[1]
DTM (Diagnostic Test Mode)	Diagnostic Test Mode[1]	DTM[1]
Dual Bed	Three Way + Oxidation Catalytic Converter[1]	TWC+OC[1]
Duty Solenoid for Purge Valve	Evaporative Emission Canister Purge Valve[1]	EVAP Canister Purge Valve[1]
E2PROM (Electrically Erasable Programmable Read Only Memory)	Electrically Erasable Programmable Read Only Memory[1]	EEPROM[1]
Early Fuel Evaporation	Early Fuel Evaporation[1]	EFE[1]
EATX (Electronic Automatic Transmission/ Transaxle)	Automatic Transmission	A/T
	Automatic Transaxle	A/T
EC (Engine Control)	Engine Control[1]	EC[1]
ECA (Electronic Control Assembly)	Powertrain Control Module[1]	PCM[1]
ECL (Engine Coolant Level)	Engine Coolant Level	ECL
ECM (Engine Control Module)	Engine Control Module[1]	ECM[1]
ECT (Engine Coolant Temperature)	Engine Coolant Temperature[1]	ECT[1]
ECT (Engine Coolant Temperature) Sender	Engine Coolant Temperature Sensor[1]	ECT Sensor[1]
ECT (Engine Coolant Temperature) Sensor	Engine Coolant Temperature Sensor[1]	ECT Sensor[1]
ECT (Engine Coolant Temperature) Switch	Engine Coolant Temperature Switch[1]	ECT Switch[1]
ECU4 (Electronic Control Unit 4)	Powertrain Control Module[1]	PCM
EDF (Electro-Drive Fan) Control	Fan Control	FC
EDIS (Electronic Distributor Ignition System)	Distributor Ignition System[1]	DI System[1]
EDIS (Electronic Distributorless Ignition System)	Electronic Ignition System[1]	EI System[1]
EDIS (Electronic Distributor Ignition System) Module	Distributor Ignition Control Module[1]	Distributor ICM
EEC (Electronic Engine Control)	Engine Control[1]	EC[1]
EEC (Electronic Engine Control) Processor	Powertrain Control Module[1]	PCM[1]
EECS (Evaporative Emission Control System)	Evaporative Emission System[1]	EVAP System[1]
EEPROM (Electrically Erasable Programmable Read Only Memory)	Electrically Erasable Programmable Read Only Memory[1]	EEPROM[1]
EFE (Early Fuel Evaporation)	Early Fuel Evaporation[1]	EFE[1]
EFI (Electronic Fuel Injection)	Multiport Fuel Injection[1]	MFI[1]
EFI (Electronic Fuel Injection)	Throttle Body Fuel Injection[1]	TBI[1]
EGO (Exhaust Gas Oxygen) Sensor	Oxygen Sensor[1]	O2S[1]
EGOS (Exhaust Gas Oxygen Sensor)	Oxygen Sensor[1]	O2S[1]
EGR (Exhaust Gas Recirculation)	Exhaust Gas Recirculation[1]	EGR[1]
EGR (Exhaust Gas Recirculation) Diagnostic Valve	Exhaust Gas Recirculation Diagnostic Valve[1]	EGR Diagnostic Valve[1]
EGR (Exhaust Gas Recirculation) System	Exhaust Gas Recirculation System[1]	EGR System[1]
EGR (Exhaust Gas Recirculation) Thermal Vacuum Valve	Exhaust Gas Recirculation Thermal Vacuum Valve[1]	EGR TVV[1]
EGR (Exhaust Gas Recirculation) Valve	Exhaust Gas Recirculation Valve[1]	EGR Valve[1]
EGR TVV (Exhaust Gas Recirculation Thermal Vacuum Valve)	Exhaust Gas Recirculation Thermal Vacuum Valve[1]	EGR TVV[1]
EGRT (Exhaust Gas Recirculation Temperature)	Exhaust Gas Recirculation Temperature[1]	EGRT

Recommended Terms and Recommended Acronyms See Figure 2

[1] Emission-Related Term

Figure 12-21D SAE J1930 names and terminology (courtesy SAE International).

SAE J1930 Revised JUN93

Existing Usage	Acceptable Usage	Acceptable Acronized Usage
EGRT (Exhaust Gas Recirculation Temperature) Sensor	Exhaust Gas Recirculation Temperature Sensor[1]	EGRT Sensor[1]
EGRV (Exhaust Gas Recirculation Valve)	Exhaust Gas Recirculation Valve[1]	EGR Valve[1]
EGRVC (Exhaust Gas Recirculation Valve Control)	Exhaust Gas Recirculation Valve Control[1]	EGR Valve Control[1]
EGS (Exhaust Gas Sensor)	Oxygen Sensor[1]	O2S[1]
EI (Electronic Ignition) (With Distributor)	Distributor Ignition[1]	DI[1]
EI (Electronic Ignition) (Without Distributor)	Electronic Ignition[1]	EI[1]
Electrically Erasable Programmable Read Only Memory	Electrically Erasable Programmable Read Only Memory[1]	EEPROM[1]
Electronic Engine Control	Electronic Engine Control[1]	Electronic EC[1]
Electronic Ignition	Electronic Ignition[1]	EI[1]
Electronic Spark Advance	Ignition Control[1]	IC[1]
Electronic Spark Timing	Ignition Control[1]	IC[1]
EM (Engine Modification)	Engine Modification[1]	EM[1]
EMR (Engine Maintenance Reminder)	Service Reminder Indicator[1]	SRI[1]
Engine Control	Engine Control[1]	EC[1]
Engine Coolant Fan Control	Fan Control	FC
Engine Coolant Level	Engine Coolant Level	ECL
Engine Coolant Level Indicator	Engine Coolant Level Indicator	ECL Indicator
Engine Coolant Temperature	Engine Coolant Temperature[1]	ECT[1]
Engine Coolant Temperature Sender	Engine Coolant Temperature Sensor[1]	ECT Sensor[1]
Engine Coolant Temperature Sensor	Engine Coolant Temperature Sensor[1]	ECT Sensor[1]
Engine Coolant Temperature Switch	Engine Coolant Temperature Switch[1]	ECT Switch[1]
Engine Modification	Engine Modification[1]	EM[1]
Engine Speed	Engine Speed[1]	RPM[1]
EOS (Exhaust Oxygen Sensor)	Oxygen Sensor[1]	O2S[1]
EPROM (Erasable Programmable Read Only Memory)	Erasable Programmable Read Only Memory[1]	EPROM[1]
Erasable Programmable Read Only Memory	Erasable Programmable Read Only Memory[1]	EPROM[1]
ESA (Electronic Spark Advance)	Ignition Control[1]	IC[1]
ESAC (Electronic Spark Advance Control)	Distributor Ignition[1]	DI[1]
EST (Electronic Spark Timing)	Ignition Control[1]	IC[1]
EVAP CANP	Evaporative Emission Canister Purge[1]	EVAP Canister Purge[1]
EVAP (Evaporative Emission)	Evaporative Emission[1]	EVAP[1]
EVAP (Evaporative Emission) Canister	Evaporative Emission Canister[1]	EVAP Canister[1]
EVAP (Evaporative Emission) Purge Valve	Evaporative Emission Canister Purge Valve[1]	EVAP Canister Purge Valve[1]
Evaporative Emission	Evaporative Emission[1]	EVAP[1]
Evaporative Emission Canister	Evaporative Emission Canister[1]	EVAP Canister[1]
EVP (Exhaust Gas Recirculation Valve Position) Sensor	Exhaust Gas Recirculation Valve Position Sensor[1]	EGR Valve Position Sensor[1]
EVR (Exhaust Gas Recirculation Vacuum Regulator) Solenoid	Exhaust Gas Recirculation Vacuum Regulator Solenoid[1]	EGR Vacuum Regulator Solenoid[1]
EVRV (Exhaust Gas Recirculation Vacuum Regulator Valve)	Exhaust Gas Recirculation Vacuum Regulator Valve[1]	EGR Vacuum Regulator Valve[1]
Exhaust Gas Recirculation	Exhaust Gas Recirculation[1]	EGR[1]
Exhaust Gas Recirculation Temperature	Exhaust Gas Recirculation Temperature[1]	EGRT[1]
Exhaust Gas Recirculation Temperature Sensor	Exhaust Gas Recirculation Temperature Sensor[1]	EGRT Sensor[1]
Exhaust Gas Recirculation Valve	Exhaust Gas Recirculation Valve[1]	EGR Valve[1]

Recommended Terms and Recommended Acronyms See Figure 2

[1] Emission-Related Term

Figure 12-21E SAE J1930 names and terminology (courtesy SAE International).

SAE J1930 Revised JUN93

Existing Usage	Acceptable Usage	Acceptable Acronized Usage
Fan Control	Fan Control	FC
Fan Control Module	Fan Control Module	FC Module
Fan Control Relay	Fan Control Relay	FC Relay
Fan Motor Control Relay	Fan Control Relay	FC Relay
Fast Idle Thermo Valve	Idle Air Control Thermal Valve[1]	IAC Thermal Valve[1]
FBC (Feed Back Carburetor)	Carburetor[1]	CARB[1]
FBC (Feed Back Control)	Mixture Control[1]	MC[1]
FC (Fan Control)	Fan Control	FC
FC (Fan Control) Relay	Fan Control Relay	FC Relay
FEEPROM (Flash Electrically Erasable Programmable Read Only Memory)	Flash Electrically Erasable Programmable Read Only Memory[1]	FEEPROM[1]
FEPROM (Flash Erasable Programmable Read Only Memory)	Flash Erasable Programmable Read Only Memory[1]	FEPROM[1]
FF (Flexible Fuel)	Flexible Fuel[1]	FF[1]
FI (Fuel Injection)	Central Multiport Fuel Injection[1]	Central MFI[1]
FI (Fuel Injection)	Continuous Fuel Injection[1]	CFI[1]
FI (Fuel Injection)	Direct Fuel Injection[1]	DFI[1]
FI (Fuel Injection)	Indirect Fuel Injection[1]	IFI[1]
FI (Fuel Injection)	Multiport Fuel Injection[1]	MFI[1]
FI (Fuel Injection)	Sequential Multiport Fuel Injection[1]	SFI[1]
FI (Fuel Injection)	Throttle Body Fuel Injection[1]	TBI[1]
Flash EEPROM (Electrically Erasable Programmable Read Only Memory)	Flash Electrically Erasable Programmable Read Only Memory[1]	FEEPROM[1]
Flash EPROM (Erasable Programmable Read Only Memory)	Flash Erasable Programmable Read Only Memory[1]	FEPROM[1]
Flexible Fuel	Flexible Fuel[1]	FF[1]
Flexible Fuel Sensor	Flexible Fuel Sensor[1]	FF Sensor[1]
Fourth Gear	Fourth Gear	4GR
FP (Fuel Pump)	Fuel Pump	FP
FP (Fuel Pump) Module	Fuel Pump Module	FP Module
FT (Fuel Trim)	Fuel Trim[1]	FT[1]
Fuel Charging Station	Throttle Body[1]	TB[1]
Fuel Concentration Sensor	Flexible Fuel Sensor[1]	FF Sensor[1]
Fuel Injection	Central Multiport Fuel Injection[1]	Central MFI[1]
Fuel Injection	Continuous Fuel Injection[1]	CFI[1]
Fuel Injection	Direct Fuel Injection[1]	DFI[1]
Fuel Injection	Indirect Fuel Injection[1]	IFI[1]
Fuel Injection	Multiport Fuel Injection[1]	MFI[1]
Fuel Injection	Sequential Multiport Fuel Injection[1]	SFI[1]
Fuel Injection	Throttle Body Fuel Injection[1]	TBI[1]
Fuel Level Sensor	Fuel Level Sensor	Fuel Level Sensor
Fuel Module	Fuel Pump Module	FP Module
Fuel Pressure	Fuel Pressure[1]	Fuel Pressure[1]
Fuel Pressure Regulator	Fuel Pressure Regulator[1]	Fuel Pressure Regulator[1]
Fuel Pump	Fuel Pump	FP
Fuel Pump Relay	Fuel Pump Relay	FP Relay
Fuel Quality Sensor	Flexible Fuel Sensor[1]	FF Sensor[1]
Fuel Regulator	Fuel Pressure Regulator[1]	Fuel Pressure Regulator[1]
Fuel Sender	Fuel Pump Module	FP Module
Fuel Sensor	Fuel Level Sensor	Fuel Level Sensor
Fuel Tank Unit	Fuel Pump Module	FP Module

Recommended Terms and Recommended Acronyms See Figure 2

[1] Emission-Related Term

Figure 12-21F SAE J1930 names and terminology (courtesy SAE International).

SAE J1930 Revised JUN93

Existing Usage	Acceptable Usage	Acceptable Acronized Usage
Fuel Trim	Fuel Trim[1]	FT[1]
Full Throttle	Wide Open Throttle[1]	WOT[1]
GCM (Governor Control Module)	Governor Control Module	GCM
GEM (Governor Electronic Module)	Governor Control Module	GCM
GEN (Generator)	Generator	GEN
Generator	Generator	GEN
Governor	Governor	Governor
Governor Control Module	Governor Control Module	GCM
Governor Electronic Module	Governor Control Module	GCM
GND (Ground)	Ground	GND
GRD (Ground)	Ground	GND
Ground	Ground	GND
Heated Oxygen Sensor	Heated Oxygen Sensor[1]	HO2S[1]
HEDF (High Electro-Drive Fan) Control	Fan Control	FC
HEGO (Heated Exhaust Gas Oxygen) Sensor	Heated Oxygen Sensor[1]	HO2S[1]
HEI (High Energy Ignition)	Distributor Ignition[1]	DI[1]
High Speed FC (Fan Control) Switch	High Speed Fan Control Switch	High Speed FC Switch
HO2S (Heated Oxygen Sensor)	Heated Oxygen Sensor[1]	HO2S[1]
HOS (Heated Oxygen Sensor)	Heated Oxygen Sensor[1]	HO2S[1]
Hot Wire Anemometer	Mass Air Flow Sensor[1]	MAF Sensor[1]
IA (Intake Air)	Intake Air	IA
IA (Intake Air) Duct	Intake Air Duct	IA Duct
IAC (Idle Air Control)	Idle Air Control[1]	IAC[1]
IAC (Idle Air Control) Thermal Valve	Idle Air Control Thermal Valve[1]	IAC Thermal Valve[1]
IAC (Idle Air Control) Valve	Idle Air Control Valve[1]	IAC Valve[1]
IACV (Idle Air Control Valve)	Idle Air Control Valve[1]	IAC Valve[1]
IAT (Intake Air Temperature)	Intake Air Temperature[1]	IAT[1]
IAT (Intake Air Temperature) Sensor	Intake Air Temperature Sensor[1]	IAT Sensor[1]
IATS (Intake Air Temperature Sensor)	Intake Air Temperature Sensor[1]	IAT Sensor[1]
IC (Ignition Control)	Ignition Control[1]	IC[1]
ICM (Ignition Control Module)	Ignition Control Module[1]	ICM[1]
IDFI (Indirect Fuel Injection)	Indirect Fuel Injection[1]	IFI[1]
IDI (Integrated Direct Ignition)	Electronic Ignition[1]	EI[1]
IDI (Indirect Diesel Injection)	Indirect Fuel Injection[1]	IFI[1]
Idle Air Bypass Control	Idle Air Control[1]	IAC[1]
Idle Air Control	Idle Air Control[1]	IAC[1]
Idle Air Control Valve	Idle Air Control Valve[1]	IAC Valve[1]
Idle Speed Control	Idle Air Control[1]	IAC[1]
Idle Speed Control	Idle Speed Control[1]	ISC[1]
Idle Speed Control Actuator	Idle Speed Control Actuator[1]	ISC Actuator[1]
IFI (Indirect Fuel Injection)	Indirect Fuel Injection[1]	IFI[1]
IFS (Inertia Fuel Shutoff)	Inertia Fuel Shutoff	IFS
Ignition Control	Ignition Control[1]	IC[1]
Ignition Control Module	Ignition Control Module[1]	ICM[1]
In Tank Module	Fuel Pump Module	FP Module
Indirect Fuel Injection	Indirect Fuel Injection[1]	IFI[1]
Inertia Fuel Shutoff	Inertia Fuel Shutoff	IFS
Inertia Fuel - Shutoff Switch	Inertia Fuel Shutoff Switch	IFS Switch
Inertia Switch	Inertia Fuel Shutoff Switch	IFS Switch
INT (Integrator)	Short Term Fuel Trim[1]	Short Term FT[1]
Intake Air	Intake Air	IA

Recommended Terms and Recommended Acronyms See Figure 2
[1] Emission-Related Term

Figure 12-21G SAE J1930 names and terminology (courtesy SAE International).

SAE J1930 Revised JUN93

Existing Usage	Acceptable Usage	Acceptable Acronized Usage
Intake Air Duct	Intake Air Duct	IA Duct
Intake Air Temperature	Intake Air Temperature[1]	IAT[1]
Intake Air Temperature Sensor	Intake Air Temperature Sensor[1]	IAT Sensor[1]
Intake Manifold Absolute Pressure Sensor	Manifold Absolute Pressure Sensor[1]	MAP Sensor[1]
Integrated Relay Module	Relay Module	RM
Integrator	Short Term Fuel Trim[1]	Short Term FT[1]
Inter Cooler	Charge Air Cooler[1]	CAC[1]
ISC (Idle Speed Control)	Idle Air Control[1]	IAC[1]
ISC (Idle Speed Control)	Idle Speed Control[1]	ISC[1]
ISC (Idle Speed Control) Actuator	Idle Speed Control Actuator[1]	ISC Actuator[1]
ISC BPA (Idle Speed Control By Pass Air)	Idle Air Control[1]	IAC[1]
ISC (Idle Speed Control) Solenoid Vacuum Valve	Idle Speed Control Solenoid Vacuum Valve[1]	ISC Solenoid Vacuum Valve[1]
K-Jetronic	Continuous Fuel Injection[1]	CFI[1]
KAM (Keep Alive Memory)	NonVolatile Random Access Memory[1]	NVRAM[1]
KAM (Keep Alive Memory)	Keep Alive Random Access Memory[1]	Keep Alive RAM[1]
KE-Jetronic	Continuous Fuel Injection[1]	CFI[1]
KE-Motronic	Continuous Fuel Injection[1]	CFI[1]
Knock Sensor	Knock Sensor[1]	KS[1]
KS (Knock Sensor)	Knock Sensor[1]	KS[1]
L-Jetronic	Multiport Fuel Injection[1]	MFI[1]
Lambda	Oxygen Sensor[1]	O2S[1]
LH-Jetronic	Multiport Fuel Injection[1]	MFI[1]
Light Off Catalyst	Warm Up Three Way Catalytic Converter[1]	WU-TWC[1]
Light Off Catalyst	Warm Up Oxidation Catalytic Converter[1]	WU-OC[1]
Lock Up Relay	Torque Converter Clutch Relay[1]	TCC Relay[1]
Long Term FT (Fuel Trim)	Long Term Fuel Trim[1]	Long Term FT[1]
Low Speed FC (Fan Control) Switch	Low Speed Fan Control Switch	Low Speed FC Switch
LUS (Lock Up Solenoid) Valve	Torque Converter Clutch Solenoid Valve[1]	TCC Solenoid Valve[1]
M/C (Mixture Control)	Mixture Control[1]	MC[1]
MAF (Mass Air Flow)	Mass Air Flow[1]	MAF[1]
MAF (Mass Air Flow) Sensor	Mass Air Flow Sensor[1]	MAF Sensor[1]
Malfunction Indicator Lamp	Malfunction Indicator Lamp[1]	MIL[1]
Manifold Absolute Pressure	Manifold Absolute Pressure[1]	MAP[1]
Manifold Absolute Pressure Sensor	Manifold Absolute Pressure Sensor	MAP Sensor[1]
Manifold Differential Pressure	Manifold Differential Pressure[1]	MDP[1]
Manifold Surface Temperature	Manifold Surface Temperature[1]	MST[1]
Manifold Vacuum Zone	Manifold Vacuum Zone[1]	MVZ[1]
Manual Lever Position Sensor	Transmission Range Sensor[1]	TR Sensor[1]
MAP (Manifold Absolute Pressure)	Manifold Absolute Pressure[1]	MAP[1]
MAP (Manifold Absolute Pressure) Sensor	Manifold Absolute Pressure Sensor[1]	MAP Sensor[1]
MAPS (Manifold Absolute Pressure Sensor)	Manifold Absolute Pressure Sensor[1]	MAP Sensor[1]
Mass Air Flow	Mass Air Flow[1]	MAF[1]
Mass Air Flow Sensor	Mass Air Flow Sensor[1]	MAF Sensor[1]
MAT (Manifold Air Temperature)	Intake Air Temperature[1]	IAT[1]
MATS (Manifold Air Temperature Sensor)	Intake Air Temperature Sensor[1]	IAT Sensor[1]
MC (Mixture Control)	Mixture Control[1]	MC[1]
MCS (Mixture Control Solenoid)	Mixture Control Solenoid[1]	MC Solenoid[1]
MCU (Microprocessor Control Unit)	Powertrain Control Module[1]	PCM[1]
MDP (Manifold Differential Pressure)	Manifold Differential Pressure[1]	MDP[1]
MFI (Multiport Fuel Injection)	Multiport Fuel Injection[1]	MFI[1]

Recommended Terms and Recommended Acronyms See Figure 2

[1] Emission-Related Term

Figure 12-21H SAE J1930 names and terminology (courtesy SAE International).

SAE J1930 Revised JUN93

Existing Usage	Acceptable Usage	Acceptable Acronized Usage
MIL (Malfunction Indicator Lamp)	Malfunction Indicator Lamp[1]	MIL[1]
Mixture Control	Mixture Control[1]	MC[1]
Modes	Diagnostic Test Mode[1]	DTM[1]
Monotronic	Throttle Body Fuel Injection[1]	TBI[1]
Motronic	Multiport Fuel Injection[1]	MFI[1]
MPI (Multipoint Injection)	Multiport Fuel Injection[1]	MFI[1]
MPI (Multiport Injection)	Multiport Fuel Injection[1]	MFI[1]
MRPS (Manual Range Position Switch)	Transmission Range Switch	TR Switch
MST (Manifold Surface Temperature)	Manifold Surface Temperature[1]	MST[1]
Multiport Fuel Injection	Multiport Fuel Injection[1]	MFI[1]
MVZ (Manifold Vacuum Zone)	Manifold Vacuum Zone[1]	MVZ[1]
NDS (Neutral Drive Switch)	Park/Neutral Position Switch[1]	PNP Switch[1]
Neutral Safety Switch	Park/Neutral Position Switch[1]	PNP Switch[1]
NGS (Neutral Gear Switch)	Park/Neutral Position Switch[1]	PNP Switch[1]
Nonvolatile Random Access Memory	Nonvolatile Random Access Memory[1]	NVRAM[1]
NPS (Neutral Position Switch)	Park/Neutral Position Switch[1]	PNP Switch[1]
NVM (Nonvolatile Memory)	Nonvolatile Random Access Memory[1]	NVRAM[1]
NVRAM (Nonvolatile Random Access Memory)	Nonvolatile Random Access Memory[1]	NVRAM[1]
O2 (Oxygen) Sensor	Oxygen Sensor[1]	O2S[1]
O2S (Oxygen Sensor)	Oxygen Sensor[1]	O2S[1]
OBD (On Board Diagnostic)	On Board Diagnostic[1]	OBD[1]
OC (Oxidation Catalyst)	Oxidation Catalytic Converter[1]	OC[1]
Oil Pressure Sender	Oil Pressure Sensor	Oil Pressure Sensor
Oil Pressure Sensor	Oil Pressure Sensor	Oil Pressure Sensor
Oil Pressure Switch	Oil Pressure Switch	Oil Pressure Switch
OL (Open Loop)	Open Loop[1]	OL[1]
On Board Diagnostic	On Board Diagnostic[1]	OBD[1]
Open Loop	Open Loop[1]	OL[1]
OS (Oxygen Sensor)	Oxygen Sensor[1]	O2S[1]
Oxidation Catalytic Converter	Oxidation Catalytic Converter[1]	OC[1]
OXS (Oxygen Sensor) Indicator	Service Reminder Indicator[1]	SRI[1]
Oxygen Sensor	Oxygen Sensor[1]	O2S[1]
P/N (Park/Neutral)	Park/Neutral Position[1]	PNP[1]
P/S (Power Steering) Pressure Switch	Power Steering Pressure Switch	PSP Switch
P- (Pressure) Sensor	Manifold Absolute Pressure Sensor[1]	MAP Sensor[1]
PAIR (Pulsed Secondary Air Injection)	Pulsed Secondary Air Injection[1]	PAIR[1]
Park/Neutral Position	Park/Neutral Position[1]	PNP[1]
PCM (Powertrain Control Module)	Powertrain Control Module[1]	PCM[1]
PCV (Positive Crankcase Ventilation)	Positive Crankcase Ventilation[1]	PCV[1]
PCV (Positive Crankcase Ventilation) Valve	Positive Crankcase Ventilation Valve[1]	PCV Valve[1]
Percent Alcohol Sensor	Flexible Fuel Sensor[1]	FF Sensor[1]
Periodic Trap Oxidizer	Periodic Trap Oxidizer[1]	PTOX[1]
PFE (Pressure Feedback Exhaust Gas Recirculation) Sensor	Feedback Pressure Exhaust Gas Recirculation Sensor[1]	Feedback Pressure EGR Sensor[1]
PFI (Port Fuel Injection)	Multiport Fuel Injection[1]	MFI[1]
PG (Pulse Generator)	Vehicle Speed Sensor[1]	VSS[1]
PGM-FI (Programmed Fuel Injection)	Multiport Fuel Injection[1]	MFI[1]
PIP (Position Indicator Pulse)	Crankshaft Position[1]	CKP[1]
PNP (Park/Neutral Position)	Park/Neutral Position[1]	PNP[1]
Positive Crankcase Ventilation	Positive Crankcase Ventilation[1]	PCV[1]
Positive Crankcase Ventilation Valve	Positive Crankcase Ventilation Valve[1]	PCV Valve[1]

Recommended Terms and Recommended Acronyms See Figure 2
[1] Emission-Related Term

Figure 12-21l SAE J1930 names and terminology (courtesy SAE International).

SAE J1930 Revised JUN93

Existing Usage	Acceptable Usage	Acceptable Acronized Usage
Power Steering Pressure	Power Steering Pressure	PSP
Power Steering Pressure Switch	Power Steering Pressure Switch	PSP Switch
Powertrain Control Module	Powertrain Control Module[1]	PCM[1]
Pressure Feedback EGR (Exhaust Gas Recirculation)	Feedback Pressure Exhaust Gas Recirculation[1]	Feedback Pressure EGR[1]
Pressure Sensor	Manifold Absolute Pressure Sensor[1]	MAP Sensor[1]
Pressure Transducer EGR (Exhaust Gas Recirculation) System	Pressure Transducer Exhaust Gas Recirculation System[1]	Pressure Transducer EGR System[1]
PRNDL (Park, Reverse, Neutral, Drive, Low)	Transmission Range	TR
Programmable Read Only Memory	Programmable Read Only Memory[1]	PROM[1]
PROM (Programmable Read Only Memory)	Programmable Read Only Memory[1]	PROM[1]
PSP (Power Steering Pressure)	Power Steering Pressure	PSP
PSP (Power Steering Pressure) Switch	Power Steering Pressure Switch	PSP Switch
PSPS (Power Steering Pressure Switch)	Power Steering Pressure Switch	PSP Switch
PTOX (Periodic Trap Oxidizer)	Periodic Trap Oxidizer[1]	PTOX[1]
Pulsair	Pulsed Secondary Air Injection[1]	PAIR[1]
Pulsed Secondary Air Injection	Pulsed Secondary Air Injection[1]	PAIR[1]
Radiator Fan Control	Fan Control	FC
Radiator Fan Relay	Fan Control Relay	FC Relay
RAM (Random Access Memory)	Random Access Memory[1]	RAM[1]
Random Access Memory	Random Access Memory[1]	RAM[1]
Read Only Memory	Read Only Memory[1]	ROM[1]
Recirculated Exhaust Gas Temperature Sensor	Exhaust Gas Recirculation Temperature Sensor[1]	EGRT Sensor[1]
Reed Valve	Pulsed Secondary Air Injection Valve[1]	PAIR Valve[1]
REGTS (Recirculated Exhaust Gas Temperature Sensor)	Exhaust Gas Recirculation Temperature Sensor[1]	EGRT Sensor[1]
Relay Module	Relay Module	RM
Remote Mount TFI (Thick Film Ignition)	Distributor Ignition[1]	DI[1]
Revolutions per Minute	Engine Speed[1]	RPM[1]
RM (Relay Module)	Relay Module	RM
ROM (Read Only Memory)	Read Only Memory[1]	ROM[1]
RPM (Revolutions per Minute)	Engine Speed[1]	RPM[1]
SABV (Secondary Air Bypass Valve)	Secondary Air Injection Bypass Valve[1]	AIR Bypass Valve[1]
SACV (Secondary Air Check Valve)	Secondary Air Injection Control Valve[1]	AIR Control Valve[1]
SASV (Secondary Air Switching Valve)	Secondary Air Injection Switching Valve[1]	AIR Switching Valve[1]
SBEC (Single Board Engine Control)	Powertrain Control Module[1]	PCM[1]
SBS (Supercharger Bypass Solenoid)	Supercharger Bypass Solenoid[1]	SCB Solenoid[1]
SC (Supercharger)	Supercharger[1]	SC[1]
Scan Tool	Scan Tool[1]	ST[1]
SCB (Supercharger Bypass)	Supercharger Bypass[1]	SCB[1]
Secondary Air Bypass Valve	Secondary Air Injection Bypass Valve[1]	AIR Bypass Valve[1]
Secondary Air Check Valve	Secondary Air Injection Check Valve[1]	AIR Check Valve[1]
Secondary Air Injection	Secondary Air Injection[1]	AIR[1]
Secondary Air Injection Bypass	Secondary Air Injection Bypass[1]	AIR Bypass[1]
Secondary Air Injection Diverter	Secondary Air Injection Diverter[1]	AIR Diverter[1]
Secondary Air Switching Valve	Secondary Air Injection Switching Valve[1]	AIR Switching Valve[1]
SEFI (Sequential Electronic Fuel Injection)	Sequential Multiport Fuel Injection[1]	SFI[1]
Self Test	On Board Diagnostic[1]	OBD[1]
Self Test Codes	Diagnostic Trouble Code[1]	DTC[1]
Self Test Connector	Data Link Connector[1]	DLC[1]

Recommended Terms and Recommended Acronyms See Figure 2

[1] Emission-Related Term

Figure 12-21J SAE J1930 names and terminology (courtesy SAE International).

SAE J1930 Revised JUN93

Existing Usage	Acceptable Usage	Acceptable Acronized Usage
Sequential Multiport Fuel Injection	Sequential Multiport Fuel Injection[1]	SFI[1]
Service Engine Soon	Service Reminder Indicator[1]	SRI[1]
Service Engine Soon	Malfunction Indicator Lamp[1]	MIL[1]
Service Reminder Indicator	Service Reminder Indicator[1]	SRI[1]
SFI (Sequential Fuel Injection)	Sequential Multiport Fuel Injection[1]	SFI[1]
Short Term FT (Fuel Trim)	Short Term Fuel Trim[1]	Short Term FT[1]
SLP (Selection Lever Position)	Transmission Range	TR
SMEC (Single Module Engine Control)	Powertrain Control Module[1]	PCM[1]
Smoke Puff Limiter	Smoke Puff Limiter[1]	SPL[1]
SPI (Single Point Injection)	Throttle Body Fuel Injection[1]	TBI[1]
SPL (Smoke Puff Limiter)	Smoke Puff Limiter[1]	SPL[1]
SRI (Service Reminder Indicator)	Service Reminder Indicator[1]	SRI[1]
SRT (System Readiness Test)	System Readiness Test[1]	SRT[1]
ST (Scan Tool)	Scan Tool[1]	ST[1]
Supercharger	Supercharger[1]	SC[1]
Supercharger Bypass	Supercharger Bypass[1]	SCB[1]
Sync Pickup	Camshaft Position[1]	CMP[1]
System Readiness Test	System Readiness Test[1]	SRT[1]
TAB (Thermactor Air Bypass)	Secondary Air Injection Bypass[1]	AIR Bypass[1]
TAD (Thermactor Air Diverter)	Secondary Air Injection Diverter[1]	AIR Diverter[1]
TB (Throttle Body)	Throttle Body[1]	TB[1]
TBI (Throttle Body Fuel Injection)	Throttle Body Fuel Injection[1]	TBI[1]
TBT (Throttle Body Temperature)	Intake Air Temperature[1]	IAT[1]
TC (Turbocharger)	Turbocharger[1]	TC[1]
TCC (Torque Converter Clutch)	Torque Converter Clutch[1]	TCC[1]
TCC (Torque Converter Clutch) Relay	Torque Converter Clutch Relay[1]	TCC Relay[1]
TCM (Transmission Control Module)	Transmission Control Module	TCM
TFI (Thick Film Ignition)	Distributor Ignition[1]	DI[1]
TFI (Thick Film Ignition) Module	Ignition Control Module[1]	ICM[1]
Thermac	Secondary Air Injection[1]	AIR[1]
Thermac Air Cleaner	Air Cleaner[1]	ACL[1]
Thermactor	Secondary Air Injection[1]	AIR[1]
Thermactor Air Bypass	Secondary Air Injection Bypass[1]	AIR Bypass[1]
Thermactor Air Diverter	Secondary Air Injection Diverter[1]	AIR Diverter[1]
Thermactor II	Pulsed Secondary Air Injection[1]	PAIR[1]
Thermal Vacuum Switch	Thermal Vacuum Valve[1]	TVV[1]
Thermal Vacuum Valve	Thermal Vacuum Valve[1]	TVV[1]
Third Gear	Third Gear	3GR
Three Way + Oxidation Catalytic Converter	Three Way + Oxidation Catalytic Converter[1]	TWC+OC[1]
Three Way Catalytic Converter	Three Way Catalytic Converter[1]	TWC[1]
Throttle Body	Throttle Body[1]	TB[1]
Throttle Body Fuel Injection	Throttle Body Fuel Injection[1]	TBI[1]
Throttle Opener	Idle Speed Control[1]	ISC[1]
Throttle Opener Vacuum Switching Valve	Idle Speed Control Solenoid Vacuum Valve[1]	ISC Solenoid Vacuum Valve[1]
Throttle Opener VSV (Vacuum Switching Valve)	Idle Speed Control Solenoid Vacuum Valve[1]	ISC Solenoid Vacuum Valve[1]
Throttle Position	Throttle Position[1]	TP
Throttle Position Sensor	Throttle Position Sensor[1]	TP Sensor[1]
Throttle Position Switch	Throttle Position Switch[1]	TP Switch[1]
Throttle Potentiometer	Throttle Position Sensor[1]	TP Sensor[1]
TOC (Trap Oxidizer - Continuous)	Continuous Trap Oxidizer[1]	CTOX[1]
TOP (Trap Oxidizer - Periodic)	Periodic Trap Oxidizer[1]	PTOX[1]

Recommended Terms and Recommended Acronyms See Figure 2

[1] Emission-Related Term

Figure 12-21K SAE J1930 names and terminology (courtesy SAE International).

SAE J1930 Revised JUN93

Existing Usage	Acceptable Usage	Acceptable Acronized Usage
Torque Converter Clutch	Torque Converter Clutch[1]	TCC[1]
Torque Converter Clutch Relay	Torque Converter Clutch Relay[1]	TCC Relay[1]
TP (Throttle Position)	Throttle Position[1]	TP[1]
TP (Throttle Position) Sensor	Throttle Position Sensor[1]	TP Sensor[1]
TP (Throttle Position) Switch	Throttle Position Switch[1]	TP Switch[1]
TPI (Tuned Port Injection)	Multiport Fuel Injection[1]	MFI[1]
TPS (Throttle Position Sensor)	Throttle Position Sensor[1]	TP Sensor[1]
TPS (Throttle Position Switch)	Throttle Position Switch[1]	TP Switch[1]
TR (Transmission Range)	Transmission Range	TR
Transmission Control Module	Transmission Control Module	TCM
Transmission Position Switch	Transmission Range Switch	TR Switch
Transmission Range Selection	Transmission Range	TR
TRS (Transmission Range Selection)	Transmission Range	TR
TRSS (Transmission Range Selection Switch)	Transmission Range Switch	TR Switch
Tuned Port Injection	Multiport Fuel Injection[1]	MFI[1]
Turbo (Turbocharger)	Turbocharger[1]	TC[1]
Turbocharger	Turbocharger[1]	TC[1]
TVS (Thermal Vacuum Switch)	Thermal Vacuum Valve[1]	TVV[1]
TVV (Thermal Vacuum Valve)	Thermal Vacuum Valve[1]	TVV[1]
TWC (Three Way Catalytic Converter)	Three Way Catalytic Converter[1]	TWC[1]
TWC + OC (Three Way + Oxidation Catalytic Converter)	Three Way + Oxidation Catalytic Converter[1]	TWC+OC[1]
VAC (Vacuum) Sensor	Manifold Differential Pressure Sensor[1]	MDP Sensor
Vacuum Switches	Manifold Vacuum Zone Switch	MVZ Switch[1]
VAF (Volume Air Flow)	Volume Air Flow[1]	VAF[1]
Vane Air Flow	Volume Air Flow[1]	VAF[1]
Variable Fuel Sensor	Flexible Fuel Sensor	FF Sensor[1]
VAT (Vane Air Temperature)	Intake Air Temperature[1]	IAT[1]
VCC (Viscous Converter Clutch)	Torque Converter Clutch[1]	TCC[1]
Vehicle Speed Sensor	Vehicle Speed Sensor[1]	VSS[1]
VIP (Vehicle In Process) Connector	Data Link Connector[1]	DLC[1]
Viscous Converter Clutch	Torque Converter Clutch[1]	TCC[1]
Voltage Regulator	Voltage Regulator	VR
Volume Air Flow	Volume Air Flow[1]	VAF[1]
VR (Voltage Regulator)	Voltage Regulator	VR
VSS (Vehicle Speed Sensor)	Vehicle Speed Sensor[1]	VSS[1]
VSV (Vacuum Solenoid Valve) (Canister)	Evaporative Emission Canister Purge Valve[1]	EVAP Canister Purge Valve[1]
VSV (Vacuum Solenoid Valve) (EVAP)	Evaporative Emission Canister Purge Valve[1]	EVAP Canister Purge Valve[1]
VSV (Vacuum Solenoid Valve) (Throttle)	Idle Speed Control Solenoid Vacuum Valve[1]	ISC Solenoid Vacuum Valve[1]
Warm Up Oxidation Catalytic Converter	Warm Up Oxidation Catalytic Converter[1]	WU-OC[1]
Warm Up Three Way Catalytic Converter	Warm Up Three Way Catalytic Converter[1]	WU-OC[1]
Wide Open Throttle	Wide Open Throttle[1]	WOT[1]
WOT (Wide Open Throttle)	Wide Open Throttle[1]	WOT[1]
WOTS (Wide Open Throttle Switch)	Wide Open Throttle Switch[1]	WOT Switch[1]
WU-OC (Warm Up Oxidation Catalytic Converter)	Warm Up Oxidation Catalytic Converter[1]	WU-OC[1]
WU-TWC (Warm Up Three Way Catalytic Converter)	Warm Up Three Way Catalytic Converter[1]	WU-TWC[1]

Recommended Terms and Recommended Acronyms See Figure 2

[1] Emission-Related Term

Figure 12-21L SAE J1930 names and terminology (courtesy SAE International).

OBD-II EMISSIONS WARRANTY

Federal	2 years/24,000	Emissions-Related Components
	8 years/80,000	Major Emissions-Related Components: 1. Catalyst 2. Electronic Control Module
California	3 years/50,000	Emissions-Related Components
	7 years/70,000	Refer to the selected high-cost components as defined in the warranty policy and procedures manual.

Figure 12-22 OBD-II emissions warranty.

WARRANTY REQUIREMENTS

Emissions-related components must remain operable for the useful life of the system. This is defined as ten years or 100,000 miles, whichever comes first, for both federally and California-certified vehicles. Each manufacturer must demonstrate that emissions systems are durable enough to meet this requirement before the government will certify these vehicles for sale. Warranty periods as required by the legislation are indicated in Figure 12-22.

CARB is monitoring emissions system repairs performed by several dealers throughout the State of California. If a specific component or system fails on more than 4% of the vehicles sampled, the manufacturer is required to research and evaluate the reasons. If the failure rate is not attributable to owner negligence or misuse, a component failure rate of 4% could lead to a vehicle recall. The federal government is considering a similar program. Manufacturers are relying on technicians throughout the country to appropriately diagnose and repair emissions-related concerns and to replace only those parts that have failed. The sloppy practice of hanging parts on a car under an emissions warranty could lead to an unnecessary and expensive fleet-wide recall.

Glossary

Acceleration Simulation Mode (ASM) — A high-load, steady-state test procedure that allows testing to be performed using a relatively simple chassis dynamometer and an NO_x analyzer.

active test — A test that takes some sort of action when performing diagnostic functions, and does not affect performance. Active tests are performed when a passive test fails.

actuator — A device which delivers motion in response to an electrical signal.

advanced engine performance specialist (L1) test — A test designed to measure the technicians' knowledge of the diagnostic skills necessary for sophisticated emissions and engine performance problems.

analog — A continuously-variable voltage signal.

analog to digital converter — A circuit that takes an analog signal and converts it to a digital signal to be used in the computer.

annually — Once per year.

arming conditions — Specific criteria that must be met before the computer will monitor an input or output circuit for a fault code.

Assembly Line Data Link (ALDL) — A connector located in the passenger or engine compartment that transmits serial data to the scan tool.

attainment — An area which is exempt from the Clean Air Act Amendment.

automotive emissions — Gaseous and particulate compounds that are emitted from a car's crankcase, exhaust, carburetor, fuel tank, etc., including hydrocarbons, nitrogen oxide, and carbon monoxide.

BAR '90 — Bureau of Automotive Repair 1990 equipment specifications.

barometric pressure — The ambient or atmospheric air pressure.

basic — A moderate ozone non-attainment area; these areas are subject to less stringent emissions controls than those areas with enhanced programs.

baud rate — The speed at which computer data is processed or transmitted.

bi-directional — A two-way transmission. For example, the scan tool can receive data and can send commands to the computer, instructing it to perform system-specific tests on the actuators.

biannually — Every two years.

binary — A number system using two digits: 0 and 1.

breakout box — Essentially a box with a series of test terminals or lugs, which can be connected to the vehicle's computer wiring harness.

carbon monoxide (CO) — A gas created when there is not enough oxygen present during combustion. When there is an oxygen deficiency in the combustion process, some carbon in the gasoline will join some oxygen in the air, on a one-to-one basis, to form carbon monoxide (CO). Carbon monoxide can be measured as a percentage (%) or in grams per mile (g/mi).

Corporate Average Fuel Economy (CAFE) — A set of federal requirements and regulations which govern fuel economy standards.

centralized system — A vehicle emissions testing facility that is run by a state-appointed contractor or state agency. In a centralized program, one entity (either the contractor or a state agency) is responsible for purchasing and constructing test sites, hiring inspection personnel, and conducting vehicle emissions testing.

chassis dynamometer — A device that applies a controlled load to the drive wheels of the vehicle. The load is placed at a right angle to the tire across the treads. To correctly simulate the road load for a given vehicle, the dynamometer must be programmed for the correct inertia weight.

chip — A miniaturized electronic circuit etched into a base of silicon.

circuit — A system through which electricity flows before it returns to its source.

closed loop — A system that feeds back its output to the input side of the computer, which monitors the output and makes corrections as necessary.

common ground — A term used to describe the ground or negative path through the chassis of a car. Circuits are completed by connecting to the common ground. The battery negative cable is connected to this ground as well.

composite standards — Standards based on the cumulative emissions levels during the full driving cycle.

comprehensive systems monitoring — A series of diagnostic tests of each emission control system that evaluates whether a component or system is operating properly.

computer — A device that takes information, processes it, performs calculations, and outputs those calculations through commands.

conductor — A substance or body capable of transmitting electricity.

Constant Volume Sampling (CVS) — A system that guarantees a constant volume and rate of ambient air and exhaust will pass through the gas analyzer.

current — The movement of free electrons along a conductor.

cycle — A complete sequence of a wave pattern that recurs at regular intervals. That is, the change of an alternating wave from zero to a positive peak, back to zero, to a negative peak, and back to zero again. The number of cycles which occur in one second is the frequency of the wave.

Data Communications Link (DCL) — A connector located in the passenger or engine compartment that transmits serial data to the scan tool.

decentralized system — A system in which testing is conducted by independent private sector facilities (i.e., repair shops).

de-energized — Having the electric current or energy source turned off.

default value — A value that is substituted by the computer for a failed sensor that has set a hard fault.

defective — An obvious condition that is noticed and is due to normal wear or deterioration. The emission control system or component is not operational, but is not the result of tampering. The component or system must still be in place.

Digital Storage Oscilloscope (DSO) — A device that converts a signal at its input to digital information and stores it in memory.

disconnect — To separate components that are normally connected. An emission control system has been disconnected if a hose, wire, belt, or component required for the operation of the system is present, but has been disconnected.

density — The number of parts or units in a given area or space. For example, air density decreases as temperature increases in a measured space.

Diagnostic Executive — A new piece of software that allows the computer to coordinate testing operations and to keep systems functioning in the most efficient manner possible.

digital — A signal that has two states: on or off.

diode — An electrical device that permits current to flow in only one direction. It is most often used as a component in electronic controls and accessories.

downstream — Occurs during closed loop when air injection flow is to the catalytic converter.

dual traces — A situation in which two traces are displayed on the CRT at once. The traces displayed can be unrelated signals or related signals that need to be compared for time or voltage.

duty-cycle — A comparison of "on" time versus "off" time, such as the operation of a fuel injector.

EGR valve position sensor — A device which provides information on EGR valve position to the electronic engine control system processor.

enable criteria — A description of the conditions necessary for a given diagnostic test to run.

enhanced — A severe and extreme ozone non-attainment area (over 12.7 ppm) with an urbanized population of 200,000 or more; or more than 100,000 in the Northeast Ozone Transport Region.

electromagnetic field — A field having a direction or force which surrounds a charged conductor or coil.

electromechanical — A device which incorporates both electronic and mechanical elements.

electronic — Pertaining to the control of systems or devices through the use of small electrical signals and various semiconductor devices and circuits.

emission control system — A group of devices and adjustments which lower the amount of pollutants exhausted by the engine into the atmosphere.

energized — Having the electrical current or electrical source turned on (activated).

EPA — United States Environmental Protection Agency.

failure records mode — A state of operation in which critical operating conditions are automatically recorded at the instant any diagnostic test fails.

fault code cycle parameter — A specific waiting period that occurs before a hard fault code can be set; the problem must remain long enough for the computer to believe the circuit is defective.

Federal Test Procedure (FTP) — A test used to determine the compliance of light-duty vehicles (LDVs) and light-duty trucks (LDTs) with federal emissions standards. As designed, the test is intended to represent typical driving patterns in primarily urban areas.

feedback carburetion — A type of carburetion system in which the power circuit is controlled by a computer.

Final Rule — An EPA rule that established standards and other requirements for basic and enhanced vehicle inspection and maintenance programs.

flags — Indications that certain required diagnostic routines have been successfully completed, and the vehicle system or component has been diagnosed.

frame — Part of a snapshot captured by a scan tool that contains data for each input and output during one loop of the data stream.

freeze frame — Allows the Diagnostic Executive to capture data the moment an emissions-related DTC is stored and the MIL is commanded on.

fusible link — A circuit protection device that consists of a short length of wire smaller in gauge than the wire in the protected circuit. It is covered with a thick, nonflammable insulation. An overload causes the link to heat and the insulation to blister. If the overload remains, the link will melt, causing an open circuit.

graticule — A grid pattern that covers the face of the Lab Scope that serves as the reference for measurements. It is partitioned into squares called major divisions, which are broken down further into minor divisions.

hard fault code — A problem noted by the computer during its self diagnosis. Hard faults are faults that exist while the computer is checking the system.

hydrocarbon (HC) — Consists of gasoline that did not burn in the combustion process. Hydrocarbons are measured in parts per million (ppm).

IM240 — The last 240 seconds of the Federal Test Procedure (FTP); the test measures the level of emissions of certain pollutants while a vehicle is in a transient state.

inspection and maintenance (I/M) — Refers to the two components of a vehicle emissions program.

inspection station — The place where emissions testing is performed; in some states, these are referred to as inspection facilities, test sites, or testing stations.

inspection technician — The person who administers the emissions test; in some states, they are referred to as inspection personnel and/or inspectors, test inspectors, or emissions testers.

intake temperature sensor — A thermistor sensor responsible for measuring the input temperature of an air stream in the air filter or intake manifold.

intermittent fault code — A problem noted by the computer during its self diagnosis, but the problem is not occurring at the time of testing. Intermittent faults are faults that are stored in memory while the computer is checking the system.

intrusive testing — Testing that is initiated when the PCM does not see the expected results after performing the passive and active test. The intrusive test will affect the vehicle's performance and emissions levels.

laboratory oscilloscope — A cathode ray tube that displays a line pattern representing voltages in relation to time. Voltage is shown along the vertical axis and time is shown along the horizontal axis.

light-off — Refers to the time that the catalytic converter begins to effectively oxidize hydrocarbons and carbon monoxide.

loaded — Refers to the chassis dynamometer; this machine simulates vehicle inertia and road load by having the vehicle's wheels rest between the dynamometer's two rolls.

misfire — The result of poor or no combustion in the cylinder due to the absence of spark, poor fuel metering, or compression.

missing — When all or part of an emission control system has been removed from the vehicle.

modified — An emission control system or component has been modified if: 1) it has been physically or functionally altered, 2) it has been replaced with a non-original equipment manufacturer (OEM) part that has been identified by the manufacturer as not legal for use, 3) a replacement part designed for one application is used on a different application for which it was not designed.

National Automotive Technicians Education Foundation (NATEF) — A sister organization of ASE. Through its certification process, it examines the structure and resources of the education programs and evaluates them against nationally-accepted standards of quality.

ohm — The practical unit for measuring electrical resistance. It is abbreviated by the Greek letter Omega (Ω).

ohm's law — The basic rule for the relationship between voltage, current, and resistance. Current flow is directly proportional to voltage and inversely proportional to resistance.

On-Board Diagnostics One (OBD-I) — Required 1988 and newer models first sold in California to be equipped with a check engine light that alerts the driver that a computer malfunction has occurred.

On-Board Diagnostics Two (OBD-II) — Required 1994 and newer California vehicles and 1996 federal vehicles to be equipped with an enhanced diagnostic system capable of monitoring system functionality and rationality.

output driver — A transistor in the output control area of the processor that is used to turn various actuators on and off.

oxides of nitrogen (NO_x) — Compounds created when combustion chamber temperatures exceed 2500°F. Under these conditions, nitrogen combines with oxygen to form a family of new compounds called oxides of nitrogen. To simplify things, we call them (NO_x). The "N" stands for the nitrogen, the "O" for oxygen, and the "x" means that the amount of oxygen combined with nitrogen varies for each of the nitrogen compounds. Oxides of nitrogen can be measured in parts per million (ppm) or grams per mile (g/mi).

parallel circuit — A circuit in which two or more loads are connected side by side and are controlled by the same switch, but operate independently of each other. This arrangement provides full electrical potential (voltage) to all portions of the circuit.

partial stream sampling — A method of testing that draws in a low volume of exhaust gases and measures them in concentration of parts per million (ppm) and percentages (%).

passive test — A diagnostic test that monitors a system or component and no action is taken.

pattern failures — Groups of vehicles that pass the Federal Test Procedure but show consistent patterns of I/M failure.

performance monitoring — A rating system designed to let vehicle owners know how well competing repair facilities are doing in terms of getting cars to pass reinspection on the first trip back to the inspection station.

Phase 1 — The first 95 seconds of the drive cycle.

Phase 2 — The time period from 96 seconds to 240 seconds during the drive cycle.

photochemical smog — A unique type of air pollution that was first recognized in the Los Angeles Basin of California during the last half of the 1940s. The atmospheric requirements for smog formation are sunshine and still air. When the concentration of HC in the atmosphere becomes sufficiently high and NO_x is present in the correct ratio, the action of sunshine causes them to react chemically, forming photochemical smog.

pickup coil — An electromagnetic device used in electronic control system sensors. It creates an electromagnetic field due to the permanent magnet that is part of the pickup assembly. The magnetic field produces an electrical signal that is transmitted to the processor.

piezoelectric — Refers to an electronic device which is capable of generating a voltage when subjected to mechanical pressure.

ping-ponging — The condition in which a motorist has to go back and forth between the repair facility and the inspection center when a vehicle continues to fail a retest after repairs have been performed.

potential — An electrical force that is measured in volts. Sometimes used interchangeably with voltage.

potentiometer — A variable resistor with three connections. Two connections are to the ends of the resistive element. The third connection is to a wiper which moves up and down the resistive element.

preconditioning — Bringing a vehicle up to operating temperature before it is driven through the IM240 drive cycle.

pressure test — A procedure used to check for leaks in the evaporative system that could allow fuel vapors to escape into the atmosphere. The vapor line from the fuel tank is disconnected and connected to the pressure test equipment; then, fuel tank vapor lines are filled with nitrogen to a specific pressure and the pressure loss is measured for two minutes.

processor — A metal housing which contains a microcomputer and other components used in providing electronic system control.

program — A set of detailed instructions which a microcomputer follows when controlling a system.

pulse-width — The length of time an actuator, such as a fuel injector, remains on.

purge test — Used to determine if fuel vapors in the evaporative canister and fuel tank are being properly drawn into the engine. Purge flow is measured during the transient test cycle by a meter at one end of the hose running between the evaporative canister and the engine.

queuing time — The length of time a vehicle is waiting in line before being driven through the IM240 drive cycle.

Random Access Memory (RAM) — A type of memory which is used to store information temporarily. Information can be written to and read from RAM.

Read-Only Memory (ROM) — A type of memory which is used to store information permanently. Information cannot be written to ROM; as the name implies, information can only be read from ROM.

real time scope — A scope that provides instantaneous updates; this means that what is at the probe tip is what is on the screen. These scopes offer the fastest screen update rate available.

reference voltage — A voltage provided by a voltage regulator to operate potentiometers and other sensors at a constant level.

refresh rate — The time it takes for the computer to update the words in the data stream.

relay — An electromagnetic switching device that uses low current to open or close a high-current circuit.

remote sensing — Involves a computer-controlled device that transmits an infrared beam across a single lane of traffic to a receiving unit.

resistance — The opposition offered by a conductor to a flow of electricity; measured in ohms.

resistor — A device made of carbon or wire installed in an electrical circuit which tends to prevent or reduce the flow of current.

response time test — A test which measures the ability of the HO_2S to detect changes quickly when exposed to different levels of oxygen.

scan tool — Also known as a diagnostic readout tool. It converts computer pulses or signals directly into a digital or number display. This device makes it easier to read electronic trouble codes and other important data.

sensor — A device that measures an operating condition and provides an input signal to a microcomputer.

serial data — Data that allows computers to communicate with each other. It is actually a series of rapidly-changing voltage signals pulsed from high voltage to low voltage.

serial data line — A connection through which voltages are transmitted from the computer through a wire to a diagnostic connector located on the vehicle.

series circuit — A circuit in which the separate components are connected successively end to end to form a single path for the current to flow.

series-parallel circuit — A combination of series and parallel circuits. It has two or more loads in parallel, plus an additional load or loads in series.

Sealed Housing for Evaporative Determination (SHED) — A test where the entire vehicle is enclosed in a shed to collect emissions from all evaporative sources.

signal — A voltage condition that transmits specific information in an electronic system.

snapshot — A record of the serial data sent to the scan tool. It helps isolate intermittent problems by storing, in memory, data stream samples immediately before, during, and after the problem occurs.

solenoid — A wire coil with a moveable core which changes position by means of electromagnetism when current flows through the coil.

solid state — Refers to circuits that use transistors, integrated circuits, and/or other semiconductors. Any electronic circuit that does not use tubes can be considered solid state.

square wave — A digital on/off type of signal that has a very fast rise and fall.

State Implementation Program (SIP) — A state's plan for how they intend to meet the requirements of the Clean Air Act of 1990. For states which have both basic and enhanced programs, they must submit a basic SIP and an enhanced SIP. SIPs must be submitted to the EPA by July 15. It should be noted that the submission of an SIP does not bind a state to implementing the program detailed.

sweep rate — The speed at which the beam moves across a Lab Scope. It is determined by the frequency of the input signal or an internal signal of a user-variable frequency.

switch — A device used to open, close, or redirect the current in an electrical circuit.

test composite — A test that incorporates Phase 1 and Phase 2 of the IM240 drive cycle.

thermistor — A resistor that changes its resistance with temperature.

Throttle Position Sensor (TPS) — A device which provides information on the throttle position to the electronic engine control system processor.

trigger — A control that determines the instant that the scope is to start drawing the signal. Equally important is that the graph of each signal should start predictably at the same point on the waveform so that the display remains stable.

uni-directional — One-way communication. For example, the scan tool can only receive data from the on-board computer.

upstream — This occurs during open loop when air injection flow is to the exhaust manifold.

vacuum — A term used to describe a pressure that is less than atmospheric pressure.

vector — A systematic diagnostic approach to assist the technician with diagnosing driveability complaints.

volt — The unit of measurement for the force or pressure of the flow of electricity.

voltage — The electrical pressure which causes current to flow in a circuit.

voltage drop — The voltage developed between the terminals of a circuit component by the flow of current through the resistance of that part.

voltmeter — An electric meter used to measure electromotive force (EMF).

volumetric efficiency — The efficiency with which an engine fills each cylinder with a sufficient amount of oxygen and fuel.

waveform — The trace formed by an oscilloscope as voltage changes over time.

ZSTEM terminal emulation software — This software is needed to connect an IBM PC or compatible to a Snap-On, MPSI, or OTC scan tool.

Acronyms

EMISSIONS-RELATED

AIR	Air Injection Reaction System
ASM	Acceleration Simulation Mode
ATP	Anti-Tampering Program
CAAA	Clean Air Act Amendment
CARB	California Air Resources Board
CO	Carbon Monoxide
CO_2	Carbon Dioxide
CST	Certification Short Test
CVS	Constant Volume Sampling
DMV	Department of Motor Vehicles
EGR	Exhaust Gas Recirculation
EPA	Environmental Protection Agency
EVP/EEC	Evaporative Emission Controls
FTP	Federal Test Procedure
G/TEST	Grams Per Test
G/MI	Grams Per Mile
H_2O	Water
HC	Hydrocarbons
I/M	Inspection and Maintenance
IM240	Inspection and Maintenance 240-Second Emissions Test
LDT	Light-Duty Trucks
LDV	Light-Duty Vehicles
N_2	Nitrogen
NAAQS	National Ambient Air Quality Standards
NO	Nitric Oxide
NO_2	Nitrogen Dioxide
NO_x	Oxides of Nitrogen
O_2	Oxygen
OC	Oxidizing Catalytic Converter
PCV	Positive Crankcase Ventilation
PPM	Parts Per Million
RG	Repair Grade
RSD	Remote Sensing Device
SCFM	Standard Cubic Feet Per Minute
SHED	Sealed Housing for Evaporative Determination
SIP	State Implementation Plan
TAC	Thermostatic Air Cleaner
THC	Total Hydrocarbons
TVS	Thermo Vacuum Switch
TWC	Reduction/Oxidizing Catalytic Converter (Three-Way Catalyst)
UDDS	Urban Dynamometer Driving Schedule
VIN	Vehicle Identification Number
VIR	Vehicle Inspection Report

COMPUTER ENGINE CONTROL SYSTEMS

A/D	Analog to Digital
AC/DC	Alternating and Direct Current
ALDL	Assembly Line Data Link
BLM	Block Learn
BMAP	Barometric Manifold Absolute Pressure
BP	Barometric Pressure
CPU	Central Processing Unit
DCL	Data Communications Link
DPFE	Differential Pressure Feedback EGR
DTC	Diagnostic Trouble Codes
DVOM	Digital multimeter
ECT	Engine Coolant Temperature
EEPROM	Electronically Erasable Programmable Read-Only Memory
EFE	Early Fuel Evaporation
EMF	Electromotive Force
EPROM	Erasable Programmable Read-Only Memory
HO_2S	Heated Oxygen Sensor
I/O	Input and Output
IAC	Idle Air Control
IBOB	Intelligent Breakout Box

INT	Block Integrator
ISC	Idle Speed Control
KOEO	Key-On, Engine Off
LED	Light Emitting Diodes
LFTRIM	Long-Term Fuel Trim
MAF	Mass Air Flow
MAP	Manifold Absolute Pressure
MC	Mixture Control
MIL	Malfunction Indicator Light
N.C.	Normally Closed
N.O.	Normally Open
NTC	Negative Temperature Coefficient
OBD-I	On-Board Diagnostics Generation One
OBD-II	On-Board Diagnostics Generation Two
PFE	Pressure Feedback EGR
PNP	Positive/Negative/Positive Transistor
PROM	Programmable Read-Only Memory
PTC	Positive Thermo Coefficient
PWM	Pulse-Width Modulation
QDM	Quad Driver Module
RAM	Random Access Memory
ROM	Read-Only Memory
SFTRIM	Short-Term Fuel Trim
TPS	Throttle Position Sensor
UV	Ultra-Violet
VAF	Vane Air Flow
Vf	Voltage Feedback
VT100	Video Display Terminal
W.O.T.	Wide-Open-Throttle

LABORATORY SCOPES

CRT	Cathode Ray Tube
DSO	Digital Storage Oscilloscope
Lab	Laboratory
TIME/DIV	Time Per Division
VOLTS/DIV	Voltage Per Division

GENERAL

ASE	Automotive Service Excellence
CD-ROM	Compact Disk Read-Only Memory
ETL	Evaluation Team Leader
GUI	Graphical User Interface
NATEF	National Automotive Technician Education Foundation
OEM	Original Equipment Manufacturer
PC	Personal Computer
SAE	Society of Automotive Engineers
TSB	Technical Service Bulletins

Index

Note: Page Numbers in **Bold type** reference non-text material.

H

I

L

M

N

O

P

Q